Information Bottleneck

Information Bottleneck: Theory and Applications in Deep Learning

Editors

Bernhard C. Geiger
Gernot Kubin

MDPI • Basel • Beijing • Wuhan • Barcelona • Belgrade • Manchester • Tokyo • Cluj • Tianjin

Editors
Bernhard C. Geiger
Know-Center GmbH
Austria

Gernot Kubin
Graz University of Technology
Austria

Editorial Office
MDPI
St. Alban-Anlage 66
4052 Basel, Switzerland

This is a reprint of articles from the Special Issue published online in the open access journal *Entropy* (ISSN 1099-4300) (available at: https://www.mdpi.com/journal/entropy/special_issues/information_theoretic_computational_intelligence).

For citation purposes, cite each article independently as indicated on the article page online and as indicated below:

LastName, A.A.; LastName, B.B.; LastName, C.C. Article Title. *Journal Name* **Year**, *Volume Number*, Page Range.

ISBN 978-3-0365-0802-3 (Hbk)
ISBN 978-3-0365-0803-0 (PDF)

© 2021 by the authors. Articles in this book are Open Access and distributed under the Creative Commons Attribution (CC BY) license, which allows users to download, copy and build upon published articles, as long as the author and publisher are properly credited, which ensures maximum dissemination and a wider impact of our publications.

The book as a whole is distributed by MDPI under the terms and conditions of the Creative Commons license CC BY-NC-ND.

Contents

About the Editors vii

Bernhard C. Geiger and Gernot Kubin
Information Bottleneck: Theory and Applications in Deep Learning
Reprinted from: *Entropy* **2020**, *22*, 1408, doi:10.3390/e22121408 1

Julius Kunze, Louis Kirsch, Hippolyt Ritter and David Barber
Gaussian Mean Field Regularizes by Limiting Learned Information
Reprinted from: *Entropy* **2019**, *21*, 758, doi:10.3390/e21080758 5

Tailin Wu, Ian Fischer, Isaac L. Chuang and Max Tegmark
Learnability for the Information Bottleneck
Reprinted from: *Entropy* **2019**, *21*, 924, doi:10.3390/e21100924 21

Thanh Tang Nguyen and Jaesik Choi
Markov Information Bottleneck to Improve Information Flow in Stochastic Neural Networks
Reprinted from: *Entropy* **2019**, *21*, 976, doi:10.3390/e21100976 55

Artemy Kolchinsky, Brendan D. Tracey and David H. Wolpert
Nonlinear Information Bottleneck
Reprinted from: *Entropy* **2019**, *21*, 1181, doi:10.3390/e21121181 77

Max Tegmark and Tailin Wu
Pareto-Optimal Data Compression for Binary Classification Tasks
Reprinted from: *Entropy* **2020**, *22*, 7, doi:10.3390/e22010007 93

Borja Rodríguez Gálvez, Ragnar Thobaben and Mikael Skoglund
The Convex Information Bottleneck Lagrangian
Reprinted from: *Entropy* **2020**, *22*, 98, doi:10.3390/e22010098 121

Giulio Franzese, Monica Visintin
Probabilistic Ensemble of Deep Information Networks
Reprinted from: *Entropy* **2020**, *22*, 100, doi:10.3390/e22010100 149

Hlynur Jónsson, Giovanni Cherubini and Evangelos Eleftheriou
Convergence Behavior of DNNs with Mutual-Information-Based Regularization
Reprinted from: *Entropy* **2020**, *22*, 727, doi:10.3390/e22070727 167

Slava Voloshynovskiy, Olga Taran, Mouad Kondah, Taras Holotyak and Danilo Rezende
Variational Information Bottleneck for Semi-Supervised Classification
Reprinted from: *Entropy* **2020**, *22*, 943, doi:10.3390/e22090943 181

Ian Fischer
The Conditional Entropy Bottleneck
Reprinted from: *Entropy* **2020**, *22*, 999, doi:10.3390/e22090999 215

Ian Fischer and Alexander A. Alemi
CEB Improves Model Robustness
Reprinted from: *Entropy* **2020**, *22*, 1081, doi:10.3390/e22101081 237

Bernhard C. Geiger and Ian S. Fischer
A Comparison of Variational Bounds for the Information Bottleneck Functional
Reprinted from: *Entropy* **2020**, *22*, 1229, doi:10.3390/e22111229 253

About the Editors

Bernhard C. Geiger (Dipl.-Ing. Dr.) received the Dipl.-Ing. degree in Electrical Engineering (with distinction) and the Dr. techn. degree in Electrical and Information Engineering (with distinction) from Graz University of Technology, Austria, in 2009 and 2014, respectively. In 2009, he joined the Signal Processing and Speech Communication Laboratory, Graz University of Technology, as a Project Assistant and later accepted a position as Research and Teaching Associate at the same lab in 2010. He was a Senior Scientist and Erwin Schrödinger Fellow at the Institute for Communications Engineering, Technical University of Munich, Germany (2014–2017) and a postdoctoral researcher at the Signal Processing and Speech Communication Laboratory, Graz University of Technology, Austria (2017–2018). He is currently a Senior Researcher at Know-Center GmbH, Graz, Austria, where he also leads the Machine Learning Group within the Knowledge Discovery Area. Dr. Geiger's research interests cover information theory for machine learning, theory-assisted machine learning, and information–theoretic model reduction for Markov chains and hidden Markov models. He is a Senior Member of the IEEE.

Gernot Kubin Gernot Kubin (Dipl.-Ing. Dr.) has been University Professor and Founding Director of the Signal Processing and Speech Communication Laboratory at TU Graz, Austria, since 2000. He received the Dipl.-Ing. degree in 1982 and Dr. techn. degree (sub auspiciis praesidentis) in Electrical Engineering in 1990 from TU Vienna, Austria. At TU Graz, he has served as Dean of Studies in Electrical and Audio Engineering (2004–2007), Coordinator of the Key Research Area Smart Systems for a Mobile Society (2004–2011), Coordinator of the Field of Expertise Information, Communications, and Computing (2013–2015), Deputy Chair of the Curricular Committee for Electrical and Audio Engineering (2004–2008 and 2013–2019), Coordinator of the Doctoral School in Information and Communications Engineering (2007–now), Member of the Commission for Scientific Integrity and Ethics (2007–2011 and 2016–now), and Chair of the Senate (2007–2010 and 2013–now). Earlier international appointments include CERN Geneva (1980), TU Vienna (1983–2000), Erwin Schrödinger Fellow at Philips Research Labs Eindhoven (1985), AT&T Bell Labs Murray Hill (1992–1993 and 1995), KTH Stockholm (1998), Global IP Sound Stockholm (2000–2001) and San Francisco (2006), UC San Diego (2006), Danang UT (2009), and TU Munich (2015, 2017, and 2018). At national level, he has demonstrated leadership in the Vienna Telecommunications Research Centre FTW (1999–2016), the Christian Doppler Laboratory for Nonlinear Signal Processing (2002–2010), the Competence Network for Advanced Speech Technologies COAST (2006–2010), the FWF Research Network on Signal and Information Processing in Science and Engineering SISE (2008–2011), the COMET Excellence Projects Advanced Audio Processing AAP (2008–2013) and Acoustic Sensing and Design ASD (2013–2017), the Higher-Education Conference HSK (2016–now), the TU Graz Excellence Project Dependable Internet of Things (2016–now), the Competence Network for Digital Humanities KONDE (2017–now), the Complexity Science Hub Vienna (2017–now), the Scientific Board of the Christian Doppler Forschungsgesellschaft (2020–now), and as Speaker of the Senate Chairs Conference (SVK) of the Universities of Austria (2019–now). His research interests are in nonlinear signal processing as well as speech and audio communication. He has co-authored over 180 peer-reviewed publications and advised over 30 PhD students. Dr. Kubin is the recipient of the 2015 Nikola Tesla medal for the highest number of patents awarded to a TU Graz scientist in 5 years. He has served as a Member of the Board, Austrian Acoustics Association (2000–2015), as an elected member of the IEEE Speech and Language Processing Technical Committee (2011–2016), and as an elected member of the Speech Acoustics and Speech Processing committees of the German Information Technology Society ITG since 2015. He was General Chair for the INTERSPEECH 2019 conference held in Graz, Austria, in September 2019.

Editorial

Information Bottleneck: Theory and Applications in Deep Learning

Bernhard C. Geiger [1,*,†] and Gernot Kubin [2,†]

1. Know-Center GmbH, Inffeldgasse 13/6, 8010 Graz, Austria
2. Signal Processing and Speech Communication Laboratory, Graz University of Technology, Inffeldgasse 16c, 8010 Graz, Austria; g.kubin@ieee.org
* Correspondence: geiger@ieee.org
† These authors contributed equally to this work.

Received: 2 December 2020; Accepted: 9 December 2020; Published: 14 December 2020

Keywords: information bottleneck; deep learning; neural networks

---

The information bottleneck (IB) framework, proposed in [1], describes the problem of representing an observation X in a lossy manner, such that its representation T is informative of a relevance variable Y. Mathematically, the IB problem aims to find a lossy compression scheme described by a conditional distribution $P_{T|X}$ that is a minimizer of the following functional:

$$\min_{P_{T|X}} \Big(I(X;T) - \beta I(Y;T) \Big) \qquad (1)$$

where the minimization is performed over a well-defined feasible set.

The IB framework has received significant attention in information theory and machine learning; cf. [2,3]. Recently, the IB framework has also gained popularity in the analysis and design of neural networks (NNs): The framework has been proposed to investigate the stochastic optimization of NN parameters with information–theoretic quantities, e.g., [4,5], and the IB functional was used as a cost function for NN training [6,7].

Based on this increased attention, this Special Issue aims to investigate the properties of the IB functional in this new context and to propose learning mechanisms inspired by the IB framework. More specifically, we invited authors to submit manuscripts that provide novel insight into the properties of the IB functional that apply the IB principle for training deep, i.e., multi-layer machine learning structures such as NNs and that investigate the learning behavior of NNs using the IB framework. To cover the breadth of the current literature, we also solicited manuscripts that discuss frameworks inspired by the IB principle, but that depart from them in a well-motivated manner.

In the remainder of this Editorial, we provide a brief summary of the papers in this Special Issue, in order of their appearance.

- Kunze et al. show that maximizing the evidence lower bound with a factorized Gaussian approximate posterior effectively limits mutual information between the available data and the learned parameters [8]. The effect of this tunable "model capacity" is validated in supervised and unsupervised settings, illustrating intuitive connections with overfitting, the NN architecture, and the dataset size;
- Wu et al. investigate the learnability within the IB framework. They show that if the parameter β in (1) falls below a certain threshold β_0, then a trivial representation T that is independent of X and Y minimizes the IB functional [9]. This threshold depends on the joint distribution of X and Y, and the authors propose an algorithm to estimate β_0 for a given dataset;

- Ngyuen and Choi argue that every layer in a feedforward NN should be optimized w.r.t. the IB functional (1) separately, with the parameter β adapted to the layer index [10]. Proposing a cost function for this multiobjective optimization problem, a computable variational bound, and a greedy optimization procedure, they achieve superior accuracy and adversarial robustness in stochastic binary NNs;
- Kolchinsky et al. propose an NN-based implementation of the IB problem, i.e., the compression scheme $P_{T|X}$ and conditional distribution $P_{Y|T}$ are parameterized by NNs [11]. Acknowledging the issues in [12], these NNs are trained to minimize an upper bound on $(I(X;T))^2 - \beta I(Y;T)$, combining variational and non-parametric approaches for bounding. Their experiments yield a better trade-off between $I(X;T)$ and $I(Y;T)$ and more meaningful latent representations in the bottleneck layer than a corresponding reformulation of [6];
- Tegmark and Wu investigate binary classification from real-valued observations [13]. They show that the observations can be compressed to a discrete representation T parameterized by β in such a way that the Pareto frontier of (1) is swept, essentially characterizing the binary classification problem. The authors further show that the corner points of this Pareto frontier, corresponding to a maximization of $I(Y;T)$ for a given alphabet size of T, can be computed without multiobjective optimization;
- Rodríguez Gálvez et al. discuss the scenario in which the target Y is a deterministic function of X in [14]. In this case, it is known that sweeping the paramter β in (1) is not sufficient to sweep the Pareto frontier of optimal $(I(X;T), I(Y;T))$ pairs [12]. The authors show that this shortcoming can be removed by optimizing $u(I(X;T)) - \beta_u I(Y;T)$ instead, where u is a strictly convex function. Furthermore, the authors demonstrate that the particular choice of the strictly convex function u helps to obtain a desired value of $I(X;T)$ over a wide range of parameters β_u;
- Franzese and Visintin propose using the IB functional as a cost function to train ensembles of decision trees for classification [15]. The authors show that these ensembles perform similarly to bagged trees, while they outperform the naive Bayes and k-nearest neighbor classifiers;
- Jónsson et al. [16] investigate the learning behavior of a high-dimensional VGG-16 convolutional NN in the information plane. Using MINE [17] to estimate $I(X;T)$ and $I(Y;T)$ throughout training, the authors observed a separate compression phase, during which the estimate of $I(X;T)$ decreases, thus aligning with [4]. The authors further show that regularizing NN training via an MINE-based estimate of the compression term $I(X;T)$ yields improved classification performance;
- Voloshynovskiy et al. propose an IB-based framework for semi-supervised classification, considering variational bounds both with learned and hand-crafted marginal distributions and achieving competitive performance [18]. A close investigation of their cost function yields improved insight into previously proposed approaches to semi-supervised classification;
- Fischer formulates the principle of minimum necessary information and derives from it the conditional entropy bottleneck functional [19]. This functional is mathematically equivalent to the IB functional, but uses the chain rule of mutual information to replace $I(X;T)$ in (1) by $I(X;T|Y)$. This results in different variational bounds, which are shown to yield better classification accuracy, improved robustness to adversarial examples, and stronger out-of-distribution detection than deterministic models or models based on variational approximations of (1), cf. [6];
- Fischer and Alemi provide additional empirical evidence for the claims in [19]. Specifically, they show that optimizing the proposed variational bounds leads to improved robustness against targeted and untargeted projected gradient descent attacks and to common corruptions (cf. [20]) of the ImageNet data [21]. Furthermore, the authors indicate that the conditional entropy bottleneck functional yields improved calibration for both clean and corrupted test data;
- Geiger and Fischer investigate the variational bounds proposed in [6,19]. While the underlying IB and conditional entropy bottleneck functionals are equivalent, the authors show that the variational bounds are not; these bounds are generally unordered, but an ordering can be enforced by restricting the feasible sets appropriately [22]. Their analysis is valid for general optimization and does not rely on the assumption that the variational bounds are implemented using NNs.

We thank all the authors for their excellent contributions and timely submission of their works. We are looking forward to many future developments that will build on the current bounty of insightful results and that will make machine learning better explainable.

Funding: The work of Bernhard C. Geiger was supported by the iDev40 project and by the COMET programs within the Know-Center and the K2 Center "Integrated Computational Material, Process and Product Engineering (IC-MPPE)" (Project No 859480). The iDev40 project has received funding from the ECSEL Joint Undertaking (JU) under Grant Agreement No 783163. The JU receives support from the European Union's Horizon 2020 research and innovation program. It is co-funded by the consortium members, grants from Austria, Germany, Belgium, Italy, Spain, and Romania. The COMET programs are supported by the Austrian Federal Ministry for Climate Action, Environment, Energy, Mobility, Innovation and Technology, the Austrian Federal Ministry of Digital and Economic Affairs, and by the States of Styria, Upper Austria, and the Tyrol. COMET is managed by the Austrian Research Promotion Agency FFG.

Acknowledgments: We would like to express our gratitude to the Editorial Assistants of Entropy for their help in organizing this Special Issue.

Conflicts of Interest: The authors declare no conflict of interest.

References

1. Tishby, N.; Pereira, F.C.; Bialek, W. The Information Bottleneck Method. In Proceedings of the Allerton Conference on Communication, Control, and Computing, Monticello, IL, USA, 22–24 September 1999; pp. 368–377.
2. Zaidi, A.; Estella-Aguerri, I.; Shamai (Shitz), S. On the Information Bottleneck Problems: Models, Connections, Applications and Information Theoretic Views. *Entropy* **2020**, *22*, 151. [CrossRef] [PubMed]
3. Goldfeld, Z.; Polyanskiy, Y. The Information Bottleneck Problem and Its Applications in Machine Learning. *IEEE J. Sel. Areas Inf. Theory* **2020**, *1*, 19–38. [CrossRef]
4. Shwartz-Ziv, R.; Tishby, N. Opening the Black Box of Deep Neural Networks via Information. *arXiv* **2017**, arXiv:1703.00810.
5. Geiger, B.C. On Information Plane Analyses of Neural Network Classifiers—A Review. *arXiv* **2020**, arXiv:2003.09671.
6. Alemi, A.A.; Fischer, I.; Dillon, J.V.; Murphy, K. Deep Variational Information Bottleneck. In Proceedings of the International Conference on Learning Representations (ICLR), Toulon, France, 24–26 April 2017.
7. Achille, A.; Soatto, S. Information Dropout: Learning Optimal Representations Through Noisy Computation. *IEEE Trans. Pattern Anal. Mach. Intell.* **2018**, *40*, 2897–2905. [CrossRef] [PubMed]
8. Kunze, J.; Kirsch, L.; Ritter, H.; Barber, D. Gaussian Mean Field Regularizes by Limiting Learned Information. *Entropy* **2019**, *21*, 758. [CrossRef] [PubMed]
9. Wu, T.; Fischer, I.; Chuang, I.L.; Tegmark, M. Learnability for the Information Bottleneck. *Entropy* **2019**, *21*, 924. [CrossRef]
10. Nguyen, T.T.; Choi, J. Markov Information Bottleneck to Improve Information Flow in Stochastic Neural Networks. *Entropy* **2019**, *21*, 976. [CrossRef]
11. Kolchinsky, A.; Tracey, B.D.; Wolpert, D.H. Nonlinear Information Bottleneck. *Entropy* **2019**, *21*, 1181. [CrossRef]
12. Kolchinsky, A.; Tracey, B.D.; Van Kuyk, S. Caveats for information bottleneck in deterministic scenarios. In Proceedings of the International Conference on Learning Representations (ICLR), New Orleans, LA, USA, 6–9 May 2019.
13. Tegmark, M.; Wu, T. Pareto-Optimal Data Compression for Binary Classification Tasks. *Entropy* **2020**, *22*, 7. [CrossRef] [PubMed]
14. Rodríguez Gálvez, B.; Thobaben, R.; Skoglund, M. The Convex Information Bottleneck Lagrangian. *Entropy* **2020**, *22*, 98. [CrossRef] [PubMed]
15. Franzese, G.; Visintin, M. Probabilistic Ensemble of Deep Information Networks. *Entropy* **2020**, *22*, 100. [CrossRef] [PubMed]
16. Jónsson, H.; Cherubini, G.; Eleftheriou, E. Convergence Behavior of DNNs with Mutual-Information-Based Regularization. *Entropy* **2020**, *22*, 727. [CrossRef] [PubMed]

17. Belghazi, M.I.; Baratin, A.; Rajeshwar, S.; Ozair, S.; Bengio, Y.; Courville, A.; Hjelm, D. Mutual Information Neural Estimation. In Proceedings of the International Conference on Machine Learning (ICML), Stockholm, Sweden, 10–15 July 2018; pp. 531–540.
18. Voloshynovskiy, S.; Taran, O.; Kondah, M.; Holotyak, T.; Rezende, D. Variational Information Bottleneck for Semi-Supervised Classification. *Entropy* **2020**, *22*, 943. [CrossRef] [PubMed]
19. Fischer, I. The Conditional Entropy Bottleneck. *Entropy* **2020**, *22*, 999. [CrossRef] [PubMed]
20. Hendrycks, D.; Dietterich, T. Benchmarking Neural Networks Robustness to Common Corruptions and Perturbations. In Proceedings of the International Conference on Learning Representations (ICLR), New Orleans, LA, USA, 6–9 May 2019.
21. Fischer, I.; Alemi, A.A. CEB Improves Model Robustness. *Entropy* **2020**, *22*, 1081. [CrossRef] [PubMed]
22. Geiger, B.C.; Fischer, I.S. A Comparison of Variational Bounds for the Information Bottleneck Functional. *Entropy* **2020**, *22*, 1229. [CrossRef] [PubMed]

Publisher's Note: MDPI stays neutral with regard to jurisdictional claims in published maps and institutional affiliations.

© 2020 by the authors. Licensee MDPI, Basel, Switzerland. This article is an open access article distributed under the terms and conditions of the Creative Commons Attribution (CC BY) license (http://creativecommons.org/licenses/by/4.0/).

Article

Gaussian Mean Field Regularizes by Limiting Learned Information

Julius Kunze [1,*], Louis Kirsch [1,2], Hippolyt Ritter [1] and David Barber [1,3]

1. Computer Science, University College London, London WC1E 6BT, UK
2. The Swiss AI Lab (IDSIA), University of Lugano (USI) & University of Applied Sciences of Southern Switzerland (SUPSI), 6928 Manno, Switzerland
3. Alan Turing Institute, London NW1 2DB, UK
* Correspondence: juliuskunze@gmail.com

Received: 14 June 2019; Accepted: 1 August 2019; Published: 3 August 2019

Abstract: Variational inference with a factorized Gaussian posterior estimate is a widely-used approach for learning parameters and hidden variables. Empirically, a regularizing effect can be observed that is poorly understood. In this work, we show how mean field inference improves generalization by limiting mutual information between learned parameters and the data through noise. We quantify a maximum capacity when the posterior variance is either fixed or learned and connect it to generalization error, even when the KL-divergence in the objective is scaled by a constant. Our experiments suggest that bounding information between parameters and data effectively regularizes neural networks on both supervised and unsupervised tasks.

Keywords: information theory; variational inference; machine learning

1. Introduction

Bayesian machine learning is a popular framework for dealing with uncertainty in a principled way by integrating over model parameters rather than finding point estimates [1–3]. Unfortunately, exact inference is usually not feasible due to the intractable normalization constant of the posterior. A popular alternative is variational inference [4], where a tractable approximate distribution is optimized to resemble the true posterior as closely as possible. Due to its amenability to stochastic gradient descent [5–8], variational inference is scalable to large models and datasets.

The most common choice for the variational posterior is a factorized Gaussian. Outside of Bayesian inference, parameter noise has been found to be an effective regularizer [9–11], e.g., for training neural networks. In combination with L2-regularization, additive Gaussian parameter noise corresponds to variational inference with a Gaussian approximate posterior with fixed variance. Interestingly, it has been observed that flexible posteriors can perform worse than simple ones [12–15].

Variational inference follows the Minimum Description Length (MDL) principle [16–18], a formalization of Occam's Razor. Loosely speaking, it states that of two models describing the data equally well, the "simpler" one should be preferred. However, MDL is only an objective for compressing the training data and the model and makes no formal statement about generalization to unseen data. Yet, generalization to new data is a key property of a machine learning algorithm.

Recent work [19–22] has proposed upper bounds on the generalization error as a function of the mutual information between model parameters and training data. It states that the gap between training and test error can be reduced by limiting the mutual information. However, to the best of our knowledge, these bounds and specific inference methods have so far not been linked.

In this work, we show that Gaussian mean field inference in models with Gaussian priors can be reinterpreted as point estimation in corresponding noisy models. This leads to an upper bound on the mutual information between model parameters and data through the data processing inequality. Our result holds for both supervised and unsupervised models. We discuss the connection to generalization bounds from Xu and Raginsky [19] and Bu et al. [20], suggesting that the Gaussian mean field aids generalization. In our experiments, we show that limiting model capacity via mutual information is an effective measure of regularization, further supporting our theoretical framework.

2. Regularization through the Mean Field

In our derivation, we denote a generic model as $p(\theta, D) = p(\theta)p(D \mid \theta)$ with unobserved variables θ and data D. We refer to θ as the model parameters; however, in latent variable models, θ can also include the per-data point latent variables. The model consists of a prior $p(\theta)$ and a likelihood $p(D \mid \theta)$. Ideally, one would like to find the posterior $p(\theta \mid D) = p(D \mid \theta)p(\theta)/Z$, where $Z = \int p(D \mid \theta)p(\theta)d\theta$ is the normalizer. However, calculating Z is typically intractable. Variational inference finds an approximation by maximizing the Evidence Lower Bound (ELBO)

$$\begin{aligned}\log p(D) &\geq \log p(D) - D_{\text{KL}}\left(q(\theta)||p(\theta \mid D)\right) \\ &= \mathbb{E}_{q(\theta)} \log p(D \mid \theta) - D_{\text{KL}}\left(q(\theta)||p(\theta)\right)\end{aligned} \quad (1)$$

w.r.t. the approximate posterior $q(\theta)$. Our focus in this work lies on Gaussian mean field inference, so q is a fully-factorized normal distribution with a learnable mean μ and variance σ^2. The prior is also chosen to be component-wise independent $p(\theta) = \mathcal{N}\left(0, \sigma_p^2 I\right)$. The generative and inference models for this setting are shown in Figure 1a.

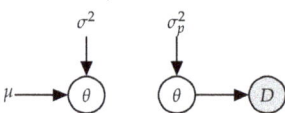

(a) Original model

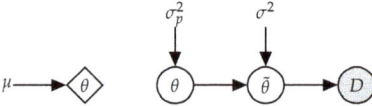

(b) Noisy model: fixed variance

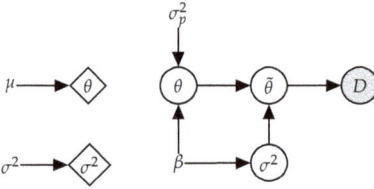

(c) Noisy model: learned variance

Figure 1. Gaussian mean field inference on model parameters θ with a Gaussian prior (a) can be reinterpreted as optimizing a point estimate on a model with injected noise, both when variance is fixed (b) and learned (c). For the latter case, we show this for the more general case where the complexity term in the objective is scaled by a constant $\beta > 0$, with $\beta = 1$ recovering variational inference.

2.1. Fixed-Variance Gaussian Mean Field Inference

When the variance σ^2 of the approximate posterior is fixed, the ELBO can be written as:

$$\mathbb{E}_{\theta \sim \mathcal{N}(\mu, \sigma^2 I)} \log p(D \mid \theta) - \frac{1}{2\sigma_p^2} \sum_i \mu_i^2 + c \tag{2}$$

which is optimized with respect to μ. We use $i \in \{1, \ldots, K\}$ to denote the parameter index and c for constant terms.

To show how the Gaussian mean field implicitly limits learned information, we extend the model with a noisy version of the parameters $\tilde{\theta} \sim p(\tilde{\theta} \mid \theta)$ and let the likelihood depend on those noisy parameters. We choose the noise distribution to be the same as the inference distribution for the original model and find a lower bound on the log-joint of the noisy model. This leads to the same objective as mean field variational inference in the original model.

Specifically, we define the noisy model $p'(\theta, \tilde{\theta}, D) = p'(\theta) p'(\tilde{\theta} \mid \theta) p'(D \mid \tilde{\theta})$ as visualized in Figure 1b. We use p' to emphasize the distinction between distributions of the modified noisy model and the original one. As in the original model, $\theta \sim \mathcal{N}\left(0, \sigma_p^2 I\right)$ represents the parameters (with the same prior), i.e., $p(\theta) = p'(\theta)$. We denote the noisy parameters as $\tilde{\theta} \sim \mathcal{N}(\theta, \sigma^2)$. The likelihood remains unchanged, i.e., $p'(D \mid \tilde{\theta}) = p(D \mid \theta)$, except that it now depends on the noisy parameters instead of the "clean" ones.

We now show that maximizing a lower bound on the log-joint probability of the noisy model results in an identical objective as for variational inference in the clean model

$$\log p'(D, \theta) \tag{3}$$

$$= \log \int p'(D \mid \tilde{\theta}) p'(\tilde{\theta} \mid \theta) \, d\tilde{\theta} + \log p'(\theta) \tag{4}$$

$$\geq \mathbb{E}_{\tilde{\theta} \sim \mathcal{N}(\theta, \sigma^2 I)} \log p'(D \mid \tilde{\theta}) - \frac{1}{2\sigma_p^2} \sum_i \theta_i^2 + c \tag{5}$$

$$= \mathbb{E}_{\tilde{\theta} \sim \mathcal{N}(\mu, \sigma^2 I)} \log p'(D \mid \tilde{\theta}) - \frac{1}{2\sigma_p^2} \sum_i \mu_i^2 + c \tag{6}$$

where Equation (5) follows from Jensen's inequality as in Equation (1). In the final equation, we have replaced θ with μ (which is simply a change of names, since we are maximizing the objective over this free variable) to emphasize that the objective functions are identical.

Since D is independent of θ given $\tilde{\theta}$, the joint $p(\theta, \tilde{\theta}, D)$ forms a Markov chain, and the data processing inequality [23] limits the mutual information $I(D, \theta)$ between learned parameters and data through:

$$I(D, \theta) \leq I(\tilde{\theta}, \theta) \tag{7}$$

The upper bound is given by:

$$I(\tilde{\theta}, \theta) = H(\tilde{\theta}) - H(\tilde{\theta} \mid \theta) = \frac{K}{2} \log \left(1 + \frac{\sigma_p^2}{\sigma^2}\right) \tag{8}$$

where K denotes the number of parameters. Here, we exploit that θ and $\tilde{\theta} \mid \theta$ are Gaussian with $H(\tilde{\theta}) = \frac{K}{2} \log 2\pi e \left(\sigma^2 + \sigma_p^2\right)$ and $H(\tilde{\theta} \mid \theta) = \frac{K}{2} \log 2\pi e \sigma^2$. This quantity is known as the capacity of channels with Gaussian noise in signal processing [23]. Intuitively, a high prior variance σ_p^2 corresponds to a large capacity, while a high noise variance σ^2 reduces it. Any desired capacity can be achieved by simply adjusting the signal-to-noise ratio σ_p^2 / σ^2.

2.2. Generalization Error vs. Limited Information

Intuitively, we characterize overfitting as learning too much information about the training data, suggesting that limiting the amount of information extracted from the training data into the hypothesis should improve generalization. This idea was recently formalized by Xu and Raginsky [19], Bu et al. [20], Bassily et al. [21], Russo and Zou [22], showing that limiting mutual information between data and learned parameters bounds the expected generalization error under certain assumptions.

Specifically, their work characterizes the following process: Assume that our training dataset is sampled from a true distribution $p_t(D)$. Based on this training set, a learning algorithm subsequently returns a distribution over hypotheses given by $p_t(\theta \mid D)$. The process defines mutual information $I_t(D, \theta)$ on the joint distribution $p_t(D, \theta) = p_t(D)p_t(\theta \mid D)$. Under certain assumptions on the loss function, Xu and Raginsky [19] derived a bound on the generalization error of the learning algorithm in expectation over this sampling process. Bu et al. [20] relaxed the condition on the loss and proved the applicability to a simple estimation algorithm involving L2-loss.

Exact Bayesian inference returns the true posterior $p(\theta \mid D)$ on a model $p(\theta, D)$. The theorem then states that a bound on $I(D, \theta)$ limits the expected generalization error as described in Bu et al. [20] if the model captures the nature of the generating process in the marginal $p(D) = \int d\theta p(\theta) p(D \mid \theta)$. This is a common assumption necessary to justify any (variational) Bayesian approach.

Exact inference is intractable on deep models, and instead, one typically learns variational or point estimates for the posterior. That is also true for the objective on the noisy model above, where we only used a point estimate as given by Equation (6). Therefore, the assumption of exact inference is not met. Yet, we believe that those bounds motivate the expectation that variational inference aids generalization by limiting the learned information. If we performed exact inference on the noisy model in the last section, the given mutual information would imply a bound on generalization error as implied by Xu and Raginsky [19] and Bu et al. [20]. Therefore, we are optimistic that the gap between variational inference and those generalization bounds can be closed either by performing more accurate inference in the noisy model or by taking the dynamics of the training algorithm into account when bounding mutual information (see Section 5.2 for further discussion).

2.3. Learned-Variance Gaussian Mean Field Inference

The variance in Gaussian mean field inference is typically learned for each parameter [8,24,25]. Similar to when the variance in the approximate posterior is fixed, one can obtain a capacity constraint. This is the case even for a generalization of the objective from Equation (1) where the divergence term $D_{\text{KL}}(q(\theta) \| p(\theta))$ is scaled by some factor $\beta > 0$. Higgins et al. [26] proposed using $\beta > 1$ to learn "disentangled" representations in variational autoencoders. Further, β is commonly annealed from 0–1 for expressive models (e.g., Blundell et al. [25], Bowman et al. [27], Sønderby et al. [28].) In the following, we quantify a general capacity depending on β, where $\beta = 1$ recovers the standard variational objective. For notational simplicity, we here assume a prior variance of $\sigma_p^2 = 1$. It is straight-forward to adapt the derivation to the general case.

In this case, the objective can be written as:

$$\mathbb{E}_{\theta \sim \mathcal{N}(\mu, \sigma^2)} \log p(D \mid \theta) + \frac{\beta}{2} \sum_i \left(\log \sigma_i^2 - \sigma_i^2 - \mu_i^2 - 1 \right) \tag{9}$$

where now, both μ and σ^2 represent learned vectors and $\mathcal{N}(\mu, \sigma^2)$ denotes a variable composed of pairwise independent Gaussian components with means and variances given by the elements of μ and σ^2.

Similar to the previous section, we show a lower bound on the log-joint of a new noisy model to be identical to Equation (9). Specifically, we define the noisy model $p'(\theta, \sigma^2, \tilde{\theta}, D) = p'(\theta)p'(\sigma^2)p'(\tilde{\theta} \mid \theta, \sigma^2)p'(D \mid \tilde{\theta})$ (Figure 1c), with independent priors $\theta_i \sim \mathcal{N}\left(0, \frac{1}{\beta}\right)$ and $\sigma_i^2 \sim \Gamma\left(\frac{\beta}{2} + 1, \frac{\beta}{2}\right)$, where

$\Gamma(\cdot, \cdot)$ denotes the Gamma distribution. As previously done Section 2.1, we define the noise-injected parameters as $\tilde{\theta} \sim \mathcal{N}(\theta, \sigma^2)$ and likelihood as $p'(D \mid \tilde{\theta}) = p(D \mid \theta)$.

The priors are chosen so that with Jensen's inequality, we find a lower bound on the log-joint probability of this model that recovers the objective from Equation (9):

$$\begin{aligned}
&\log p'(D, \theta, \sigma^2) \\
&= \log \int p'(D \mid \tilde{\theta}) p'(\tilde{\theta} \mid \theta, \sigma^2) \, d\tilde{\theta} + \log p'(\theta) + \log p'(\sigma^2) \\
&\geq \mathbb{E}_{\tilde{\theta} \sim \mathcal{N}(\theta, \sigma^2)} \log p'(D \mid \tilde{\theta}) + \sum_i \left(\log p'(\theta_i) + \log p'(\sigma_i^2) \right) \\
&= \mathbb{E}_{\tilde{\theta} \sim \mathcal{N}(\mu, \sigma^2)} \log p'(D \mid \tilde{\theta}) + \frac{\beta}{2} \sum_i \left(\log \sigma_i^2 - \sigma_i^2 - \mu_i^2 \right) + c
\end{aligned} \quad (10)$$

In the noisy model, the data processing inequality and the independence of dimensions implies a bound:

$$I(D, (\theta, \sigma^2)) \leq I(\tilde{\theta}, (\theta, \sigma^2)) = \sum_i I(\tilde{\theta}_i, (\theta_i, \sigma_i^2)) \quad (11)$$

where the capacity $I(\theta_i, (\theta_i, \sigma_i^2))$ per dimension is derived in Appendix A.

Figure 2 shows numerical results for various values of β. Standard variational inference ($\beta = 1$) results in a capacity of 0.45 bits per dimension. We observe that higher β corresponds to smaller capacity, which is given by the mutual information $I(\tilde{\theta}_i, (\theta_i, \sigma_i^2))$ between our new latent (θ_i, σ_i^2) and $\tilde{\theta}_i$. This formalizes the intuition that a higher weight of the complexity term in our objective increases regularization by decreasing a limit on the capacity.

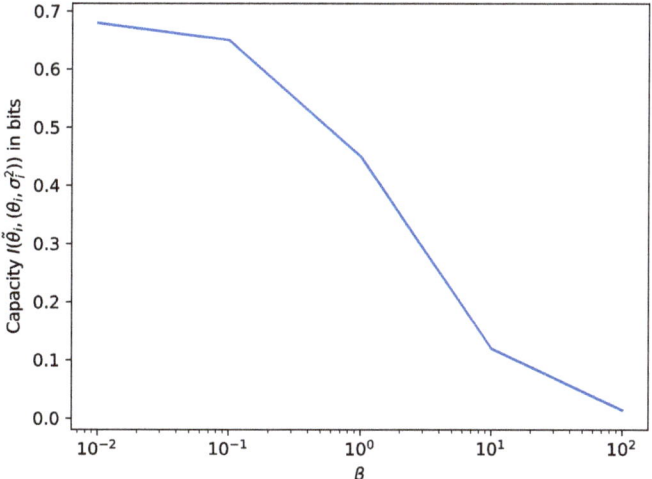

Figure 2. Relationship between β and capacity $I(\tilde{\theta}_i, (\theta_i, \sigma_i^2))$ per parameter dimension in Gaussian mean field inference with learned variance and the complexity term scaled by $\beta > 0$.

2.4. Supervised and Unsupervised Learning

The above derivations apply to any learning algorithm that is purely trained with Gaussian mean field inference. This covers supervised and unsupervised tasks.

In supervised learning, the training data typically consist of pairs of inputs and labels, and a loss is assigned to each pair that depends on the trained model, e.g., neural network parameters. When all parameters are learned with one of the discussed mean field methods, the given bounds apply.

The derivation also comprises unsupervised methods with per-data latent variables and even amortized inference such as variational autoencoders [8,24], again as long as all learned variables are learned via Gaussian mean field inference. While this might be helpful to find generalizing representations, the focus of the experiments is on validating the generalizing behaviour of the Bayesian mean field variational approach on neural network parameters for overfitting regimes, namely a small dataset and complex models.

2.5. Flexible Variational Distributions

The objective function for variational inference is maximized when the approximate posterior is equal to the true one. This motivates the development of flexible families of posterior distributions [8,29–36]. In the case of exact inference, a bound on generalization as discussed in Section 2.2 only applies if the model itself has finite mutual information between data and parameters. However, estimating mutual information is generally a hard problem, particularly in high-dimensional, non-linear models. This makes it hard to state a generic bound, which is why we focus on the case of Gaussian mean field inference.

3. Related Work

3.1. Regularization in Neural Networks

The Gaussian mean field is intimately related to other popular regularization approaches in deep learning: As is apparent from Equation (6), the fixed-variance Gaussian mean field applied to training neural network weights is equivalent to L2-regularization (weight decay) combined with Gaussian parameter noise [9–11] on all network weights. Molchanov et al. [37] showed that additive parameter noise results in multiplicative noise on the unit activations. The resulting dependencies between noise components on the layer output can be ignored without significantly changing empirical results [38]. This is in turn equivalent to scaled Gaussian dropout [24].

3.2. Information Bottlenecks

The information bottleneck principle by Tishby et al. [39], Shamir et al. [40] aims to find a representation Z of some input X that is most useful to predict an output Y. For this purpose, the objective is to maximize the amount of information $I(Y, Z)$ the representation contains about the output under a bounded amount of information $I(X, Z)$ about the input:

$$\max_{I(X,Z)<C} I(Y,Z) \tag{12}$$

They described a training procedure using the softly-constrained objective:

$$\min \mathcal{L}_{IB} = \min I(X,Z) - \beta I(Y,Z) \tag{13}$$

where $\beta > 0$ controls the trade-off.

Alemi et al. [41] suggested a variational approximation for this objective. For the task of reconstruction, where labels Y are identical to inputs X, this results exactly in the β-VAE objective [42,43]. This is in accordance with our result from Section 2.3 that there is a maximum capacity per latent dimension that decreases for higher β. Setting $\beta > 1$, as suggested by Higgins et al. [26], for obtaining disentangled representations, corresponds to lower capacity per latent component than achieved by standard variational inference.

Both Tishby et al. [39] and Higgins et al. [26] introduced β as a trade-off parameter without a quantitative interpretation. With our information-theoretic perspective, we quantify the implied capacity and provide a link to the generalization error. Further, both methods are concerned with the information in the latent representation. They do not limit the mutual information with the model parameters, leaving them vulnerable to model overfitting under our theoretical assumptions. We experimentally validated this vulnerability and explore the effect of filling this gap by applying Gaussian mean field inference to the model parameters.

3.3. Information Estimation with Neural Networks

Multiple recent techniques [44–46] proposed the use of neural networks for obtaining a lower bound on the mutual information. This is useful in settings when we want to maximize mutual information, e.g., between the data and a lower-dimensional representation. In contrast, we show that Gaussian variational inference on variables with a Gaussian prior implicitly places an upper bound on the mutual information between these variables and the data and explore its regularizing effect.

4. Experiments

In this section, we analyse the implications of applying Gaussian mean field inference of a fixed scale to the model parameters in the supervised and unsupervised context. Our theoretical results suggest that varying the capacity will affect the generalization capability, and we show this effect on small data regimes and how it changes with the training set size. Furthermore, we investigate whether capacity is the only predictor for generalization or whether varying priors and architectures also have an effect. Finally, we demonstrate qualitatively how the capacity bounds are reflected in Fashion MNIST reconstruction.

4.1. Supervised Learning

We begin with a supervised classification task on the CIFAR10 dataset, training only on a subset of the first 5000 samples. We used 63×3 convolutional layers with 128 channels each followed by a ReLU activation function, every second of which implemented striding of 2 to reduce the input dimensionality. Finally, the last layer was a linear projection, which parameterized a categorical distribution. The capacity of each parameter in this network was set to specific values given by Equation (8).

Figure 3 shows that decreasing the model capacity per dimension (by increasing the noise) reduced the training log-likelihood and increased the test log-likelihood until both of them meet at an optimal capacity. One can observe that very small capacities led to a signal that was too noisy, and good predictions were no longer possible. In short, regimes of underfitting and overfitting were generated depending on the capacity.

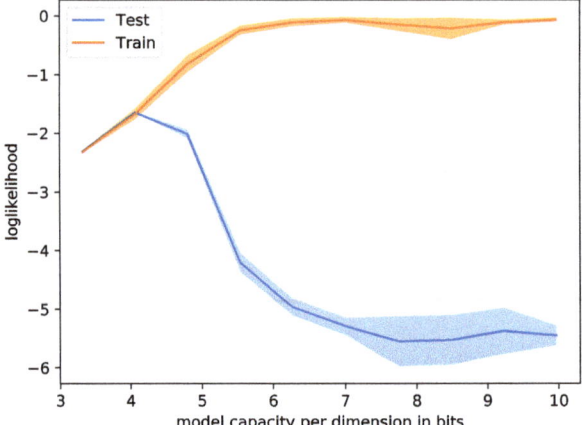

Figure 3. Classifying CIFAR10 with varying model capacities. Large capacities lead to overfitting, while small capacities drown the signal in noise. Each configuration was evaluated 5 times; the mean and standard deviation are displayed.

4.2. Unsupervised Learning

We now evaluate the regularizing effect of fixed-scale Gaussian mean field inference in an unsupervised setting for image reconstruction on the MNIST dataset. Therefore, we used a VAE [6] with 2 latent dimensions and a 3-layer neural network parameterizing the conditional factorized Gaussian distribution. As usual, it was trained using the free energy objective, but different from the original work, we also used Gaussian mean field inference for the model parameters. Again, we used a small training set of 200 examples for the following experiments if not denoted otherwise.

4.2.1. Varying Model Capacity and Priors

In our first experiment, we analysed generalization by inspecting the test evidence lower bound (ELBO) when varying the model capacity, which can be seen in Figure 4a. Similar to the supervised case, we can observe that there was a certain model capacity range that explained the data very well, while less or more capacity resulted in noise drowning and overfitting, respectively. In the same figure, we also investigated whether the information-theoretic model capacity can predict generalization independently of the specific prior distribution. Since we merely state an upper bound on mutual information in Section 2.1, the prior may have an effect in practice, which cannot be explained only by the capacity. Figure 4a shows that indeed, while the general behaviour remained the same for different model priors, the generalization error was not entirely independent. Furthermore, the observation that all curves descended with larger capacities, for all priors, suggests that weight decay [47] of fixed scale without parameters noise was not sufficient to regularize arbitrarily large networks. In Figure 4b, we investigated the extreme case of dropping the prior entirely and switching to maximum-likelihood learning instead by using an improper uniform prior. This approach recovered Gaussian dropout [24,48]. Dropping the prior set the bottleneck capacity to infinity and should lead to worse generalization. Comparing the test ELBO of this Gaussian dropout variant to the original Gaussian mean field inference in Figure 4b confirmed this result for larger capacities. For larger noise scales, generalization was still working well, a result that was not explained in our information-theoretic framework, but plausible due to the deployed limited architecture.

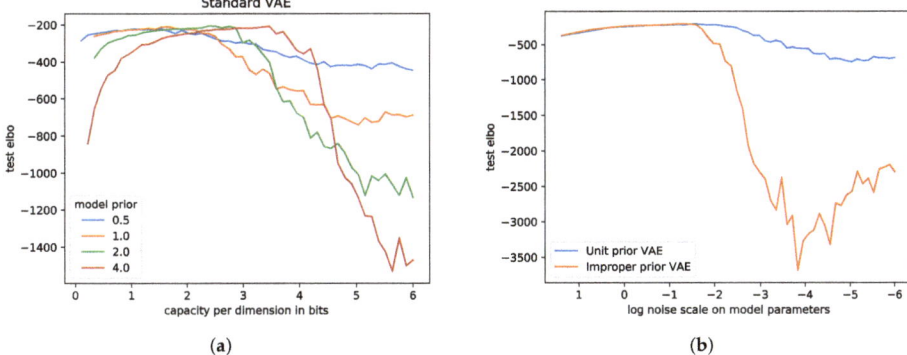

Figure 4. MNIST test reconstruction with a variational autoencoder training on 200 samples for various capacities and Gaussian priors. (**a**) The test Evidence Lower Bound (ELBO) is not invariant when varying the prior on the model parameters. Nevertheless, the first increasing and then decreasing trend when changing the capacity remains; (**b**) Using an improper prior, similar to just using Gaussian dropout on the weights, leads to an accelerated decrease of generalization for smaller noise scales.

4.2.2. Varying Training Set Size

Figure 5a shows how limiting the capacity affects the test ELBO for varying amounts of training data. Models with very small capacity extracted less information from the data into the model, thus yielding a good test ELBO somewhat independent of the dataset size. This is visible as a graph that ascends very little with more training data (e.g., total model capacity of 330 kbits). Note that we here report the capacity of the entire model, which is the sum of the capacities for each parameter. In order to improve the test ELBO, more information from the data had to be extracted into the model. However, clearly, this led to non-generalizing information being extracted when the dataset was small, leading to overfitting. Only for larger datasets, the extracted information generalized. This is visible as a strongly-ascending test ELBO with larger dataset sizes and bad generalization for small datasets. We can therefore conclude that the information bottleneck needs to be chosen based on the amount of data that are available. Intuitively, when more information is available, the more information should be extracted into the model.

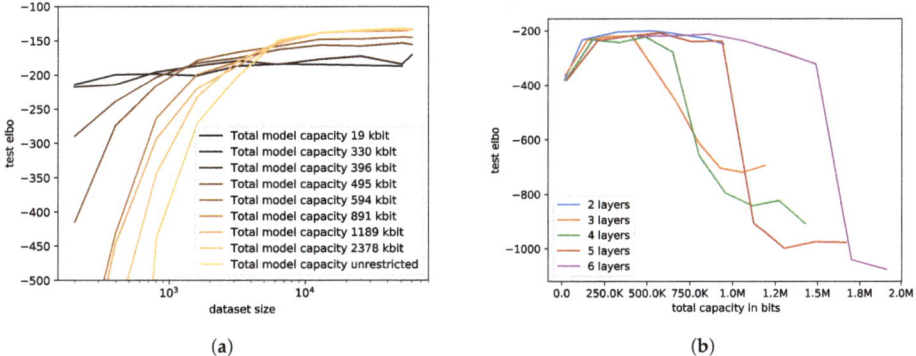

Figure 5. MNIST test reconstruction with a VAE; training on varying dataset sizes, architectures, and model capacities. (**a**) Varying the number of samples. Depending on the size of the dataset, higher capacities of the model are required to fit all the data points; (**b**) Varying architecture. Overfitting is not getting worse for more layers if the capacity is low enough. More layers do overfit only for higher capacities.

4.2.3. Varying Model Size

Furthermore, we inspected how the size of the model (here, in terms of the number of layers) affected generalization in Figure 5b. Similar to varying the prior distribution, we were interested in how well the total capacity predicted generalization and the role the architecture plays. It can be observed that larger networks were more resilient to larger total capacities before they started overfitting. This indicates that the total capacity was less important than the individual capacity (i.e., noise) per parameter. Nevertheless, larger networks were more prone to overfitting for very large model capacities. This makes sense as their functional form was less constrained, an aspect that was not captured by our theory.

4.2.4. Qualitative Reconstruction

Finally, we plot test reconstruction means for the binarized Fashion MNIST dataset under the same setup for various capacities in Figure 6. In accordance with the previous experiments, we observed that if the capacity was chosen too small, the model was not learning anything useful, while too large capacities resulted in overconfidence. This can be observed in most means being close to either 0–1. An intermediate capacity, on the other hand, made sensible predictions (given that it was trained only on 200 samples) with sensible uncertainty, visible through grey pixels that correspond to high entropy.

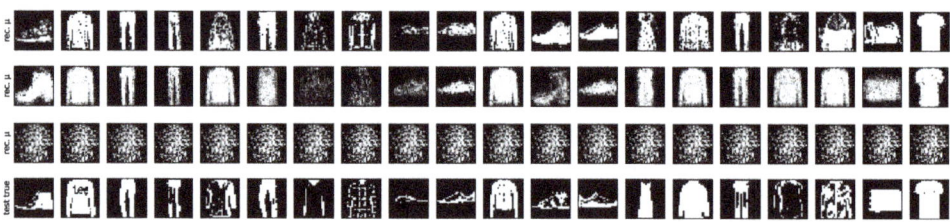

Figure 6. Test reconstruction means for binarized Fashion MNIST trained on 200 samples with per-parameter capacities of 5, 2, and 1 bits (**top**) compared to the true data (**bottom**).

5. Discussion

In this section, we discuss how the capacity can be set, as well as the effect of the model architecture and learning dynamics.

5.1. Choosing the Capacity

We have obtained a new trade-off parameter, the capacity, that has a simple quantitative interpretation: It determines how many bits to extract maximally from the training set. In contrast, for the β parameter introduced in Tishby et al. [39] and Higgins et al. [26], a clear interpretation is not known. Yet, it may still be hard to set the capacity optimally. Simple mechanisms such as evaluation of a validation set to determine its value may be used. We expect that more theoretically-rigorous methods could be developed.

Furthermore, in this paper, we focused on the regularization that Gaussian mean field inference implies on the model parameters. The same concept is valid for data-dependent latent variables, for instance in VAEs, as discussed in Section 2.4. In VAEs, Gaussian mean field inference on the latent variables leads to a restricted latent capacity, but leaves the capacity of the model unbounded. This leaves VAEs vulnerable to model overfitting, as demonstrated in the experiments, and setting β as done in Higgins et al. [26] is not sufficient to control complexity. This motivates the limitation of capacity between the data and both per-data point latent variables and model parameters. The interaction between the two is an interesting future research direction.

5.2. Role of Learning Dynamics and Architecture

As discussed in Section 2.2, it is necessary to perform exact inference in the noisy model for the bounds on the generalization error to hold. However, this assumption was not met. In practice, $p_t(\theta \mid D)$ encodes the complete learning algorithm, which in deep learning typically includes parameter initialization and the dynamics of the stochastic gradient descent optimization.

Our experiments confirmed the relevance of the aforementioned factors: L2-regularization works in practice, even though no noise was added to the parameters. This could be explained by the fact that noise was already implicitly added through stochastic gradient descent [49] or through the output distribution of the network. Similarly, Gaussian dropout [9–11] without a prior on the parameters helped generalization. Again, early stopping combined with a finite reach of gradient descent steps effectively shaped a prior of finite variance in the parameter space. This could also formalize why the annealing schedule employed by Blundell et al. [25], Bowman et al. [27] and Sønderby et al. [28] was effective.

Since these other factors affect generalization, quantifying mutual information $I_t(\theta, D)$ of the actual distribution due to the learning dynamics might be a promising approach to explain why neural networks often generalize well on their own. This idea is in accordance with recent work that links the learning dynamics of small neural networks to generalization behaviour [50].

On the other hand, the architecture choice also had an influence on generalization. This does not contradict our theory, we since we only formulated an upper bound on mutual information. Tightening this bound based on the model architecture and output distribution is usually hard, as discussed in Section 2.5, but might be possible.

Another promising direction would be to sample approximately from the exact posterior on network parameters (i.e., as done by Marceau-Caron and Ollivier [51]), on a capacity-limited architecture, instead of the usual approach of point estimation. In the limit of infinite training time, this would fully realize the discussed bound on the expected generalization error.

6. Conclusions

We explored an information-theoretic perspective on the regularizing effects observed in Gaussian mean field approaches. The derivation featured a capacity that can be naturally interpreted as a limit on the amount of information extracted about the given data by the inferred model. We validated its practicality for both supervised and unsupervised learning.

How this capacity should be set for parameters and latent variables depending on the task and data is an interesting direction of research. We exploited a theoretical link of mutual information and generalization error. While this work is restricted to the Gaussian mean field, incorporating the effect of learning dynamics on mutual information in the future work might allow understanding why overparameterized neural networks still generalize well to unseen data.

Author Contributions: Conceptualization, methodology, software, validation, formal analysis, investigation, visualization, and writing of the original draft were performed by J.K. and L.K. with supervision by D.B., and review and editing was done by H.R.

Funding: This research received no external funding.

Conflicts of Interest: The authors declare no conflict of interest.

Appendix A. Capacity in Learned-Variance Gaussian Mean Field Inference

The capacity per dimension for the model discussed in Section 2.3 is given by:

$$\begin{aligned}
&I(\tilde{\theta}_i, (\theta_i, \sigma_i^2)) \\
&= H(\tilde{\theta}_i) - H(\tilde{\theta}_i \mid \theta_i, \sigma_i^2) \\
&= -\int_{-\infty}^{\infty} p'(\tilde{\theta}_i) \log p'(\tilde{\theta}_i) \, d\tilde{\theta}_i - \int_{0}^{\infty} p'(\sigma_i^2) \frac{1}{2} \log 2\pi e \sigma_i^2 \, d\sigma_i^2
\end{aligned} \quad \text{(A1)}$$

$\tilde{\theta}_i \mid \theta_i, \sigma_i^2 \sim \mathcal{N}\left(\theta_i, \sigma_i^2\right)$ with $\theta_i \sim \mathcal{N}\left(0, \frac{1}{\beta}\right)$ implies $\tilde{\theta}_i \mid \sigma_i^2 \sim \mathcal{N}\left(0, \sigma_i^2 + \frac{1}{\beta}\right)$. Together with $\sigma_i^2 \sim \Gamma\left(\frac{\beta}{2} + 1, \frac{\beta}{2}\right)$, this implies:

$$
\begin{aligned}
p'(\tilde{\theta}_i) &= \int_0^\infty d\sigma_i^2 \, p'(\sigma_i^2) p'(\tilde{\theta}_i \mid \sigma_i^2) \\
&= \int_0^\infty d\sigma_i^2 \frac{1}{\Gamma\left(\frac{\beta}{2}\right)} \left(\frac{\beta}{2}\sigma_i^2 e^{-\sigma_i^2}\right)^{\frac{\beta}{2}} \cdot \left(2\pi\left(\sigma_i^2 + \frac{1}{\beta}\right)\right)^{-\frac{1}{2}} e^{-\frac{1}{2\left(\sigma_i^2 + \frac{1}{\beta}\right)}\tilde{\theta}_i^2}
\end{aligned}
\qquad (A2)
$$

Numerical results for the capacity $I(\tilde{\theta}_i, (\theta_i, \sigma_i^2))$ with varying β are given below (see Table A1) and plotted in Figure 2.

Table A1. Capacity $I(\tilde{\theta}_i, (\theta_i, \sigma_i^2))$ in learned-variance Gaussian mean field inference with varying factor β on the divergence term.

β	$I(\tilde{\theta}_i, (\theta_i, \sigma_i^2))$
0.01	0.68 bits
0.1	0.65 bits
1	0.45 bits
10	0.12 bits
100	0.014 bits

References

1. Bishop, C.M. *Pattern Recognition and Machine Learning*; Springer: Berlin, Germany, 2006.
2. Barber, D. *Bayesian Reasoning and Machine Learning*; Cambridge University Press: Cambridge, UK, 2012.
3. Ghahramani, Z. Probabilistic machine learning and artificial intelligence. *Nature* **2015**, *521*, 452. [CrossRef] [PubMed]
4. Wainwright, M.J.; Jordan, M.I. Graphical models, exponential families, and variational inference. *Found. Trends Mach. Learn.* **2008**, *1*, 1–305. [CrossRef]
5. Hoffman, M.D.; Blei, D.M.; Wang, C.; Paisley, J. Stochastic variational inference. *J. Mach. Learn. Res.* **2013**, *14*, 1303–1347.
6. Kingma, D.P.; Welling, M. Auto-encoding variational Bayes. *arXiv* **2013**, arXiv:1312.6114.
7. Titsias, M.; Lázaro-Gredilla, M. Doubly stochastic variational Bayes for non-conjugate inference. In Proceedings of the International Conference on Machine Learning, Beijing, China, 21–26 June 2014.
8. Rezende, D.J.; Mohamed, S. Variational inference with normalizing flows. In Proceedings of the International Conference on Machine Learning, Lille, France, 6–11 July 2015.
9. Graves, A.; Mohamed, A.R.; Hinton, G. Speech recognition with deep recurrent neural networks. In Proceedings of the IEEE International Conference on Acoustics, Speech and Signal Processing, Vancouver, BC, Canada, 26–31 May 2013.
10. Plappert, M.; Houthooft, R.; Dhariwal, P.; Sidor, S.; Chen, R.Y.; Chen, X.; Asfour, T.; Abbeel, P.; Andrychowicz, M. Parameter Space Noise for Exploration. In Proceedings of the International Conference on Learning Representations, Vancouver, BC, Canada, 30 April–3 May 2018.
11. Fortunato, M.; Azar, M.G.; Piot, B.; Menick, J.; Hessel, M.; Osband, I.; Graves, A.; Mnih, V.; Munos, R.; Hassabis, D.; et al. Noisy Networks For Exploration. In Proceedings of the International Conference on Learning Representations, Vancouver, BC, Canada, 30 April–3 May 2018.
12. Turner, R.; Sahani, M. Two problems with variational expectation maximisation for time-series models. In *Bayesian Time Series Models*; Cambridge University Press: Cambridge, UK, 2011.
13. Trippe, B.; Turner, R. Overpruning in variational Bayesian neural networks. *arXiv* **2018**, arXiv:1801.06230.
14. Braithwaite, D.; Kleijn, W.B. Bounded Information Rate Variational Autoencoders. *arXiv* **2018**, arXiv:1807.07306.
15. Shu, R.; Bui, H.H.; Zhao, S.; Kochenderfer, M.J.; Ermon, S. Amortized Inference Regularization. *arXiv* **2018**, arXiv:1805.08913.
16. Rissanen, J. Modeling by shortest data description. *Automatica* **1978**, *14*, 465–471. [CrossRef]

17. Rissanen, J. A universal prior for integers and estimation by minimum description length. *Ann. Stat.* **1983**, *11*, 416–431. [CrossRef]
18. Hinton, G.; van Camp, D. Keeping neural networks simple by minimising the description length of weights. In *Computational Learning Theory*; ACM Press: New York, NY, USA, 1993.
19. Xu, A.; Raginsky, M. Information-theoretic analysis of generalization capability of learning algorithms. In *Advances in Neural Information Processing Systems*; NIPS: Grenada, Spain, 2017.
20. Bu, Y.; Zou, S.; Veeravalli, V.V. Tightening Mutual Information Based Bounds on Generalization Error. *arXiv* **2019**, arXiv:1901.04609.
21. Bassily, R.; Moran, S.; Nachum, I.; Shafer, J.; Yehudayoff, A. Learners that use little information. In Proceedings of the Algorithmic Learning Theory, Lanzarote, Spain, 7–9 April 2018.
22. Russo, D.; Zou, J. How much does your data exploration overfit? Controlling bias via information usage. *arXiv* **2015**, arXiv:1511.05219.
23. Cover, T.M.; Thomas, J.A. *Elements of Information Theory*; John Wiley & Sons: New York, NY, USA, 2012.
24. Kingma, D.P.; Salimans, T.; Welling, M. Variational dropout and the local reparameterization trick. In *Advances in Neural Information Processing Systems*; NIPS: Grenada, Spain, 2015.
25. Blundell, C.; Cornebise, J.; Kavukcuoglu, K.; Wierstra, D. Weight uncertainty in neural networks. In Proceedings of the International Conference on Machine Learning, Beijing, China, 21–26 June 2015.
26. Higgins, I.; Matthey, L.; Pal, A.; Burgess, C.; Glorot, X.; Botvinick, M.; Mohamed, S.; Lerchner, A. β-VAE: Learning basic visual concepts with a constrained variational framework. In Proceedings of the International Conference on Learning Representations, Toulon, France, 24–26 April 2017.
27. Bowman, S.R.; Vilnis, L.; Vinyals, O.; Dai, A.M.; Jozefowicz, R.; Bengio, S. Generating sentences from a continuous space. *arXiv* **2015**, arXiv:1511.06349.
28. Sønderby, C.K.; Raiko, T.; Maaløe, L.; Sønderby, S.K.; Winther, O. Ladder variational autoencoders. In *Advances in Neural Information Processing Systems*; NIPS: Grenada, Spain, 2016.
29. Kingma, D.P.; Salimans, T.; Jozefowicz, R.; Chen, X.; Sutskever, I.; Welling, M. Improved variational inference with inverse autoregressive flow. In *Advances in Neural Information Processing Systems*; NIPS: Grenada, Spain, 2016.
30. Salimans, T.; Kingma, D.; Welling, M. Markov chain Monte Carlo and variational inference: Bridging the gap. In Proceedings of the International Conference on Machine Learning, Beijing, China, 21–26 June 2015; pp. 1218–1226.
31. Ranganath, R.; Tran, D.; Altosaar, J.; Blei, D. Operator variational inference. In *Advances in Neural Information Processing Systems*; NIPS: Grenada, Spain, 2016.
32. Huszár, F. Variational inference using implicit distributions. *arXiv* **2017**, arXiv:1702.08235.
33. Chen, T.Q.; Rubanova, Y.; Bettencourt, J.; Duvenaud, D. Neural Ordinary Differential Equations. *arXiv* **2018**, arXiv:1806.07366.
34. Vertes, E.; Sahani, M. Flexible and accurate inference and learning for deep generative models. *arXiv* **2018**, arXiv:1805.11051.
35. Burda, Y.; Grosse, R.; Salakhutdinov, R. Importance weighted autoencoders. *arXiv* **2015**, arXiv:1509.00519.
36. Cremer, C.; Morris, Q.; Duvenaud, D. Reinterpreting importance-weighted autoencoders. In Proceedings of the International Conference on Learning Representations Workshop, Toulon, France, 24–26 April 2017.
37. Molchanov, D.; Ashukha, A.; Vetrov, D. Variational dropout sparsifies deep neural networks. In Proceedings of the International Conference on Machine Learning, Sydney, Australia, 6–11 August 2017; pp. 2498–2507.
38. Wang, S.; Manning, C. Fast dropout training. In Proceedings of the International Conference on Machine Learning, Atlanta, GA, USA, 16–21 June 2013.
39. Tishby, N.; Pereira, F.C.; Bialek, W. The information bottleneck method. *arXiv* **2000**, arXiv:physics/0004057.
40. Shamir, O.; Sabato, S.; Tishby, N. Learning and generalization with the information bottleneck. *Theor. Comput. Sci.* **2010**, *411*, 2696–2711. [CrossRef]
41. Alemi, A.A.; Fischer, I.; Dillon, J.V.; Murphy, K. Deep variational information bottleneck. In Proceedings of the International Conference on Learning Representations, San Juan, Puerto Rico, 2–4 May 2016.
42. Achille, A.; Soatto, S. Emergence of invariance and disentangling in deep representations. *arXiv* **2017**, arXiv:1706.01350.
43. Alemi, A.; Poole, B.; Fischer, I.; Dillon, J.; Saurous, R.A.; Murphy, K. Fixing a broken ELBO. In Proceedings of the International Conference on Machine Learning, Stockholm, Sweden, 10–15 July 2018.

44. Belghazi, I.; Rajeswar, S.; Baratin, A.; Hjelm, R.D.; Courville, A. MINE: Mutual information neural estimation. In Proceedings of the International Conference on Machine Learning, Stockholm, Sweden, 10–15 July 2018.
45. Van den Oord, A.; Li, Y.; Vinyals, O. Representation learning with contrastive predictive coding. *arXiv* **2018**, arXiv:1807.03748.
46. Hjelm, R.D.; Fedorov, A.; Lavoie-Marchildon, S.; Grewal, K.; Trischler, A.; Bengio, Y. Learning deep representations by mutual information estimation and maximization. *arXiv* **2018**, arXiv:1808.06670.
47. Krogh, A.; Hertz, J.A. A simple weight decay can improve generalization. In *Advances in Neural Information Processing Systems*; NIPS: Grenada, Spain, 1992.
48. Srivastava, N.; Hinton, G.; Krizhevsky, A.; Sutskever, I.; Salakhutdinov, R. Dropout: A simple way to prevent neural networks from overfitting. *J. Mach. Learn. Res.* **2014**, *15*, 1929–1958.
49. Lei, D.; Sun, Z.; Xiao, Y.; Wang, W.Y. Implicit Regularization of Stochastic Gradient Descent in Natural Language Processing: Observations and Implications. *arXiv* **2018**, arXiv:1811.00659.

50. Li, Y.; Liang, Y. Learning overparameterized neural networks via stochastic gradient descent on structured data. In *Advances in Neural Information Processing Systems*; NIPS: Grenada, Spain, 2018.
51. Marceau-Caron, G.; Ollivier, Y. Natural Langevin dynamics for neural networks. In *International Conference on Geometric Science of Information*; Springer: London, UK, 2017.

© 2019 by the authors. Licensee MDPI, Basel, Switzerland. This article is an open access article distributed under the terms and conditions of the Creative Commons Attribution (CC BY) license (http://creativecommons.org/licenses/by/4.0/).

Article
Learnability for the Information Bottleneck

Tailin Wu [1,*], Ian Fischer [2], Isaac L. Chuang [1] and Max Tegmark [1]

1 Department of Physics, MIT, 77 Massachusetts Ave, Cambridge, MA 02139, USA; ichuang@mit.edu (I.L.C.); tegmark@mit.edu (M.T.)
2 Google Research, 1600 Amphitheatre Parkway, Mountain View, CA 94043, USA; iansf@google.com
* Correspondence: tailin@mit.edu

Received: 1 August 2019; Accepted: 12 September 2019; Published: 23 September 2019

Abstract: The Information Bottleneck (IB) method provides an insightful and principled approach for balancing compression and prediction for representation learning. The IB objective $I(X;Z) - \beta I(Y;Z)$ employs a Lagrange multiplier β to tune this trade-off. However, in practice, not only is β chosen empirically without theoretical guidance, there is also a lack of theoretical understanding between β, learnability, the intrinsic nature of the dataset and model capacity. In this paper, we show that if β is improperly chosen, learning cannot happen—the trivial representation $P(Z|X) = P(Z)$ becomes the global minimum of the IB objective. We show how this can be avoided, by identifying a sharp phase transition between the unlearnable and the learnable which arises as β is varied. This phase transition defines the concept of IB-Learnability. We prove several sufficient conditions for IB-Learnability, which provides theoretical guidance for choosing a good β. We further show that IB-learnability is determined by the largest *confident*, *typical* and *imbalanced subset* of the examples (the *conspicuous subset*), and discuss its relation with model capacity. We give practical algorithms to estimate the minimum β for a given dataset. We also empirically demonstrate our theoretical conditions with analyses of synthetic datasets, MNIST and CIFAR10.

Keywords: learnability; information bottleneck; representation learning; conspicuous subset

1. Introduction

Tishby et al. [1] introduced the *Information Bottleneck* (IB) objective function which learns a representation Z of observed variables (X, Y) that retains as little information about X as possible but simultaneously captures as much information about Y as possible:

$$\min \text{IB}_\beta(X, Y; Z) = \min[I(X;Z) - \beta I(Y;Z)] \qquad (1)$$

$I(\cdot)$ is the mutual information. The hyperparameter β controls the trade-off between compression and prediction, in the same spirit as Rate-Distortion Theory [2] but with a learned representation function $P(Z|X)$ that automatically captures some part of the "semantically meaningful" information, where the semantics are determined by the observed relationship between X and Y. The IB framework has been extended to and extensively studied in a variety of scenarios, including Gaussian variables [3], meta-Gaussians [4], continuous variables via variational methods [5–7], deterministic scenarios [8,9], geometric clustering [10] and is used for learning invariant and disentangled representations in deep neural nets [11,12].

From the IB objective (Equation (1)) we see that when $\beta \to 0$ it will encourage $I(X;Z) = 0$ which leads to a trivial representation Z that is independent of X, while when $\beta \to +\infty$, it reduces to a maximum likelihood objective (e.g., in classification, it reduces to cross-entropy loss). Therefore, as we vary β from 0 to $+\infty$, there must exist a point β_0 at which IB starts to learn a nontrivial representation where Z contains information about X.

As an example, we train multiple variational information bottleneck (VIB) models on binary classification of MNIST [13] digits 0 and 1 with 20% label noise at different β. The accuracy vs. β is shown in Figure 1. We see that when $\beta < 3.25$, no learning happens and the accuracy is the same as random guessing. Beginning with $\beta > 3.25$, there is a clear phase transition where the accuracy sharply increases, indicating the objective is able to learn a nontrivial representation. In general, we observe that different datasets and model capacity will result in different β_0 at which IB starts to learn a nontrivial representation. How does β_0 depend on the aspects of the dataset and model capacity and how can we estimate it? What does an IB model learn at the onset of learning? Answering these questions may provide a deeper understanding of IB in particular and learning on two observed variables in general.

In this work, we begin to answer the above questions. Specifically:

- We introduce the concept of *IB-Learnability* and show that when we vary β, the IB objective will undergo a phase transition from the inability to learn to the ability to learn (Section 3).
- Using the second-order variation, we derive sufficient conditions for IB-Learnability, which provide upper bounds for the learnability threshold β_0 (Section 4).
- We show that IB-Learnability is determined by the largest *confident, typical* and *imbalanced subset* of the examples (the *conspicuous subset*), reveal its relationship with the slope of the Pareto frontier at the origin on the information plane $I(X;Z)$ vs. $I(Y;Z)$ and discuss its relation to model capacity (Section 5).
- We prove a deep relationship between IB-Learnability, our upper bounds on β_0, the hypercontractivity coefficient, the contraction coefficient and the maximum correlation (Section 5).

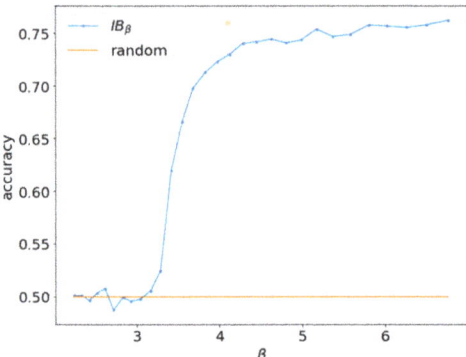

Figure 1. Accuracy for binary classification of MNIST digits 0 and 1 with 20% label noise and varying β. No learning happens for models trained at $\beta < 3.25$.

We also present an algorithm for estimating the onset of IB-Learnability and the conspicuous subset, which provide us with a tool for understanding a key aspect of the learning problem (X, Y) (Section 6). Finally, we use our main results to demonstrate on synthetic datasets, MNIST [13] and CIFAR10 [14] that the theoretical prediction for IB-Learnability closely matches experiment, and show the conspicuous subset our algorithm discovers (Section 7).

2. Related Work

The seminal IB work [1] provides a tabular method for exactly computing the optimal encoder distribution $P(Z|X)$ for a given β and cardinality of the discrete representation, $|Z|$. They did not consider the IB learnability problem as addressed in this work. Chechik et al. [3] presents the Gaussian Information Bottleneck (GIB) for learning a multivariate Gaussian representation Z of (X, Y), assuming that both X and Y are also multivariate Gaussians. Under GIB, they derive analytic formula for the optimal representation as a noisy linear projection to eigenvectors of the normalized regression matrix $\Sigma_{x|y}\Sigma_x^{-1}$ and the learnability threshold β_0 is then given by $\beta_0 = \frac{1}{1-\lambda_1}$ where λ_1 is the largest eigenvalue of the matrix $\Sigma_{x|y}\Sigma_x^{-1}$. This work provides deep insights about relations between the dataset, β_0 and optimal representations in the Gaussian scenario but the restriction to multivariate Gaussian datasets limits the generality of the analysis Another analytic treatment of IB is given in [4], which reformulates the objective in terms of the copula functions. As with the GIB approach, this formulation restricts the form of the data distributions—the copula functions for the joint distribution (X, Y) are assumed to be known, which is unlikely in practice.

Strouse and Schwab [8] present the Deterministic Information Bottleneck (DIB), which minimizes the coding cost of the representation, $H(Z)$, rather than the transmission cost, $I(X; Z)$ as in IB. This approach learns hard clusterings with different code entropies that vary with β. In this case, it is clear that a hard clustering with minimal $H(Z)$ will result in a single cluster for all of the data, which is the DIB trivial solution. No analysis is given beyond this fact to predict the actual onset of learnability, however.

The first amortized IB objective is in the Variational Information Bottleneck (VIB) of Alemi et al. [5]. VIB replaces the exact, tabular approach of IB with variational approximations of the classifier distribution ($P(Y|Z)$) and marginal distribution ($P(Z)$). This approach cleanly permits learning a stochastic encoder, $P(Z|X)$, that is applicable to any $x \in \mathcal{X}$, rather than just the particular X seen at training time. The cost of this flexibility is the use of variational approximations that may be less expressive than the tabular method. Nevertheless, in practice, VIB learns easily and is simple to implement, so we rely on VIB models for our experimental confirmation.

Closely related to IB is the recently proposed Conditional Entropy Bottleneck (CEB) [7]. CEB attempts to explicitly learn the Minimum Necessary Information (MNI), defined as the point in the information plane where $I(X;Y) = I(X;Z) = I(Y;Z)$. The MNI point may not be achievable even in principle for a particular dataset. However, the CEB objective provides an explicit estimate of how closely the model is approaching the MNI point by observing that a necessary condition for reaching the MNI point occurs when $I(X;Z|Y) = 0$. The CEB objective $I(X;Z|Y) - \gamma I(Y;Z)$ is equivalent to IB at $\gamma = \beta + 1$, so our analysis of IB-Learnability applies equally to CEB.

Kolchinsky et al. [9] show that when Y is a deterministic function of X, the "corner point" of the IB curve (where $I(X;Y) = I(X;Z) = I(Y;Z)$) is the unique optimizer of the IB objective for all $0 < \beta' < 1$ (with the parameterization of Kolchinsky et al. [9], $\beta' = 1/\beta$), which they consider to be a "trivial solution". However, their use of the term "trivial solution" is distinct from ours. They are referring to the observation that all points on the IB curve contain uninteresting interpolations between two different but valid solutions on the optimal frontier, rather than demonstrating a non-trivial trade-off between compression and prediction as expected when varying the IB Lagrangian. Our use of "trivial" refers to whether IB is capable of learning at all given a certain dataset and value of β.

Achille and Soatto [12] apply the IB Lagrangian to the weights of a neural network, yielding InfoDropout. In Achille and Soatto [11], the authors give a deep and compelling analysis of how the IB Lagrangian can yield invariant and disentangled representations. They do not, however, consider the question of the onset of learning, although they are aware that not all models will learn a non-trivial representation. More recently, Achille et al. [15] repurpose the InfoDropout IB Lagrangian as a Kolmogorov Structure Function to analyze the ease with which a previously-trained network can be fine-tuned for a new task. While that work is tangentially related to learnability, the question it addresses is substantially different from our investigation of the onset of learning.

Our work is also closely related to the hypercontractivity coefficient [16,17], defined as $\sup_{Z-X-Y} \frac{I(Y;Z)}{I(X;Z)}$, which by definition equals the inverse of β_0, our IB-learnability threshold. In [16], the authors prove that the hypercontractivity coefficient equals the contraction coefficient $\eta_{\mathrm{KL}}(P_{Y|X}, P_X)$ and Kim et al. [18] propose a practical algorithm to estimate $\eta_{\mathrm{KL}}(P_{Y|X}, P_X)$, which provides a measure for potential influence in the data. Although our goal is different, the sufficient conditions we provide for IB-Learnability are also lower bounds for the hypercontractivity coefficient.

3. IB-Learnability

We are given instances of (x, y) drawn from a distribution with probability (density) $P(X, Y)$ with support of $\mathcal{X} \times \mathcal{Y}$, where unless otherwise stated, both X and Y can be discrete or continuous variables. We use capital letters X, Y, Z for random variables and lowercase x, y, z to denote the instance of variables, with $P(\cdot)$ and $p(\cdot)$ denoting their probability or probability density, respectively. (X, Y) is our *training data* and may be characterized by different types of noise. The nature of this training data and the choice of β will be sufficient to predict the transition from unlearnable to learnable.

We can learn a representation Z of X with conditional probability $p(z|x)$, such that X, Y, Z obey the Markov chain $Z \leftarrow X \leftrightarrow Y$. Equation (1) above gives the IB objective with Lagrange multiplier β, $\text{IB}_\beta(X, Y; Z)$, which is a functional of $p(z|x)$: $\text{IB}_\beta(X, Y; Z) = \text{IB}_\beta[p(z|x)]$. The IB learning task is to find a conditional probability $p(z|x)$ that minimizes $\text{IB}_\beta(X, Y; Z)$. The larger β, the more the objective favors making a good prediction for Y. Conversely, the smaller β, the more the objective favors learning a concise representation.

How can we select β such that the IB objective learns a useful representation? In practice, the selection of β is done empirically. Indeed, Tishby et al. [1] recommends "sweeping β". In this paper, we provide theoretical guidance for choosing β by introducing the concept of IB-Learnability and providing a series of IB-learnable conditions.

Definition 1. *(X, Y) is IB_β-learnable if there exists a Z given by some $p_1(z|x)$, such that $\text{IB}_\beta(X, Y; Z)|_{p_1(z|x)} < \text{IB}_\beta(X, Y; Z)|_{p(z|x) = p(z)}$, where $p(z|x) = p(z)$ characterizes the trivial representation where $Z = Z_{trivial}$ is independent of X.*

If $(X; Y)$ is IB_β-learnable, then when $\text{IB}_\beta(X, Y; Z)$ is globally minimized, it will *not* learn a trivial representation. On the other hand, if $(X; Y)$ is not IB_β-learnable, then when $\text{IB}_\beta(X, Y; Z)$ is globally minimized, it may learn a trivial representation.

3.1. Trivial Solutions

Definition 1 defines trivial solutions in terms of representations where $I(X; Z) = I(Y; Z) = 0$. Another type of trivial solution occurs when $I(X; Z) > 0$ but $I(Y; Z) = 0$. This type of trivial solution is not directly achievable by the IB objective, as $I(X; Z)$ is minimized but it can be achieved by construction or by chance. It is possible that starting learning from $I(X; Z) > 0, I(Y; Z) = 0$ could result in access to non-trivial solutions not available from $I(X; Z) = 0$. We do not attempt to investigate this type of trivial solution in this work.

3.2. Necessary Condition for IB-Learnability

From Definition 1, we can see that IB_β-Learnability for any dataset $(X; Y)$ requires $\beta > 1$. In fact, from the Markov chain $Z \leftarrow X \leftrightarrow Y$, we have $I(Y; Z) \leq I(X; Z)$ via the data-processing inequality. If $\beta \leq 1$, then since $I(X; Z) \geq 0$ and $I(Y; Z) \geq 0$, we have that $\min(I(X; Z) - \beta I(Y; Z)) = 0 = \text{IB}_\beta(X, Y; Z_{trivial})$. Hence (X, Y) is not IB_β-learnable for $\beta \leq 1$.

Due to the reparameterization invariance of mutual information, we have the following theorem for IB_β-Learnability:

Lemma 1. *Let $X' = g(X)$ be an invertible map (if X is a continuous variable, g is additionally required to be continuous). Then (X, Y) and (X', Y) have the same IB_β-Learnability.*

The proof for Lemma 1 is in Appendix A.2. Lemma 1 implies a favorable property for any condition for IB_β-Learnability: the condition should be invariant to invertible mappings of X. We will inspect this invariance in the conditions we derive in the following sections.

4. Sufficient Conditions for IB-Learnability

Given (X, Y), how can we determine whether it is IB_β-learnable? To answer this question, we derive a series of sufficient conditions for IB_β-Learnability, starting from its definition. The conditions are in increasing order of practicality, while sacrificing as little generality as possible.

Firstly, Theorem 1 characterizes the IB_β-Learnability range for β, with proof in Appendix A.3:

Theorem 1. *If (X, Y) is IB_{β_1}-learnable, then for any $\beta_2 > \beta_1$, it is IB_{β_2}-learnable.*

Based on Theorem 1, the range of β such that (X, Y) is IB_β-learnable has the form $\beta \in (\beta_0, +\infty)$. Thus, β_0 is the *threshold* of IB-Learnability.

Lemma 2. $p(z|x) = p(z)$ *is a stationary solution for $IB_\beta(X, Y; Z)$.*

The proof in Appendix A.6 shows that both first-order variations $\delta I(X; Z) = 0$ and $\delta I(Y; Z) = 0$ vanish at the trivial representation $p(z|x) = p(z)$, so $\delta IB_\beta[p(z|x)] = 0$ at the trivial representation.

Lemma 2 yields our strategy for finding sufficient conditions for learnability: find conditions such that $p(z|x) = p(z)$ is not a local minimum for the functional $IB_\beta[p(z|x)]$. Based on the necessary condition for the minimum (Appendix A.4), we have the following theorem (The theorems in this paper deal with learnability w.r.t. true mutual information. If parameterized models are used to approximate the mutual information, the limitation of the model capacity will translate into more uncertainty of Y given X, viewed through the lens of the model.):

Theorem 2 (Suff. Cond. 1). *A sufficient condition for (X, Y) to be IB_β-learnable is that there exists a perturbation function $h(z|x)$ (so that the perturbed probability (density) is $p'(z|x) = p(z|x) + \epsilon \cdot h(z|x)$) with $\int h(z|x)dz = 0$, such that the second-order variation $\delta^2 IB_\beta[p(z|x)] < 0$ at the trivial representation $p(z|x) = p(z)$.*

The proof for Theorem 2 is given in Appendix A.4. Intuitively, if $\delta^2 IB_\beta[p(z|x)]\big|_{p(z|x)=p(z)} < 0$, we can always find a $p'(z|x) = p(z|x) + \epsilon \cdot h(z|x)$ in the neighborhood of the trivial representation $p(z|x) = p(z)$, such that $IB_\beta[p'(z|x)] < IB_\beta[p(z|x)]$, thus satisfying the definition for IB_β-Learnability.

To make Theorem 2 more practical, we perturb $p(z|x)$ around the trivial solution $p'(z|x) = p(z|x) + \epsilon \cdot h(z|x)$ and expand $IB_\beta[p(z|x) + \epsilon \cdot h(z|x)] - IB_\beta[p(z|x)]$ to the second order of ϵ. We can then prove Theorem 3:

Theorem 3 (Suff. Cond. 2). *A sufficient condition for (X, Y) to be IB_β-learnable is X and Y are not independent and*

$$\beta > \inf_{h(x)} \beta_0[h(x)] \qquad (2)$$

where the functional $\beta_0[h(x)]$ is given by

$$\beta_0[h(x)] = \frac{\mathbb{E}_{x \sim p(x)}[h(x)^2] - \left(\mathbb{E}_{x \sim p(x)}[h(x)]\right)^2}{\mathbb{E}_{y \sim p(y)}\left[\left(\mathbb{E}_{x \sim p(x|y)}[h(x)]\right)^2\right] - \left(\mathbb{E}_{x \sim p(x)}[h(x)]\right)^2}$$

Moreover, we have that $\left(\inf_{h(x)} \beta[h(x)]\right)^{-1}$ is a lower bound of the slope of the Pareto frontier in the information plane $I(Y; Z)$ vs. $I(X; Z)$ at the origin.

The proof is given in Appendix A.7, which also shows that if $\beta > \inf_{h(x)} \beta_0[h(x)]$ in Theorem 3 is satisfied, we can construct a perturbation function $h(z|x) = h^*(x)h_2(z)$ with $h^*(x) = \arg\min_{h(x)} \beta_0[h(x)]$, $\int h_2(z)dz = 0$, $\int \frac{h_2^2(z)}{p(z)} dz > 0$ for some $h_2(z)$, such that $h(z|x)$ satisfies Theorem 2. It also shows that the converse is true: if there exists $h(z|x)$ such that the condition in Theorem 2 is true, then Theorem 3 is satisfied, that is, $\beta > \inf_{h(x)} \beta_0[h(x)]$. (We do not claim that any $h(z|x)$ satisfying Theorem 2 can be decomposed to $h^*(x)h_2(z)$ at the onset of learning. But from the equivalence of Theorems 2 and 3 as explained above, when there exists an $h(z|x)$ such that Theorem 2 is satisfied, we can always construct an $h'(z|x) = h^*(x)h_2(z)$ that also satisfies Theorem 2.) Moreover, letting the perturbation function $h(z|x) = h^*(x)h_2(z)$ at the trivial solution, we have

$$p_\beta(y|x) = p(y) + \epsilon^2 C_z (h^*(x) - \overline{h}_x^*) \int p(x,y)(h^*(x) - \overline{h}_x^*) dx \tag{3}$$

where $p_\beta(y|x)$ is the estimated $p(y|x)$ by IB for a certain β, $\overline{h}_x^* = \int h^*(x)p(x)dx$ and $C_z = \int \frac{h_2^2(z)}{p(z)} dz > 0$ is a constant. This shows how the $p_\beta(y|x)$ by IB explicitly depends on $h^*(x)$ at the onset of learning. The proof is provided in Appendix A.8.

Theorem 3 suggests a method to estimate β_0: we can parameterize $h(x)$ for example, by a neural network, with the objective of minimizing $\beta_0[h(x)]$. At its minimization, $\beta_0[h(x)]$ provides an upper bound for β_0, and $h(x)$ provides a *soft clustering* of the examples corresponding to a nontrivial perturbation of $p(z|x)$ at $p(z|x) = p(z)$ that minimizes $IB_\beta[p(z|x)]$.

Alternatively, based on the property of $\beta_0[h(x)]$, we can also use a specific functional form for $h(x)$ in Equation (2) and obtain a stronger sufficient condition for IB_β-Learnability. But we want to choose $h(x)$ as near to the infimum as possible. To do this, we note the following characteristics for the R.H.S of Equation (2):

- We can set $h(x)$ to be nonzero if $x \in \Omega_x$ for some region $\Omega_x \subset \mathcal{X}$ and 0 otherwise. Then we obtain the following sufficient condition:

$$\beta > \inf_{h(x), \Omega_x \subset \mathcal{X}} \frac{\frac{\mathbb{E}_{x \sim p(x), x \in \Omega_x}[h(x)^2]}{\left(\mathbb{E}_{x \sim p(x), x \in \Omega_x}[h(x)]\right)^2} - 1}{\int \frac{dy}{p(y)} \left(\frac{\mathbb{E}_{x \sim p(x), x \in \Omega_x}[p(y|x)h(x)]}{\mathbb{E}_{x \sim p(x), x \in \Omega_x}[h(x)]}\right)^2 - 1} \tag{4}$$

- The numerator of the R.H.S. of Equation (4) attains its minimum when $h(x)$ is a constant within Ω_x. This can be proved using the Cauchy-Schwarz inequality: $\langle u, u \rangle \langle v, v \rangle \geq \langle u, v \rangle^2$, setting $u(x) = h(x)\sqrt{p(x)}$, $v(x) = \sqrt{p(x)}$ and defining the inner product as $\langle u, v \rangle = \int u(x)v(x)dx$. Therefore, the numerator of the R.H.S. of Equation (4) $\geq \frac{1}{\int_{x \in \Omega_x} p(x)} - 1$ and attains equality when $\frac{u(x)}{v(x)} = h(x)$ is constant.

Based on these observations, we can let $h(x)$ be a nonzero constant inside some region $\Omega_x \subset \mathcal{X}$ and 0 otherwise and the infimum over an arbitrary function $h(x)$ is simplified to infimum over $\Omega_x \subset \mathcal{X}$ and we obtain a sufficient condition for IB_β-Learnability, which is a key result of this paper:

Theorem 4 (Conspicuous Subset Suff. Cond.). *A sufficient condition for (X, Y) to be IB_β-learnable is X and Y are not independent and*

$$\beta > \inf_{\Omega_x \subset \mathcal{X}} \beta_0(\Omega_x) \tag{5}$$

where

$$\beta_0(\Omega_x) = \frac{\frac{1}{p(\Omega_x)} - 1}{\mathbb{E}_{y \sim p(y|\Omega_x)} \left[\frac{p(y|\Omega_x)}{p(y)} - 1\right]}$$

Ω_x *denotes the event that $x \in \Omega_x$, with probability $p(\Omega_x)$.*

$(\inf_{\Omega_x \subset \mathcal{X}} \beta_0(\Omega_x))^{-1}$ gives a lower bound of the slope of the Pareto frontier in the information plane $I(Y; Z)$ vs. $I(X; Z)$ at the origin.

The proof is given in Appendix A.9. In the proof we also show that this condition is invariant to invertible mappings of X.

5. Discussion

5.1. The Conspicuous Subset Determines β_0

From Equation (5), we see that three characteristics of the subset $\Omega_x \subset \mathcal{X}$ lead to low β_0: **(1) confidence:** $p(y|\Omega_x)$ is large; **(2) typicality and size:** the number of elements in Ω_x is large or the elements in Ω_x are typical, leading to a large probability of $p(\Omega_x)$; **(3) imbalance:** $p(y)$ is small for the subset Ω_x but large for its complement. In summary, β_0 will be determined by the largest *confident*, *typical* and *imbalanced* subset of examples or an equilibrium of those characteristics. We term Ω_x at the minimization of $\beta_0(\Omega_x)$ the *conspicuous subset*.

5.2. Multiple Phase Transitions

Based on this characterization of Ω_x, we can hypothesize datasets with multiple learnability phase transitions. Specifically, consider a region Ω_{x0} that is small but "typical", consists of all elements confidently predicted as y_0 by $p(y|x)$ and where y_0 is the least common class. By construction, this Ω_{x0} will dominate the infimum in Equation (5), resulting in a small value of β_0. However, the remaining $\mathcal{X} - \Omega_{x0}$ effectively form a new dataset, $\mathcal{X}_1$. At exactly β_0, we may have that the current encoder, $p_0(z|x)$, has no mutual information with the remaining classes in $\mathcal{X}_1$; that is, $I(Y_1; Z_0) = 0$. In this case, Definition 1 applies to $p_0(z|x)$ with respect to $I(X_1; Z_1)$. We might expect to see that, at β_0, learning will plateau until we get to some $\beta_1 > \beta_0$ that defines the phase transition for $\mathcal{X}_1$. Clearly this process could repeat many times, with each new dataset $\mathcal{X}_i$ being distinctly more difficult to learn than $\mathcal{X}_{i-1}$.

5.3. Similarity to Information Measures

The denominator of $\beta_0(\Omega_x)$ in Equation (5) is closely related to mutual information. Using the inequality $x - 1 \geq \log(x)$ for $x > 0$, it becomes:

$$\mathbb{E}_{y \sim p(y|\Omega_x)} \left[\frac{p(y|\Omega_x)}{p(y)} - 1 \right] \geq \mathbb{E}_{y \sim p(y|\Omega_x)} \left[\log \frac{p(y|\Omega_x)}{p(y)} \right] = \tilde{I}(\Omega_x; Y)$$

where $\tilde{I}(\Omega_x; Y)$ is the mutual information "density" at $\Omega_x \subset \mathcal{X}$. Of course, this quantity is also $\mathbb{D}_{\text{KL}}[p(y|\Omega_x)||p(y)]$, so we know that the denominator of Equation (5) is non-negative. Incidentally, $\mathbb{E}_{y \sim p(y|\Omega_x)} \left[\frac{p(y|\Omega_x)}{p(y)} - 1 \right]$ is the density of "rational mutual information" [19] at Ω_x.

Similarly, the numerator of $\beta_0(\Omega_x)$ is related to the self-information of Ω_x:

$$\frac{1}{p(\Omega_x)} - 1 \geq \log \frac{1}{p(\Omega_x)} = -\log p(\Omega_x) = h(\Omega_x)$$

so we can estimate β_0 as:

$$\beta_0 \simeq \inf_{\Omega_x \subset \mathcal{X}} \frac{h(\Omega_x)}{\tilde{I}(\Omega_x; Y)} \qquad (6)$$

Since Equation (6) uses upper bounds on both the numerator and the denominator, it does not give us a bound on β_0, only an estimate.

5.4. Estimating Model Capacity

The observation that a model cannot distinguish between cluster overlap in the data and its own lack of capacity gives an interesting way to use IB-Learnability to measure the capacity of a set of models relative to the task they are being used to solve. For example, for a classification task, we can use different model classes to estimate $p(y|x)$. For each such trained model, we can estimate the corresponding IB-learnability threshold β_0. A model with smaller capacity than the task needs will translate to more uncertainty in $p(y|\Omega_x)$, resulting in a larger β_0. On the other hand, models that give the same β_0 as each other all have the same capacity relative to the task, even if we would otherwise expect them to have very different capacities. For example, if two deep models have the same core architecture but one has twice the number of parameters at each layer and they both yield the same β_0, their capacities are equivalent with respect to the task. Thus, β_0 provides a way to measure model capacity in a task-specific manner.

5.5. Learnability and the Information Plane

Many of our results can be interpreted in terms of the geometry of the Pareto frontier illustrated in Figure 2, which describes the trade-off between increasing $I(Y;Z)$ and decreasing $I(X;Z)$. At any point on this frontier that minimizes $\text{IB}_\beta^{\min} \equiv \min I(X;Z) - \beta I(Y;Z)$, the frontier will have slope β^{-1} if it is differentiable. If the frontier is also concave (has negative second derivative), then this slope β^{-1} will take its maximum β_0^{-1} at the origin, which implies IB_β-Learnability for $\beta > \beta_0$, so that the threshold for IB_β-Learnability is simply the inverse slope of the frontier at the origin. More generally, as long as the Pareto frontier is differentiable, the threshold for IB_β-learnability is the inverse of its maximum slope. Indeed, Theorem 3 and Theorem 4 give lower bounds of the slope of the Pareto frontier at the origin.

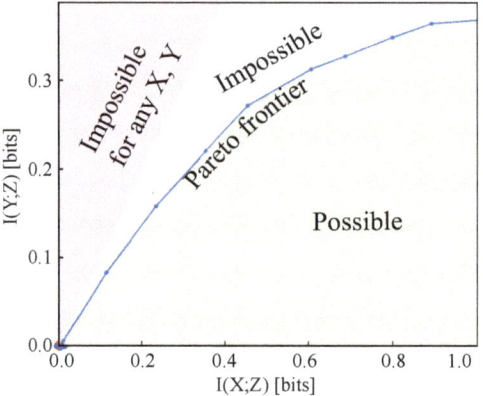

Figure 2. The Pareto frontier of the information plane, $I(X;Z)$ vs. $I(Y;Z)$, for the binary classification of MNIST digits 0 and 1 with 20% label noise described in Section 1 and Figure 1. For this problem, learning happens for models trained at $\beta > 3.25$. $H(Y) = 1$ bit since only two of ten digits are used and $I(Y;Z) \leq I(X;Y) \approx 0.5$ bits $< H(Y)$ because of the 20% label noise. The true frontier is differentiable; the figure shows a variational approximation that places an upper bound on both informations, horizontally offset to pass through the origin.

5.6. IB-Learnability, Hypercontractivity and Maximum Correlation

IB-Learnability and its sufficient conditions we provide harbor a deep connection with hypercontractivity and maximum correlation:

$$\frac{1}{\beta_0} = \xi(X;Y) = \eta_{\text{KL}} \geq \sup_{h(x)} \frac{1}{\beta_0[h(x)]} = \rho_m^2(X;Y) \tag{7}$$

which we prove in Appendix A.11. Here $\rho_m(X;Y) \equiv \max_{f,g} \mathbb{E}[f(X)g(Y)]$ s.t. $\mathbb{E}[f(X)] = \mathbb{E}[g(Y)] = 0$ and $\mathbb{E}[f^2(X)] = \mathbb{E}[g^2(Y)] = 1$ is the *maximum correlation* [20,21], $\xi(X;Y) \equiv \sup_{Z-X-Y} \frac{I(Y;Z)}{I(X;Z)}$ is the *hypercontractivity coefficient* and $\eta_{KL}(p(y|x), p(x)) \equiv \sup_{r(x) \neq p(x)} \frac{\mathbb{D}_{KL}(r(y)||p(y))}{\mathbb{D}_{KL}(r(x)||p(x))}$ is the *contraction coefficient*. Our proof relies on Anantharam et al. [16]'s proof $\xi(X;Y) = \eta_{KL}$. Our work reveals the deep relationship between IB-Learnability and these earlier concepts and provides additional insights about what aspects of a dataset give rise to high maximum correlation and hypercontractivity: the most confident, typical, imbalanced subset of (X, Y).

6. Estimating the IB-Learnability Condition

Theorem 4 not only reveals the relationship between the learnability threshold for β and the least noisy region of $P(Y|X)$ but also provides a way to practically estimate β_0, both in the general classification case and in more structured settings.

6.1. Estimation Algorithm

Based on Theorem 4, for general classification tasks we suggest Algorithm 1 to empirically estimate an upper-bound $\tilde{\beta}_0 \geq \beta_0$, as well as discovering the conspicuous subset that determines β_0.

We approximate the probability of each example $p(x_i)$ by its empirical probability, $\hat{p}(x_i)$. For example, for MNIST, $p(x_i) = \frac{1}{N}$, where N is the number of examples in the dataset. The algorithm starts by first learning a maximum likelihood model of $p_\theta(y|x)$, using for example, feed-forward neural networks. It then constructs a matrix $P_{y|x}$ and a vector p_y to store the estimated $p(y|x)$ and $p(y)$ for all the examples in the dataset. To find the subset Ω such that the $\tilde{\beta}_0$ is as small as possible, by previous analysis we want to find a *conspicuous* subset such that its $p(y|x)$ is large for a certain class j (to make the denominator of Equation (5) large) and containing as many elements as possible (to make the numerator small).

We suggest the following heuristics to discover such a conspicuous subset. For each class j, we sort the rows of $(P_{y|x})$ according to its probability for the pivot class j by decreasing order and then perform a search over $i_{\text{left}}, i_{\text{right}}$ for $\Omega = \{i_{\text{left}}, i_{\text{left}} + 1, ..., i_{\text{right}}\}$. Since $\tilde{\beta}_0$ is large when Ω contains too few or too many elements, the minimum of $\tilde{\beta}_0^{(j)}$ for class j will typically be reached with some intermediate-sized subset and we can use binary search or other discrete search algorithm for the optimization. The algorithm stops when $\tilde{\beta}_0^{(j)}$ does not improve by tolerance ε. The algorithm then returns the $\tilde{\beta}_0$ as the minimum over all the classes $\tilde{\beta}_0^{(1)}, ... \tilde{\beta}_0^{(N)}$, as well as the conspicuous subset that determines this $\tilde{\beta}_0$.

After estimating $\tilde{\beta}_0$, we can then use it for learning with IB, either directly or as an anchor for a region where we can perform a much smaller sweep than we otherwise would have. This may be particularly important for very noisy datasets, where β_0 can be very large.

Algorithm 1 Estimating the upper bound for β_0 and identifying the conspicuous subset

Require: Dataset $\mathcal{D} = \{(x_i, y_i)\}, i = 1, 2, ...N$. The number of classes is C.
Require ε: tolerance for estimating β_0
1: Learn a maximum likelihood model $p_\theta(y|x)$ using the dataset $\mathcal{D}$.
2: Construct matrix $(P_{y|x})$ such that $(P_{y|x})_{ij} = p_\theta(y = y_j|x = x_i)$.
3: Construct vector $p_y = (p_{y1}, ..., p_{yC})$ such that $p_{yj} = \frac{1}{N}\sum_{i=1}^{N}(P_{y|x})_{ij}$.
4: **for** j in $\{1, 2, ...C\}$:
5: $\quad P_{y|x}^{(\text{sort} j)} \leftarrow$ Sort the rows of $P_{y|x}$ in decreasing values of $(P_{y|x})_{ij}$.
6: $\quad \tilde{\beta}_0^{(j)}, \Omega^{(j)} \leftarrow$ Search $i_{\text{left}}, i_{\text{right}}$ until $\tilde{\beta}_0^{(j)} = \text{Get}\beta(P_{y|x}, p_y, \Omega)$ is minimal with tolerance ε,
$\quad\quad$ where $\Omega = \{i_{\text{left}}, i_{\text{left}} + 1, ...i_{\text{right}}\}$.
7: **end for**
8: $j^* \leftarrow \arg\min_j\{\tilde{\beta}_0^{(j)}\}, j = 1, 2, ...N$.
9: $\tilde{\beta}_0 \leftarrow \tilde{\beta}_0^{(j^*)}$.
10: $P_{y|x}^{(\tilde{\beta}_0)} \leftarrow$ the rows of $P_{y|x}^{(\text{sort} j^*)}$ indexed by $\Omega^{(j^*)}$.
11: **return** $\tilde{\beta}_0, P_{y|x}^{(\tilde{\beta}_0)}$

subroutine Get$\beta(P_{y|x}, p_y, \Omega)$:
s1: $N \leftarrow$ number of rows of $P_{y|x}$.
s2: $C \leftarrow$ number of columns of $P_{y|x}$.
s3: $n \leftarrow$ number of elements of Ω.
s4: $(p_{y|\Omega})_j \leftarrow \frac{1}{n}\sum_{i\in\Omega}(P_{y|x})_{ij}, j = 1, 2, ..., C$.
s5: $\tilde{\beta}_0 \leftarrow \dfrac{\frac{N}{n} - 1}{\sum_j \left[\frac{(p_{y|\Omega_x})_j^2}{p_{yj}} - 1\right]}$
s6: **return** $\tilde{\beta}_0$

6.2. Special Cases for Estimating β_0

Theorem 4 may still be challenging to estimate, due to the difficulty of making accurate estimates of $p(\Omega_x)$ and searching over $\Omega_x \subset \mathcal{X}$. However, if the learning problem is more structured, we may be able to obtain a simpler formula for the sufficient condition.

6.2.1. Class-Conditional Label Noise

Classification with noisy labels is a common practical scenario. An important noise model is that the labels are randomly flipped with some hidden class-conditional probabilities and we only observe the corrupted labels. This problem has been studied extensively [22–26]. If IB is applied to this scenario, how large β do we need? The following corollary provides a simple formula.

Corollary 1. *Suppose that the true class labels are y^* and the input space belonging to each y^* has no overlap. We only observe the corrupted labels y with class-conditional noise $p(y|x, y^*) = p(y|y^*)$ and Y is not independent of X. We have that a sufficient condition for IB_β-Learnability is:*

$$\beta > \inf_{y^*} \frac{\frac{1}{p(y^*)} - 1}{\sum_y \frac{p(y|y^*)^2}{p(y)} - 1} \tag{8}$$

We see that under class-conditional noise, the sufficient condition reduces to a discrete formula which only depends on the noise rates $p(y|y^*)$ and the true class probability $p(y^*)$, which can be accurately estimated via, for example, Northcutt et al. [26]. Additionally, if we know that the noise is class-conditional but the observed β_0 is greater than the R.H.S. of Equation (8), we can deduce that there is overlap between the true classes. The proof of Corollary 1 is provided in Appendix A.10.

6.2.2. Deterministic Relationships

Theorem 4 also reveals that β_0 relates closely to whether Y is a deterministic function of X, as shown by Corollary 2:

Corollary 2. *Assume that Y contains at least one value y such that its probability $p(y) > 0$. If Y is a deterministic function of X and not independent of X, then a sufficient condition for IB_β-Learnability is $\beta > 1$.*

The assumption in the Corollary 2 is satisfied by classification and certain regression problems. (The following scenario does not satisfy this assumption: for certain regression problems where Y is a continuous random variable and the probability density function $p_Y(y)$ is bounded, then for any y, the *probability* $P(Y = y)$ has measure 0.)This corollary generalizes the result in Reference [9] which only proves it for classification problems. Combined with the necessary condition $\beta > 1$ for any dataset (X, Y) to be IB_β-learnable (Section 3), we have that under the assumption, if Y is a deterministic function of X, then a necessary and sufficient condition for IB_β-learnability is $\beta > 1$; that is, its β_0 is 1. The proof of Corollary 2 is provided in Appendix A.10.

Therefore, in practice, if we find that $\beta_0 > 1$, we may infer that Y is not a deterministic function of X. For a classification task, we may infer that either some classes have overlap or the labels are noisy. However, recall that finite models may add effective class overlap if they have insufficient capacity for the learning task, as mentioned in Section 4. This may translate into a higher observed β_0, even when learning deterministic functions.

7. Experiments

To test how the theoretical conditions for IB_β-learnability match with experiment, we apply them to synthetic data with varying noise rates and class overlap, MNIST binary classification with varying noise rates and CIFAR10 classification, comparing with the β_0 found experimentally. We also compare with the algorithm in Kim et al. [18] for estimating the hypercontractivity coefficient (=$1/\beta_0$) via the contraction coefficient η_{KL}. Experiment details are in Section A.12.

7.1. Synthetic Dataset Experiments

We construct a set of datasets from 2D mixtures of 2 Gaussians as X and the identity of the mixture component as Y. We simulate two practical scenarios with these datasets: (1) noisy labels with class-conditional noise and (2) class overlap. For (1), we vary the class-conditional noise rates. For (2), we vary class overlap by tuning the distance between the Gaussians. For each experiment, we sweep β with exponential steps and observe $I(X; Z)$ and $I(Y; Z)$. We then compare the empirical β_0 indicated by the onset of above-zero $I(X; Z)$ with predicted values for β_0.

7.1.1. Classification with Class-Conditional Noise

In this experiment, we have a mixture of Gaussian distribution with 2 components, each of which is a 2D Gaussian with diagonal covariance matrix $\Sigma = \text{diag}(0.25, 0.25)$. The two components have distance 16 (hence virtually no overlap) and equal mixture weight. For each x, the label $y \in \{0, 1\}$ is the identity of which component it belongs to. We create multiple datasets by randomly flipping the labels y with a certain noise rate $\rho = P(y = 0|y^* = 1) = P(y = 1|y^* = 0)$. For each dataset, we train VIB models across a range of β and observe the onset of learning via random $I(X; Z)$ (Observed). To test how different methods perform in estimating β_0, we apply the following methods: **(1)** Corollary 1, since this is classification with class-conditional noise and the two true classes have virtually no overlap; **(2)** Algorithm 1 with true $p(y|x)$; **(3)** The algorithm in Kim et al. [18] that estimates $\hat{\eta}_{KL}$, provided with true $p(y|x)$; **(4)** $\beta_0[h(x)]$ in Equation (2); **(2′)** Algorithm 1 with $p(y|x)$ estimated by a neural net; **(3′)** $\hat{\eta}_{KL}$ with the same $p(y|x)$ as in (2′). The results are shown in Figure 3 and in Table 1.

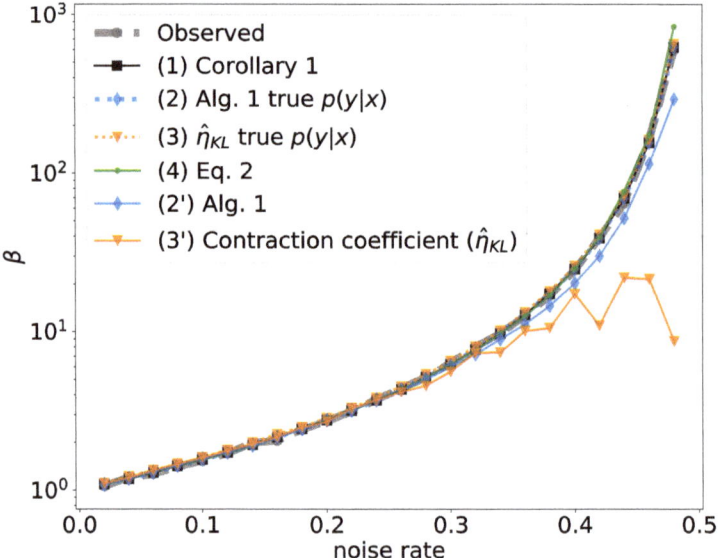

Figure 3. Predicted vs. experimentally identified β_0, for mixture of Gaussians with varying class-conditional noise rates.

Table 1. Full table of values used to generate Figure 3.

			(2) Algorithm 1	(3) $\hat{\eta}_{KL}$			
Noise Rate	Observed	(1) Corollary 1	True $p(y|x)$	True $p(y|x)$	(4) Equation (2)	(2′) Algorithm 1	(3′) $\hat{\eta}_{KL}$
0.02	1.06	1.09	1.09	1.10	1.08	1.08	1.10
0.04	1.20	1.18	1.18	1.21	1.18	1.19	1.20
0.06	1.26	1.29	1.29	1.33	1.30	1.31	1.33
0.08	1.40	1.42	1.42	1.45	1.42	1.43	1.46
0.10	1.52	1.56	1.56	1.60	1.55	1.58	1.60
0.12	1.70	1.73	1.73	1.78	1.71	1.73	1.77
0.14	1.99	1.93	1.93	1.99	1.90	1.91	1.95
0.16	2.04	2.16	2.16	2.24	2.15	2.15	2.16
0.18	2.41	2.44	2.44	2.49	2.43	2.42	2.49
0.20	2.74	2.78	2.78	2.86	2.76	2.77	2.71
0.22	3.15	3.19	3.19	3.29	3.19	3.21	3.29
0.24	3.75	3.70	3.70	3.83	3.71	3.75	3.72
0.26	4.40	4.34	4.34	4.48	4.35	4.31	4.17
0.28	5.16	5.17	5.17	5.37	5.12	4.98	4.55
0.30	6.34	6.25	6.25	6.49	6.24	6.03	5.58
0.32	8.06	7.72	7.72	8.02	7.63	7.19	7.33
0.34	9.77	9.77	9.77	10.13	9.74	8.95	7.37
0.36	12.58	12.76	12.76	13.21	12.51	11.11	10.09
0.38	16.91	17.36	17.36	17.96	16.97	14.55	10.49
0.40	24.66	25.00	25.00	25.99	25.01	20.36	17.27
0.42	39.08	39.06	39.06	40.85	39.48	30.12	10.89
0.44	64.82	69.44	69.44	71.80	76.48	51.95	21.95
0.46	163.07	156.25	156.26	161.88	173.15	114.57	21.47
0.48	599.45	625.00	625.00	651.47	838.90	293.90	8.69

From Figure 3 and Table 1 we see the following. **(A)** When using the true $p(y|x)$, both Algorithm 1 and $\hat{\eta}_{KL}$ generally upper bound the empirical β_0 and Algorithm 1 is generally tighter. **(B)** When using the true $p(y|x)$, Algorithm 1 and Corollary 1 give the same result. **(C)** Comparing Algorithm 1 and $\hat{\eta}_{KL}$ both of which use the same empirically estimated $p(y|x)$, both approaches provide good estimation in the low-noise region; however, in the high-noise region, Algorithm 1 gives more precise values than $\hat{\eta}_{KL}$, indicating that Algorithm 1 is more robust to the estimation error of $p(y|x)$. **(D)** Equation (2) empirically upper bounds the experimentally observed β_0 and gives almost the same result as theoretical estimation in Corollary 1 and Algorithm 1 with the true $p(y|x)$. In the classification setting, this approach does not require any learned estimate of $p(y|x)$, as we can directly use the empirical $p(y)$ and $p(x|y)$ from SGD mini-batches.

This experiment also shows that for dataset where the signal-to-noise is small, β_0 can be very high. Instead of blindly sweeping β, our result can provide guidance for setting β so learning can happen.

7.1.2. Classification with Class Overlap

In this experiment, we test how different amounts of overlap among classes influence β_0. We use the mixture of Gaussians with two components, each of which is a 2D Gaussian with diagonal covariance matrix $\Sigma = \text{diag}(0.25, 0.25)$. The two components have weights 0.6 and 0.4. We vary the distance between the Gaussians from 8.0 down to 0.8 and observe the $\beta_{0,exp}$. Since we do not add noise to the labels, if there were no overlap and a deterministic map from X to Y, we would have $\beta_0 = 1$ by Corollary 2. The more overlap between the two classes, the more uncertain Y is given X. By Equation (5) we expect β_0 to be larger, which is corroborated in Figure 4.

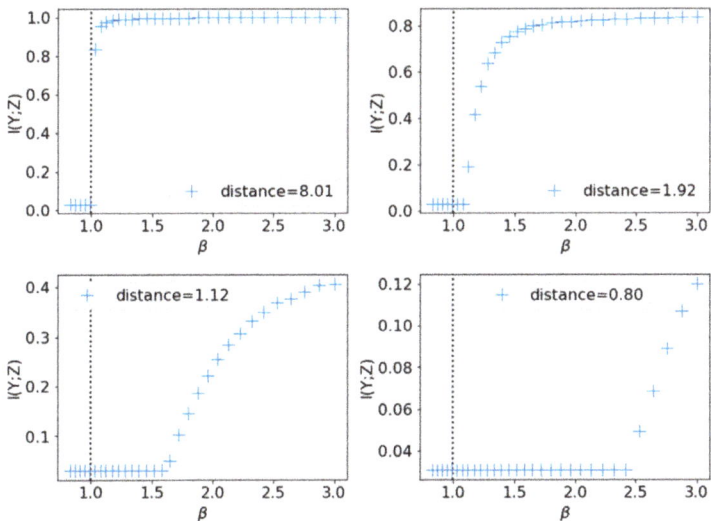

Figure 4. $I(Y; Z)$ vs. β, for mixture of Gaussian datasets with different distances between the two mixture components. The vertical lines are $\beta_{0,\text{predicted}}$ computed by the R.H.S. of Equation (8). As Equation (8) does not make predictions w.r.t. class overlap, the vertical lines are always just above $\beta_{0,\text{predicted}} = 1$. However, as expected, decreasing the distance between the classes in X space also increases the true β_0.

7.2. MNIST Experiments

We perform binary classification with digits 0 and 1 and as before, add class-conditional noise to the labels with varying noise rates ρ. To explore how the model capacity influences the onset of learning, for each dataset we train two sets of VIB models differing only by the number of neurons in their hidden layers of the encoder: one with $n = 512$ neurons, the other with $n = 128$ neurons. As we describe in Section 4, insufficient capacity will result in more uncertainty of Y given X from the point of view of the model, so we expect the observed β_0 for the $n = 128$ model to be larger. This result is confirmed by the experiment (Figure 5). Also, in Figure 5 we plot β_0 given by different estimation methods. We see that the observations (A), (B), (C) and (D) in Section 7.1 still hold.

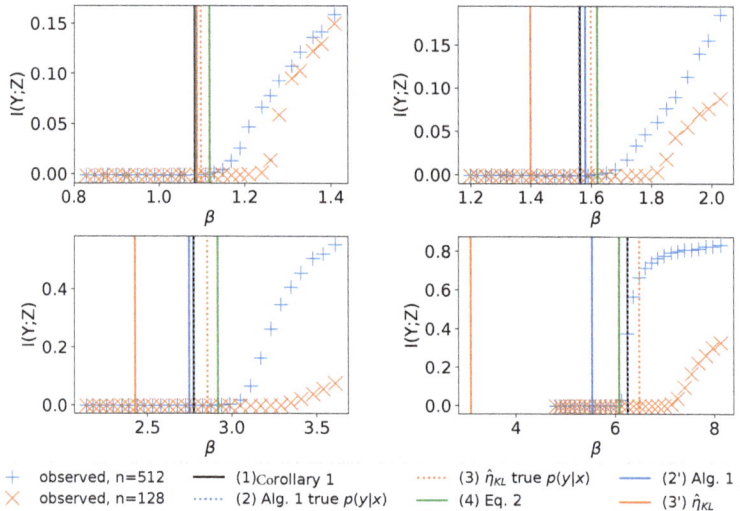

Figure 5. $I(Y;Z)$ vs. β for the MNIST binary classification with different hidden units per layer n and noise rates ρ: (upper left) $\rho = 0.02$, (upper right) $\rho = 0.1$, (lower left) $\rho = 0.2$, (lower right) $\rho = 0.3$. The vertical lines are β_0 estimated by different methods. $n = 128$ has insufficient capacity for the problem, so its observed learnability onset is pushed higher, similar to the class overlap case.

7.3. MNIST Experiments Using Equation (2)

To see what IB learns at its onset of learning for the full MNIST dataset, we optimize Equation (2) w.r.t. the full MNIST dataset and visualize the clustering of digits by $h(x)$. Equation (2) can be optimized using SGD using any differentiable parameterized mapping $h(x) : \mathcal{X} \to \mathbb{R}$. In this case, we chose to parameterize $h(x)$ with a PixelCNN++ architecture [27,28], as PixelCNN++ is a powerful autoregressive model for images that gives a scalar output (normally interpreted as $\log p(x)$). Equation (2) should generally give two clusters in the output space, as discussed in Section 4. In this setup, smaller values of $h(x)$ correspond to the subset of the data that is easiest to learn. Figure 6 shows two strongly separated clusters, as well as the threshold we choose to divide them. Figure 7 shows the first 5776 MNIST training examples as sorted by our learned $h(x)$, with the examples above the threshold highlighted in red. We can clearly see that our learned $h(x)$ has separated the "easy" one (1) digits from the rest of the MNIST training set.

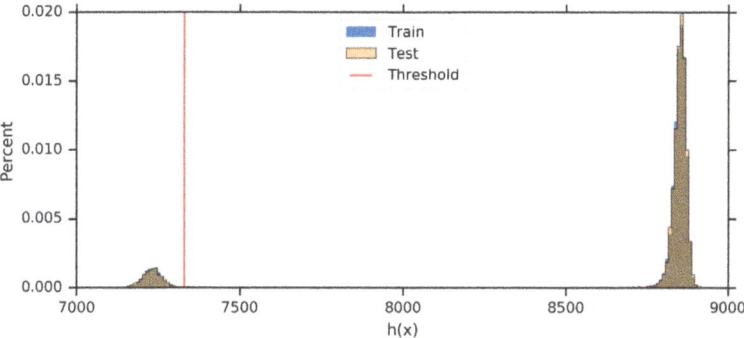

Figure 6. Histograms of the full MNIST training and validation sets according to $h(X)$. Note that both are bimodal and the histograms are indistinguishable. In both cases, $h(x)$ has learned to separate most of the ones into the smaller mode but difficult ones are in the wide valley between the two modes. See Figure 7 for all of the training images to the left of the red threshold line, as well as the first few images to the right of the threshold.

Figure 7. The first 5776 MNIST training set digits when sorted by $h(x)$. The digits highlighted in red are above the threshold drawn in Figure 6.

7.4. CIFAR10 Forgetting Experiments

For CIFAR10 [14], we study how *forgetting* varies with β. In other words, given a VIB model trained at some high β_2, if we anneal it down to some much lower β_1, what $I(Y;Z)$ does the model converge to? Using Algorithm 1, we estimated $\beta_0 = 1.0483$ on a version of CIFAR10 with 20% label noise, where the $P_{y|x}$ is estimated by maximum likelihood training with the same encoder and classifier architectures as used for VIB. For the VIB models, the lowest β with performance above chance was $\beta = 1.048$ (Figure 8), a very tight match with the estimate from Algorithm 1. See Appendix A.12 for details.

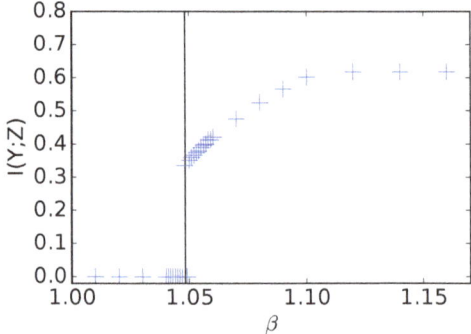

Figure 8. Plot of $I(Y;Z)$ vs. β for CIFAR10 training set with 20% label noise. Each blue cross corresponds to a fully-converged model starting with independent initialization. The vertical black line corresponds to the predicted $\beta_0 = 1.0483$ using Algorithm 1. The empirical $\beta_0 = 1.048$.

8. Conclusions

In this paper, we have presented theoretical results for predicting the onset of learning and have shown that it is determined by the conspicuous subset of the training examples. We gave a practical algorithm for predicting the transition as well as discovering this subset and showed that those predictions are accurate, even in cases of extreme label noise. We proved a deep connection between IB-learnability, our upper bounds on β_0, the hypercontractivity coefficient, the contraction coefficient and the maximum correlation. We believe that these results provide a deeper understanding of IB, as well as a tool for analyzing a dataset by discovering its conspicuous subset and a tool for measuring model capacity in a task-specific manner. Our work also raises other questions, such as whether there are other phase transitions in learnability that might be identified. We hope to address some of those questions in future work.

Author Contributions: Conceptualization, T.W. and I.F.; methodology, T.W., I.F., I.L.C. and M.T.; software, T.W. and I.F.; validation, T.W. and I.F.; formal analysis, T.W. and I.F.; investigation, T.W. and I.F.; resources, T.W., I.F., I.L.C. and M.T.; data curation, T.W. and I.F.; writing–original draft preparation, T.W., I.F., I.L.C. and M.T.; writing–review and editing, T.W., I.F., I.L.C. and M.T.; visualization, T.W. and I.F.; supervision, I.F., I.L.C. and M.T.; project administration, I.F., I.L.C. and M.T.; funding acquisition, M.T.

Funding: T.W.'s work was supported by the The Casey and Family Foundation, the Foundational Questions Institute and the Rothberg Family Fund for Cognitive Science. He thanks the Center for Brains, Minds and Machines (CBMM) for hospitality.

Acknowledgments: The authors would like to thank the anonymous reviewers for their constructive comments that contributed to improving the paper.

Conflicts of Interest: The authors declare no conflict of interest. The funders had no role in the design of the study; in the collection, analyses, or interpretation of data; in the writing of the manuscript, or in the decision to publish the results.

Appendix A

The structure of the Appendix is as follows. In Appendix A.1, we provide preliminaries for the first-order and second-order variations on functionals. We prove Theorem 1 and Theorem 1 in Appendixes A.2 and A.3, respectively. In Appendix A.4, we prove Theorem 2, the sufficient condition

1 for IB-Learnability. In Appendix A.5, we calculate the first and second variations of $\text{IB}_\beta[p(z|x)]$ at the trivial representation $p(z|x) = p(z)$, which is used in proving Lemma 2 (Appendix A.6) and the Sufficient Condition 2 for IB_β-learnability (Appendix A.7). In Appendix A.8, we prove Equation (3) at the onset of learning. After these preparations, we prove the key result of this paper, Theorem 4, in Section A.9. Then two important Corollaries 1, 2 are proved in Appendix A.10. In Appendix A.11 we explore the deep relation between β_0, $\beta_0[h(x)]$, the hypercontractivity coefficient, contraction coefficient and maximum correlation. Finally in Appendix A.12, we provide details for the experiments.

Below are some implicit conventions of the paper: for integrals, whenever a variable W is discrete, we can simply replace the integral ($\int \cdot dw$) by summation ($\sum_w \cdot$).

Appendix A.1. Preliminaries: First-Order and Second-Order Variations

Let functional $F[f(x)]$ be defined on some normed linear space $\mathscr{R}$. Let us add a perturbative function $\epsilon \cdot h(x)$ to $f(x)$, and now the functional $F[f(x) + \epsilon \cdot h(x)]$ can be expanded as

$$\Delta F[f(x)] = F[f(x) + \epsilon \cdot h(x)] - F[f(x)]$$
$$= \varphi_1[f(x)] + \varphi_2[f(x)] + \mathcal{O}(\epsilon^3 ||h||^2)$$

where $||h||$ denotes the norm of h, $\varphi_1[f(x)] = \epsilon \frac{dF[f(x)]}{d\epsilon}$ is a linear functional of $\epsilon \cdot h(x)$, and is called the *first-order variation*, denoted as $\delta F[f(x)]$. $\varphi_2[f(x)] = \frac{1}{2}\epsilon^2 \frac{d^2F[f(x)]}{d\epsilon^2}$ is a quadratic functional of $\epsilon \cdot h(x)$, and is called the *second-order variation*, denoted as $\delta^2 F[f(x)]$.

If $\delta F[f(x)] = 0$, we call $f(x)$ a stationary solution for the functional $F[\cdot]$.

If $\Delta F[f(x)] \geq 0$ for all $h(x)$ such that $f(x) + \epsilon \cdot h(x)$ is at the neighborhood of $f(x)$, we call $f(x)$ a (local) minimum of $F[\cdot]$.

Appendix A.2. Proof of Lemma 1

Proof. If (X, Y) is IB_β-learnable, then there exists $Z \in \mathcal{Z}$ given by some $p_1(z|x)$ such that $\text{IB}_\beta(X, Y; Z) < \text{IB}(X, Y; Z_{trivial}) = 0$, where $Z_{trivial}$ satisfies $p(z|x) = p(z)$. Since $X' = g(X)$ is a invertible map (if X is continuous variable, g is additionally required to be continuous), and mutual information is invariant under such an invertible map [29], we have that $\text{IB}_\beta(X', Y; Z) = I(X'; Z) - \beta I(Y; Z) = I(X; Z) - \beta I(Y; Z) = \text{IB}_\beta(X, Y; Z) < 0 = \text{IB}(X', Y; Z_{trivial})$, so (X', Y) is IB_β-learnable. On the other hand, if (X, Y) is not IB_β-learnable, then $\forall Z$, we have $\text{IB}_\beta(X, Y; Z) \geq \text{IB}(X, Y; Z_{trivial}) = 0$. Again using mutual information's invariance under g, we have for all Z, $\text{IB}_\beta(X', Y; Z) = \text{IB}_\beta(X, Y; Z) \geq \text{IB}(X, Y; Z_{trivial}) = 0$, leading to that (X', Y) is not IB_β-learnable. Therefore, we have that (X, Y) and (X', Y) have the same IB_β-learnability. □

Appendix A.3. Proof of Theorem 1

Proof. At the trivial representation $p(z|x) = p(z)$, we have $I(X; Z) = 0$, and $I(Y; Z) = 0$ due to the Markov chain, so $\text{IB}_\beta(X, Y; Z)|_{p(z|x)=p(z)} = 0$ for any β. Since (X, Y) is IB_{β_1}-learnable, there exists a Z given by a $p_1(z|x)$ such that $\text{IB}_{\beta_1}(X, Y; Z)|_{p_1(z|x)} < 0$. Since $\beta_2 > \beta_1$, and $I(Y; Z) \geq 0$, we have $\text{IB}_{\beta_2}(X, Y; Z)|_{p_1(z|x)} \leq \text{IB}_{\beta_1}(X, Y; Z)|_{p_1(z|x)} < 0 = \text{IB}_{\beta_2}(X, Y; Z)|_{p(z|x)=p(z)}$. Therefore, (X, Y) is IB_{β_2}-learnable. □

Appendix A.4. Proof of Theorem 2

Proof. To prove Theorem 2, we use the Theorem 1 of Chapter 5 of Gelfand et al. [30] which gives a necessary condition for $F[f(x)]$ to have a minimum at $f_0(x)$. Adapting to our notation, we have:

Theorem A1 ([30]). *A necessary condition for the functional $F[f(x)]$ to have a minimum at $f(x) = f_0(x)$ is that for $f(x) = f_0(x)$ and all admissible $\epsilon \cdot h(x)$,*

$$\delta^2 F[f(x)] \geq 0.$$

Applying to our functional $\text{IB}_\beta[p(z|x)]$, an immediate result of Theorem A1 is that, if at $p(z|x) = p(z)$, there exists an $\epsilon \cdot h(z|x)$ such that $\delta^2 \text{IB}_\beta[p(z|x)] < 0$, then $p(z|x) = p(z)$ is not a minimum for $\text{IB}_\beta[p(z|x)]$. Using the definition of IB_β learnability, we have that (X, Y) is IB_β-learnable. □

Appendix A.5. First- and Second-Order Variations of $IB_\beta[p(z|x)]$

In this section, we derive the first- and second-order variations of $\text{IB}_\beta[p(z|x)]$, which are needed for proving Lemma 2 and Theorem 3.

Lemma A1. *Using perturbative function $h(z|x)$, we have*

$$\delta \text{IB}_\beta[p(z|x)] = \int dxdz\, p(x)h(z|x)\log\frac{p(z|x)}{p(z)} - \beta \int dxdydz\, p(x,y)h(z|x)\log\frac{p(z|y)}{p(z)}$$

$$\delta^2 \text{IB}_\beta[p(z|x)] = \frac{1}{2}\Bigg[\int dxdz\,\frac{p(x)^2}{p(x,z)}h(z|x)^2 - \beta\int dxdx'dydz\,\frac{p(x,y)p(x',y)}{p(y,z)}h(z|x)h(z|x')$$

$$+ (\beta - 1)\int dxdx'dz\,\frac{p(x)p(x')}{p(z)}h(z|x)h(z|x')\Bigg]$$

Proof. Since $\text{IB}_\beta[p(z|x)] = I(X;Z) - \beta I(Y;Z)$, let us calculate the first and second-order variation of $I(X;Z)$ and $I(Y;Z)$ w.r.t. $p(z|x)$, respectively. Through this derivation, we use $\epsilon \cdot h(z|x)$ as a perturbative function, for ease of deciding different orders of variations. We assume that $h(z|x)$ is continuous, and there exists a constant M such that $\left|\frac{h(z|x)}{p(z|x)}\right| < M$, $\forall (x, z) \in \mathcal{X} \times \mathcal{Z}$. We will finally absorb ϵ into $h(z|x)$.

Denote $I(X;Z) = F_1[p(z|x)]$. We have

$$F_1[p(z|x)] = I(X;Z) = \int dxdz\, p(z|x)p(x)\log\frac{p(z|x)}{p(z)}$$

In this paper, we implicitly assume that the integral (or summing) are only on the support of $p(x, y, z)$.

Since

$$p(z) = \int p(z|x)p(x)dx$$

We have

$$p(z)|_{p(z|x)+\epsilon h(z|x)} = p(z)|_{p(z|x)} + \epsilon \int h(z|x)p(x)dx$$

Expanding $F_1[p(z|x) + \epsilon h(z|x)]$ to the second order of ϵ, we have

$F_1[p(z|x) + \epsilon h(z|x)]$

$= \int dxdz p(x)[p(z|x) + \epsilon h(z|x)] \log \frac{p(z|x) + \epsilon h(z|x)}{p(z) + \epsilon \int h(z|x')p(x')dx'}$

$= \int dxdz p(x)p(z|x)\left(1 + \epsilon \frac{h(z|x)}{p(z|x)}\right) \log \frac{p(z|x)\left(1 + \epsilon \frac{h(z|x)}{p(z|x)}\right)}{p(z)\left(1 + \epsilon \frac{\int h(z|x')p(x')dx'}{p(z)}\right)}$

$= \int dxdz p(x)p(z|x)\left(1 + \epsilon \frac{h(z|x)}{p(z|x)}\right) \log \left[\frac{p(z|x)}{p(z)}\left(1 + \epsilon \frac{h(z|x)}{p(z|x)}\right)\left(1 - \epsilon \frac{\int h(z|x')p(x')dx'}{p(z)}\right.\right.$

$\left.\left. + \epsilon^2 \left(\frac{\int h(z|x')p(x')dx'}{p(z)}\right)^2\right)\right] + \mathcal{O}(\epsilon^3)$

$= \int dxdz p(x)p(z|x)\left(1 + \epsilon \frac{h(z|x)}{p(z|x)}\right) \log \left[\frac{p(z|x)}{p(z)}\left(1 + \epsilon\left(\frac{h(z|x)}{p(z|x)} - \frac{\int h(z|x')p(x')dx'}{p(z)}\right)\right.\right.$

$\left.\left. + \epsilon^2 \left(\frac{\int h(z|x')p(x')dx'}{p(z)}\right)^2 - \epsilon^2 \frac{h(z|x)}{p(z|x)}\frac{\int h(z|x')p(x')dx'}{p(z)}\right)\right] + \mathcal{O}(\epsilon^3)$

$= \int dxdz p(x)p(z|x)\left(1 + \epsilon \frac{h(z|x)}{p(z|x)}\right) \left[\log \frac{p(z|x)}{p(z)} + \epsilon\left(\frac{h(z|x)}{p(z|x)} - \frac{\int h(z|x')p(x')dx'}{p(z)}\right)\right.$

$\left. + \epsilon^2 \left(\frac{\int h(z|x')p(x')dx'}{p(z)}\right)^2 - \epsilon^2 \frac{h(z|x)}{p(z|x)}\frac{\int h(z|x')p(x')dx'}{p(z)} - \frac{1}{2}\epsilon^2\left(\frac{h(z|x)}{p(z|x)} - \frac{\int h(z|x')p(x')dx'}{p(z)}\right)^2\right] + \mathcal{O}(\epsilon^3)$

Collecting the first order terms of ϵ, we have

$\delta F_1[p(z|x)]$

$= \epsilon \int dxdz p(x)p(z|x)\left(\frac{h(z|x)}{p(z|x)} - \frac{\int h(z|x')p(x')dx'}{p(z)}\right) + \epsilon \int dxdz p(x)p(z|x)\frac{h(z|x)}{p(z|x)}\log\frac{p(z|x)}{p(z)}$

$= \epsilon \int dxdz p(x)h(z|x) - \epsilon \int dx'dz p(x')h(z|x') + \epsilon \int dxdz p(x)h(z|x)\log\frac{p(z|x)}{p(z)}$

$= \epsilon \int dxdz p(x)h(z|x)\log\frac{p(z|x)}{p(z)}$

Collecting the second order terms of ϵ^2, we have

$\delta^2 F_1[p(z|x)]$

$= \epsilon^2 \int dxdz p(x)p(z|x)\left[\left(\frac{\int h(z|x')p(x')dx'}{p(z)}\right)^2 - \frac{h(z|x)}{p(z|x)}\frac{\int h(z|x')p(x')dx'}{p(z)} - \frac{1}{2}\left(\frac{h(z|x)}{p(z|x)} - \frac{\int h(z|x')p(x')dx'}{p(z)}\right)^2\right]$

$+ \epsilon^2 \int dxdz p(x)p(z|x)\frac{h(z|x)}{p(z|x)}\left(\frac{h(z|x)}{p(z|x)} - \frac{\int h(z|x')p(x')dx'}{p(z)}\right)$

$= \frac{\epsilon^2}{2}\int dxdz \frac{p(x)^2}{p(x,z)}h(z|x)^2 - \frac{\epsilon^2}{2}\int dxdx'dz \frac{p(x)p(x')}{p(z)}h(z|x)h(z|x')$

Now let us calculate the first and second-order variation of $F_2[p(z|x)] = I(Z;Y)$. We have

$F_2[p(z|x)] = I(Y;Z) = \int dydz p(z|y)p(y)\log\frac{p(y,z)}{p(y)p(z)} = \int dxdydz p(z|y)p(x,y)\log\frac{p(y,z)}{p(y)p(z)}$

Using the Markov chain $Z \leftarrow X \leftrightarrow Y$, we have

$$p(y,z) = \int p(z|x)p(x,y)dx$$

Hence

$$p(y,z)|_{p(z|x)+\epsilon h(z|x)} = p(y,z)|_{p(z|x)} + \epsilon \int h(z|x)p(x,y)dx$$

Then expanding $F_2[p(z|x) + \epsilon h(z|x)]$ to the second order of ϵ, we have

$F_2[p(z|x) + \epsilon h(z|x)]$

$= \int dxdydz\, p(x,y)p(z|x)\left(1 + \epsilon\frac{h(z|x)}{p(z|x)}\right)\log\frac{p(y,z)\left(1 + \epsilon\frac{\int h(z|x')p(x',y)dx'}{p(y,z)}\right)}{p(y)p(z)(1 + \epsilon\frac{\int h(z|x'')p(x'')dx''}{p(z)})}$

$= \int dxdydz\, p(x,y)p(z|x)\left(1 + \epsilon\frac{h(z|x)}{p(z|x)}\right)\left[\log\frac{p(y,z)}{p(y)p(z)} + \epsilon\left(\frac{\int h(z|x')p(x',y)dx'}{p(y,z)} - \frac{\int h(z|x')p(x')dx'}{p(z)}\right)\right.$

$\left. + \epsilon^2\left[\left(\frac{\int h(z|x')p(x')dx'}{p(z)}\right)^2 - \frac{\int h(z|x')p(x',y)dx'\int h(z|x'')p(x'')dx''}{p(y,z)\,p(z)} - \frac{1}{2}\left(\frac{\int h(z|x')p(x',y)dx'}{p(y,z)} - \frac{\int h(z|x')p(x')dx'}{p(z)}\right)^2\right]\right]$

$+ \mathcal{O}(\epsilon^3)$

Collecting the first order terms of ϵ, we have

$\delta F_2[p(z|x)]$

$= \epsilon\int dxdydz\, p(x,y)h(z|x)\log\frac{p(y,z)}{p(y)p(z)} + \epsilon\int dxdydz\, p(x,y)p(z|x)\frac{\int h(z|x')p(x',y)dx'}{p(y,z)}$

$- \epsilon\int dxdydz\, p(x,y)p(z|x)\frac{\int h(z|x')p(x')dx'}{p(z)}$

$= \epsilon\int dxdydz\, p(x,y)h(z|x)\log\frac{p(y,z)}{p(y)p(z)} + \epsilon\int dx'dydz\, h(z|x')p(x',y) - \epsilon\int dz\, h(z|x')p(x')dx'$

$= \epsilon\int dxdydz\, p(x,y)h(z|x)\log\frac{p(z|y)}{p(z)}$

Collecting the second order terms of ϵ, we have

$\delta^2 F_2[p(z|x)]$

$= \epsilon^2\int dxdydz\, p(x,y)p(z|x)\left[\left(\frac{\int h(z|x')p(x')dx'}{p(z)}\right)^2 - \frac{\int h(z|x')p(x',y)dx'\int h(z|x'')p(x'')dx''}{p(y,z)\,p(z)}\right]$

$- \frac{\epsilon^2}{2}\int dxdydz\, p(x,y)p(z|x)\left(\frac{\int h(z|x')p(x',y)dx'}{p(y,z)} - \frac{\int h(z|x')p(x')dx'}{p(z)}\right)^2$

$+ \epsilon^2\int dxdydz\, p(x,y)p(z|x)\frac{h(z|x)}{p(z|x)}\left(\frac{\int h(z|x')p(x',y)dx'}{p(y,z)} - \frac{\int h(z|x')p(x')dx'}{p(z)}\right)$

$= \frac{\epsilon^2}{2}\int dxdx'dydz\frac{p(x,y)p(x',y)}{p(y,z)}h(z|x)h(z|x') - \frac{\epsilon^2}{2}\int dxdx'dz\frac{p(x)p(x')}{p(z)}h(z|x)h(z|x')$

Finally, we have

$$\delta IB_\beta[p(z|x)] = \delta F_1[p(z|x)] - \beta\cdot\delta F_2[p(z|x)]$$
$$= \epsilon\left(\int dxdz\, p(x)h(z|x)\log\frac{p(z|x)}{p(z)} - \beta\int dxdydz\, p(x,y)h(z|x)\log\frac{p(z|y)}{p(z)}\right) \quad (A1)$$

$$\delta^2 \mathrm{IB}_\beta[p(z|x)] = \delta^2 F_1[p(z|x)] - \beta \cdot \delta^2 F_2[p(z|x)]$$
$$= \frac{\epsilon^2}{2} \int dxdz \frac{p(x)^2}{p(x,z)} h(z|x)^2 - \frac{\epsilon^2}{2} \int dxdx'dz \frac{p(x)p(x')}{p(z)} h(z|x)h(z|x')$$
$$- \beta \epsilon^2 \left[\frac{1}{2} \int dxdx'dydz \frac{p(x,y)p(x',y)}{p(y,z)} h(z|x)h(z|x') - \frac{1}{2} \int dxdx'dz \frac{p(x)p(x')}{p(z)} h(z|x)h(z|x') \right]$$
$$= \frac{\epsilon^2}{2} \left[\int dxdz \frac{p(x)^2}{p(x,z)} h(z|x)^2 \right.$$
$$\left. - \beta \int dxdx'dydz \frac{p(x,y)p(x',y)}{p(y,z)} h(z|x)h(z|x') + (\beta - 1) \int dxdx'dz \frac{p(x)p(x')}{p(z)} h(z|x)h(z|x') \right]$$

Absorb ϵ into $h(z|x)$, we get rid of the ϵ factor and obtain the final expression in Lemma A1. □

Appendix A.6. Proof of Lemma 2

Proof. Using Lemma A1, we have

$$\delta \mathrm{IB}_\beta[p(z|x)] = \int dxdz p(x) h(z|x) \log \frac{p(z|x)}{p(z)} - \beta \int dxdydz p(x,y) h(z|x) \log \frac{p(z|y)}{p(z)}$$

Let $p(z|x) = p(z)$ (the trivial representation), we have that $\log \frac{p(z|x)}{p(z)} \equiv 0$. Therefore, the two integrals are both 0. Hence,

$$\delta \mathrm{IB}_\beta[p(z|x)]\big|_{p(z|x)=p(z)} \equiv 0$$

Therefore, the $p(z|x) = p(z)$ is a stationary solution for $\mathrm{IB}_\beta[p(z|x)]$. □

Appendix A.7. Proof of Theorem 3

Proof. Firstly, from the necessary condition of $\beta > 1$ in Section 3, we have that any sufficient condition for IB_β-learnability should be able to deduce $\beta > 1$.

Now using Theorem 2, a sufficient condition for (X, Y) to be IB_β-learnable is that there exists $h(z|x)$ with $\int h(z|x)dx = 0$ such that $\delta^2 \mathrm{IB}_\beta[p(z|x)] < 0$ at $p(z|x) = p(x)$.

At the trivial representation, $p(z|x) = p(z)$ and hence $p(x,z) = p(x)p(z)$. Due to the Markov chain $Z \leftarrow X \leftrightarrow Y$, we have $p(y,z) = p(y)p(z)$. Substituting them into the $\delta^2 \mathrm{IB}_\beta[p(z|x)]$ in Lemma A1, the condition becomes: there exists $h(z|x)$ with $\int h(z|x)dz = 0$, such that

$$0 > \delta^2 \mathrm{IB}_\beta[p(z|x)] = \qquad\qquad\qquad\qquad\qquad\qquad\qquad\qquad\qquad\qquad\qquad\qquad\qquad\qquad\text{(A2)}$$
$$\frac{1}{2} \left[\int dxdz \frac{p(x)^2}{p(x)p(z)} h(z|x)^2 - \beta \int dxdx'dydz \frac{p(x,y)p(x',y)}{p(y)p(z)} h(z|x)h(z|x') + (\beta - 1) \int dxdx'dz \frac{p(x)p(x')}{p(z)} h(z|x)h(z|x') \right]$$

Rearranging terms and simplifying, we have

$$\int \frac{dz}{p(z)} G[h(z|x)] = \int \frac{dz}{p(z)} \left[\int dx h(z|x)^2 p(x) - \beta \int \frac{dy}{p(y)} \left(\int dx h(z|x) p(x) p(y|x) \right)^2 + (\beta - 1) \left(\int dx h(z|x) p(x) \right)^2 \right] < 0$$

where

$$G[h(x)] = \int dx h(x)^2 p(x) - \beta \int \frac{dy}{p(y)} \left(\int dx h(x) p(x) p(y|x) \right)^2 + (\beta - 1) \left(\int dx h(x) p(x) \right)^2$$

Now we prove that the condition that $\exists h(z|x)$ s.t. $\int \frac{dz}{p(z)} G[h(z|x)] < 0$ is equivalent to the condition that $\exists h(x)$ s.t. $G[h(x)] < 0$.

If $\forall h(z|x)$, $G[h(z|x)] \geq 0$, then we have $\forall h(z|x)$, $\int \frac{dz}{p(z)} G[h(z|x)] \geq 0$. Therefore, if $\exists h(z|x)$ s.t. $\int \frac{dz}{p(z)} G[h(z|x)] < 0$, we have that $\exists h(z|x)$ s.t. $G[h(z|x)] < 0$. Since the functional $G[h(z|x)]$ does not contain integration over z, we can treat the z in $G[h(z|x)]$ as a parameter and we have that $\exists h(x)$ s.t. $G[h(x)] < 0$.

Conversely, if there exists an certain function $h(x)$ such that $G[h(x)] < 0$, we can find some $h_2(z)$ such that $\int h_2(z)dz = 0$ and $\int \frac{h_2^2(z)}{p(z)} dz > 0$, and let $h_1(z|x) = h(x)h_2(z)$. Now we have

$$\int \frac{dz}{p(z)} G[h(z|x)] = \int \frac{h_2^2(z) dz}{p(z)} G[h(x)] = G[h(x)] \int \frac{h_2^2(z) dz}{p(z)} < 0$$

In other words, the condition Equation (A2) is equivalent to requiring that there exists an $h(x)$ such that $G[h(x)] < 0$. Hence, a sufficient condition for IB_β-learnability is that there exists an $h(x)$ such that

$$G[h(x)] = \int dx h(x)^2 p(x) - \beta \int \frac{dy}{p(y)} \left(\int dx h(x) p(x) p(y|x) \right)^2 + (\beta - 1) \left(\int dx h(x) p(x) \right)^2 < 0 \quad (A3)$$

When $h(x) = C = $ constant in the entire input space $\mathcal{X}$, Equation (A3) becomes:

$$C^2 - \beta C^2 + (\beta - 1)C^2 < 0$$

which cannot be true. Therefore, $h(x) = $ constant cannot satisfy Equation (A3).

Rearranging terms and simplifying, we have

$$\beta \left[\int \frac{dy}{p(y)} \left(\int dx h(x) p(x) p(y|x) \right)^2 - \left(\int dx h(x) p(x) \right)^2 \right] > \int dx h(x)^2 p(x) - \left(\int dx h(x) p(x) \right)^2 \quad (A4)$$

Written in the form of expectations, we have

$$\beta \cdot \left(\mathbb{E}_{y \sim p(y)} \left[\left(\mathbb{E}_{x \sim p(x|y)}[h(x)] \right)^2 \right] - \left(\mathbb{E}_{x \sim p(x)}[h(x)] \right)^2 \right) > \mathbb{E}_{x \sim p(x)}[h(x)^2] - \left(\mathbb{E}_{x \sim p(x)}[h(x)] \right)^2 \quad (A5)$$

Since the square function is convex, using Jensen's inequality on the L.H.S. of Equation (A5), we have

$$\mathbb{E}_{y \sim p(y)} \left[\left(\mathbb{E}_{x \sim p(x|y)}[h(x)] \right)^2 \right] \geq \left(\mathbb{E}_{y \sim p(y)} \left[\mathbb{E}_{x \sim p(x|y)}[h(x)] \right] \right)^2 = \left(\mathbb{E}_{x \sim p(x)}[h(x)] \right)^2$$

The equality holds iff $\mathbb{E}_{x \sim p(x|y)}[h(x)]$ is constant w.r.t. y, i.e., Y is independent of X. Therefore, in order for Equation (A5) to hold, we require that Y is not independent of X.

Using Jensen's inequality on the innter expectation on the L.H.S. of Equation (A5), we have

$$\mathbb{E}_{y \sim p(y)} \left[\left(\mathbb{E}_{x \sim p(x|y)}[h(x)] \right)^2 \right] \leq \mathbb{E}_{y \sim p(y)} \left[\mathbb{E}_{x \sim p(x|y)}[h(x)^2] \right] = \mathbb{E}_{x \sim p(x)}[h(x)^2] \quad (A6)$$

The equality holds when $h(x)$ is a constant. Since we require that $h(x)$ is not a constant, we have that the equality cannot be reached.

Similarly, using Jensen's inequality on the R.H.S. of Equation (A5), we have that

$$\mathbb{E}_{x \sim p(x)}[h(x)^2] > \left(\mathbb{E}_{x \sim p(x)}[h(x)] \right)^2$$

where we have used the requirement that $h(x)$ cannot be constant.

Under the constraint that Y is not independent of X, we can divide both sides of Equation (A5), and obtain the condition: there exists an $h(x)$ such that

$$\beta > \frac{\mathbb{E}_{x \sim p(x)}[h(x)^2] - \left(\mathbb{E}_{x \sim p(x)}[h(x)]\right)^2}{\mathbb{E}_{y \sim p(y)}\left[\left(\mathbb{E}_{x \sim p(x|y)}[h(x)]\right)^2\right] - \left(\mathbb{E}_{x \sim p(x)}[h(x)]\right)^2}$$

i.e.,

$$\beta > \inf_{h(x)} \frac{\mathbb{E}_{x \sim p(x)}[h(x)^2] - \left(\mathbb{E}_{x \sim p(x)}[h(x)]\right)^2}{\mathbb{E}_{y \sim p(y)}\left[\left(\mathbb{E}_{x \sim p(x|y)}[h(x)]\right)^2\right] - \left(\mathbb{E}_{x \sim p(x)}[h(x)]\right)^2}$$

which proves the condition of Theorem 3.

Furthermore, from Equation (A6) we have

$$\beta_0[h(x)] > 1$$

for $h(x) \not\equiv \text{const}$, which satisfies the necessary condition of $\beta > 1$ in Section 3.

Proof of lower bound of slope of the Pareto frontier at the origin: Now we prove the second statement of Theorem 3. Since $\delta I(X;Z) = 0$ and $\delta I(Y;Z) = 0$ according to Lemma 2, we have $\left(\frac{\Delta I(Y;Z)}{\Delta I(X;Z)}\right)^{-1} = \left(\frac{\delta^2 I(Y;Z)}{\delta^2 I(X;Z)}\right)^{-1}$. Substituting into the expression of $\delta^2 I(Y;Z)$ and $\delta^2 I(X;Z)$ from Lemma A1, we have

$$\left(\frac{\Delta I(Y;Z)}{\Delta I(X;Z)}\right)^{-1}$$

$$= \left(\frac{\delta^2 I(Y;Z)}{\delta^2 I(X;Z)}\right)^{-1}$$

$$= \frac{\frac{\epsilon^2}{2} \int dxdz \frac{p(x)^2}{p(x)p(z)} h(z|x)^2 - \frac{\epsilon^2}{2} \int dxdx'dz \frac{p(x)p(x')}{p(z)} h(z|x)h(z|x')}{\frac{\epsilon^2}{2} \int dxdx'dydz \frac{p(x,y)p(x',y)}{p(y)p(z)} h(z|x)h(z|x') - \frac{\epsilon^2}{2} \int dxdx'dz \frac{p(x)p(x')}{p(z)} h(z|x)h(z|x')}$$

$$= \frac{\left(\int dx p(x)h(x)^2 - \int dxdx' p(x)p(x')h(x)h(z|x')\right) \int \frac{h_2(z)^2}{p(z)} dz}{\left(\int dxdx'dy \frac{p(x,y)p(x',y)}{p(y)} h(x)h(z|x') - \int dxdx' p(x)p(x')h(x)h(z|x')\right) \int \frac{h_2(z)^2}{p(z)} dz}$$

$$= \frac{\int dx p(x)h(x)^2 - \int dxdx' p(x)p(x')h(x)h(z|x')}{\int dxdx'dy \frac{p(x,y)p(x',y)}{p(y)} h(x)h(z|x') - \int dxdx' p(x)p(x')h(x)h(z|x')}$$

$$= \frac{\mathbb{E}_{x \sim p(x)}[h(x)^2] - \left(\mathbb{E}_{x \sim p(x)}[h(x)]\right)^2}{\mathbb{E}_{y \sim p(y)}\left[\left(\mathbb{E}_{x \sim p(x|y)}[h(x)]\right)^2\right] - \left(\mathbb{E}_{x \sim p(x)}[h(x)]\right)^2}$$

$$= \frac{\frac{\mathbb{E}_{x \sim p(x)}[h(x)^2]}{\left(\mathbb{E}_{x \sim p(x)}[h(x)]\right)^2} - 1}{\mathbb{E}_{y \sim p(y)}\left[\left(\frac{\mathbb{E}_{x \sim p(x|y)}[h(x)]}{\mathbb{E}_{x \sim p(x)}[h(x)]}\right)^2\right] - 1}$$

$$= \beta_0[h(x)]$$

Therefore, $\left(\inf_{h(x)} \beta_0[h(x)]\right)^{-1}$ gives the largest slope of $\Delta I(Y;Z)$ vs. $\Delta I(X;Z)$ for perturbation function of the form $h_1(z|x) = h(x)h_2(z)$ satisfying $\int h_2(z) dz = 0$ and $\int \frac{h_2^2(z)}{p(z)} dz > 0$, which is a lower bound of slope of $\Delta I(Y;Z)$ vs. $\Delta I(X;Z)$ for all possible perturbation function $h_1(z|x)$. The latter is the slope of the Pareto frontier of the $I(Y;Z)$ vs. $I(X;Z)$ curve at the origin.

Inflection point for general Z: If we *do not* assume that Z is at the origin of the information plane, but at some general stationary solution Z^* with $p(z|x)$, we define

$$\beta^{(2)}[h(x)] = \left(\frac{\delta^2 I(Y;Z)}{\delta^2 I(X;Z)}\right)^{-1}$$

$$= \frac{\frac{\epsilon^2}{2}\int dxdz \frac{p(x)^2}{p(x,z)}h(z|x)^2 - \frac{\epsilon^2}{2}\int dxdx'dz \frac{p(x)p(x')}{p(z)}h(z|x)h(z|x')}{\frac{\epsilon^2}{2}\int dxdx'dydz \frac{p(x,y)p(x',y)}{p(y,z)}h(z|x)h(z|x') - \frac{\epsilon^2}{2}\int dxdx'dz \frac{p(x)p(x')}{p(z)}h(z|x)h(z|x')}$$

$$= \frac{\int dxdz \frac{p(x)^2}{p(x,z)}h(z|x)^2 - \int dxdx'dz \frac{p(x)p(x')}{p(z)}h(z|x)h(z|x')}{\int dxdx'dydz \frac{p(x,y)p(x',y)}{p(y,z)}h(z|x)h(z|x') - \int dxdx'dz \frac{p(x)p(x')}{p(z)}h(z|x)h(z|x')}$$

$$= \frac{\int \frac{dz}{p(z)}\left[\int dx \frac{p(x)^2}{p(x|z)}h(z|x)^2 - \left(\int dxp(x)h(z|x)\right)^2\right]}{\int \frac{dz}{p(z)}\left[\int \frac{dy}{p(y|z)}\left(\int dxp(x,y)h(z|x)\right)^2 - \left(\int dxp(x)h(z|x)\right)^2\right]}$$

$$= \frac{\int \frac{dz}{p(z)}\left[\frac{\int dx \frac{p(x)^2}{p(x|z)}h(z|x)^2}{\left(\int dxp(x)h(z|x)\right)^2} - 1\right]}{\int \frac{dz}{p(z)}\left[\frac{\int \frac{dy}{p(y|z)}\left(\int dxp(x,y)h(z|x)\right)^2}{\left(\int dxp(x)h(z|x)\right)^2} - 1\right]}$$

$$= \frac{\int dz \left[\frac{\int dx \frac{p(x)}{p(z|x)}h(z|x)^2}{\left(\int dxp(x)h(z|x)\right)^2} - \frac{1}{p(z)}\right]}{\int dz \left[\frac{\int \frac{dy}{p(z|y)p(y)}\left(\int dxp(x,y)h(z|x)\right)^2}{\left(\int dxp(x)h(z|x)\right)^2} - \frac{1}{p(z)}\right]}$$

$$= \frac{\int dz \left[\int dx \frac{p(x)}{p(z|x)}h(z|x)^2 - \frac{1}{p(z)}\left(\int dxp(x)h(z|x)\right)^2\right]}{\int dz \left[\int \frac{dy}{p(z|y)p(y)}\left(\int dxp(x,y)h(z|x)\right)^2 - \frac{1}{p(z)}\left(\int dxp(x)h(z|x)\right)^2\right]}$$

which reduces to $\beta_0[h(x)]$ when $p(z|x) = p(z)$. When

$$\beta > \inf_{h(z|x)} \beta^{(2)}[h(z|x)] \tag{A7}$$

it becomes a non-stable solution (non-minimum), and we will have other Z that achieves a better $IB_\beta(X,Y;Z)$ than the current Z^*. □

Appendix A.8. What IB First Learns at Its Onset of Learning

In this section, we prove that at the onset of learning, if letting $h(z|x) = h^*(x)h_2(z)$, we have

$$p_\beta(y|x) = p(y) + \epsilon^2 C_z(h^*(x) - \overline{h}_x^*)\int p(x,y)(h^*(x) - \overline{h}_x^*)dx \tag{A8}$$

where $p_\beta(y|x)$ is the estimated $p(y|x)$ by IB for a certain β, $h^*(x) = \inf_{h(x)} \beta_0[h(x)]$, $\overline{h}_x^* = \int h^*(x)p(x)dx$, $C_z = \int \frac{h_2^2(z)}{p(z)}dz$ is a constant.

Proof. In IB, we use $p_\beta(z|x)$ to obtain Z from X, then obtain the prediction of Y from Z using $p_\beta(y|z)$. Here we use subscript β to denote the probability (density) at the optimum of $IB_\beta[p(z|x)]$ at a specific β. We have

$$p_\beta(y|x) = \int p_\beta(y|z) p_\beta(z|x) dz$$

$$= \int dz \frac{p_\beta(y,z) p_\beta(z|x)}{p_\beta(z)}$$

$$= \int dz \frac{p_\beta(z|x)}{p_\beta(z)} \int p(x',y) p_\beta(z|x') dx'$$

When we have a small perturbation $\epsilon \cdot h(z|x)$ at the trivial representation, $p_\beta(z|x) = p_{\beta_0}(z) + \epsilon \cdot h(z|x)$, we have $p_\beta(z) = p_{\beta_0}(z) + \epsilon \cdot \int h(z|x'') p(x'') dx''$. Substituting, we have

$$p_\beta(y|x) = \int dz \frac{p_{\beta_0}(z) \left(1 + \epsilon \cdot \frac{h(z|x)}{p_{\beta_0}(z)}\right)}{p_{\beta_0}(z) \left(1 + \epsilon \cdot \frac{\int h(z|x'') p(x'') dx''}{p_{\beta_0}(z)}\right)} \int p(x',y) p_{\beta_0}(z) \left(1 + \epsilon \cdot \frac{h(z|x')}{p_{\beta_0}(z)}\right) dx'$$

$$= \int dz \frac{1 + \epsilon \cdot \frac{h(z|x)}{p_{\beta_0}(z)}}{1 + \epsilon \cdot \frac{\int h(z|x'') p(x'') dx''}{p_{\beta_0}(z)}} \int p(x',y) p_{\beta_0}(z) \left(1 + \epsilon \cdot \frac{h(z|x')}{p_{\beta_0}(z)}\right) dx'$$

The 0th-order term is $\int dz dx' p(x',y) p_{\beta_0}(z) = p(y)$. The first-order term is

$$\delta p_\beta(z|x) = \epsilon \cdot \int dz dx' \left(h(z|x) + h(z|x') - \int h(z|x'') p(x'') dx''\right) p(x',y)$$

$$= \epsilon \cdot \int dx' \left(\int dz h(z|x) + \int dz h(z|x')\right) - \epsilon \cdot \int dx' dx'' p(x',y) p(x'') \int dz h(z|x'')$$

$$= 0 - 0$$

$$= 0$$

since we have $\int h(z|x) dz = 0$ for any x.

For the second-order term, using $h(z|x) = h^*(x) h_2(z)$ and $C_z = \int \frac{dz}{p_{\beta_0}(z)} h_2^2(z)$, it is

$$\delta^2 p_\beta(y|x) = \epsilon^2 \cdot \int dz \left(\frac{\int h(z|x'') p(x'') dx''}{p_{\beta_0}(z)}\right)^2 \int p(x',y) p_{\beta_0}(z) dx'$$

$$- \epsilon^2 \cdot \int dz \frac{h(z|x) \int h(z|x'') p(x'') dx''}{(p_{\beta_0}(z))^2} \int p(x',y) p_{\beta_0}(z) dx'$$

$$+ \epsilon^2 \int dz \left(h(z|x) - \int h(z|x'') p(x'') dx\right) \int p(x',y) \frac{h(z|x')}{p_{\beta_0}(z)} dx'$$

$$= \epsilon^2 C_z \cdot \left(\int h^*(x'') p(x'') dx''\right)^2 p(y)$$

$$- \epsilon^2 C_z \cdot h^*(x) \int h^*(x'') p(x'') dx'' p(y)$$

$$+ \epsilon^2 C_z \cdot h^*(x) \int p(x',y) h^*(x') dx'$$

$$- \epsilon^2 C_z \cdot \int h^*(x'') p(x'') dx \int p(x',y) h^*(x') dx'$$

$$= \epsilon^2 C_z (h^*(x) - \overline{h}_x^*) \left[\left(\int p(x',y) h^*(x') dx'\right) - \overline{h}_x^* p(y)\right]$$

$$= \epsilon^2 C_z (h^*(x) - \overline{h}_x^*) \int p(x',y) \left(h^*(x') - \overline{h}_x^*\right) dx'$$

where $\overline{h}_x^* = \int h^*(x)p(x)dx$. Combining everything, we have up to the second order,

$$p_\beta(y|x) = p(y) + \epsilon^2 C_z(h^*(x) - \overline{h}_x^*) \int p(x,y)(h^*(x) - \overline{h}_x^*)dx$$

□

Appendix A.9. Proof of Theorem 4

Proof. According to Theorem 3, a sufficient condition for (X, Y) to be IB_β-learnable is that X and Y are not independent, and

$$\beta > \inf_{h(x)} \frac{\frac{\mathbb{E}_{x\sim p(x)}[h(x)^2]}{(\mathbb{E}_{x\sim p(x)}[h(x)])^2} - 1}{\mathbb{E}_{y\sim p(y)}\left[\left(\frac{\mathbb{E}_{x\sim p(x|y)}[h(x)]}{\mathbb{E}_{x\sim p(x)}[h(x)]}\right)^2\right] - 1} \quad (A9)$$

We can assume a specific form of $h(x)$, and obtain a (potentially stronger) sufficient condition. Specifically, we let

$$h(x) = \begin{cases} 1, & x \in \Omega_x \\ 0, & \text{otherwise} \end{cases} \quad (A10)$$

for certain $\Omega_x \subset \mathcal{X}$. Substituting into Equation (A10), we have that a sufficient condition for (X, Y) to be IB_β-learnable is

$$\beta > \inf_{\Omega_x \subset \mathcal{X}} \frac{\frac{p(\Omega_x)}{p(\Omega_x)^2} - 1}{\int dy\, p(y) \left(\frac{\int_{x \in \Omega_x} dx\, p(x|y)}{p(\Omega_x)}\right)^2 - 1} > 0 \quad (A11)$$

where $p(\Omega_x) = \int_{x \in \Omega_x} p(x)dx$.

The denominator of Equation (A11) is

$$\int dy\, p(y) \left(\frac{\int_{x \in \Omega_x} dx\, p(x|y)}{p(\Omega_x)}\right)^2 - 1$$

$$= \int dy\, p(y) \left(\frac{p(\Omega_x|y)}{p(\Omega_x)}\right)^2 - 1$$

$$= \int dy\, \frac{p(y|\Omega_x)^2}{p(y)} - 1$$

$$= \mathbb{E}_{y \sim p(y|\Omega_x)}\left[\frac{p(y|\Omega_x)}{p(y)} - 1\right]$$

Using the inequality $x - 1 \geq \log x$, we have

$$\mathbb{E}_{y \sim p(y|\Omega_x)}\left[\frac{p(y|\Omega_x)}{p(y)} - 1\right] \geq \mathbb{E}_{y \sim p(y|\Omega_x)}\left[\log \frac{p(y|\Omega_x)}{p(y)}\right] \geq 0$$

Both equalities hold iff $p(y|\Omega_x) \equiv p(y)$, at which the denominator of Equation (A11) is equal to 0 and the expression inside the infimum diverge, which will not contribute to the infimum. Except this scenario, the denominator is greater than 0. Substituting into Equation (A11), we have that a sufficient condition for (X, Y) to be IB_β-learnable is

$$\beta > \inf_{\Omega_x \subset \mathcal{X}} \frac{\frac{p(\Omega_x)}{p(\Omega_x)^2} - 1}{\mathbb{E}_{y \sim p(y|\Omega_x)}\left[\frac{p(y|\Omega_x)}{p(y)} - 1\right]} \quad (A12)$$

Since Ω_x is a subset of $\mathcal{X}$, by the definition of $h(x)$ in Equation (A10), $h(x)$ is not a constant in the entire $\mathcal{X}$. Hence the numerator of Equation (A12) is positive. Since its denominator is also positive, we can then neglect the "> 0", and obtain the condition in Theorem 4.

Since the $h(x)$ used in this theorem is a subset of the $h(x)$ used in Theorem 3, the infimum for Equation (5) is greater than or equal to the infimum in Equation (2). Therefore, according to the second statement of Theorem 3, we have that the $(\inf_{\Omega_x \subset \mathcal{X}} \beta_0(\Omega_x))^{-1}$ is also a lower bound of the slope for the Pareto frontier of $I(Y; Z)$ vs. $I(X; Z)$ curve.

Now we prove that the condition Equation (5) is invariant to invertible mappings of X. In fact, if $X' = g(X)$ is a uniquely invertible map (if X is continuous, g is additionally required to be continuous), let $\mathcal{X}' = \{g(x) | x \in \Omega_x\}$, and denote $g(\Omega_x) \equiv \{g(x) | x \in \Omega_x\}$ for any $\Omega_x \subset \mathcal{X}$, we have $p(g(\Omega_x)) = p(\Omega_x)$, and $p(y|g(\Omega_x)) = p(y|\Omega_x)$. Then for dataset (X, Y), let $\Omega'_x = g(\Omega_x)$, we have

$$\frac{\frac{1}{p(\Omega'_x)} - 1}{\mathbb{E}_{y \sim p(y|\Omega'_x)} \left[\frac{p(y|\Omega'_x)}{p(y)} - 1 \right]} = \frac{\frac{1}{p(\Omega_x)} - 1}{\mathbb{E}_{y \sim p(y|\Omega_x)} \left[\frac{p(y|\Omega_x)}{p(y)} - 1 \right]} \tag{A13}$$

Additionally we have $\mathcal{X}' = g(\mathcal{X})$. Then

$$\inf_{\Omega'_x \subset \mathcal{X}'} \frac{\frac{1}{p(\Omega'_x)} - 1}{\mathbb{E}_{y \sim p(y|\Omega'_x)} \left[\frac{p(y|\Omega'_x)}{p(y)} - 1 \right]} = \inf_{\Omega_x \subset \mathcal{X}} \frac{\frac{1}{p(\Omega_x)} - 1}{\mathbb{E}_{y \sim p(y|\Omega_x)} \left[\frac{p(y|\Omega_x)}{p(y)} - 1 \right]} \tag{A14}$$

For dataset $(X', Y) = (g(X), Y)$, applying Theorem 4 we have that a sufficient condition for it to be IB_β-learnable is

$$\beta > \inf_{\Omega'_x \subset \mathcal{X}'} \frac{\frac{1}{p(\Omega'_x)} - 1}{\mathbb{E}_{y \sim p(y|\Omega'_x)} \left[\frac{p(y|\Omega'_x)}{p(y)} - 1 \right]} = \inf_{\Omega_x \subset \mathcal{X}} \frac{\frac{1}{p(\Omega_x)} - 1}{\mathbb{E}_{y \sim p(y|\Omega_x)} \left[\frac{p(y|\Omega_x)}{p(y)} - 1 \right]} \tag{A15}$$

where the equality is due to Equation (A14). Comparing with the condition for IB_β-learnability for (X, Y) (Equation (5)), we see that they are the same. Therefore, the condition given by Theorem 4 is invariant to invertible mapping of X. □

Appendix A.10. Proof of Corollary 1 and Corollary 2

Appendix A.10.1. Proof of Corollary 1

Proof. We use Theorem 4. Let Ω_x contain all elements x whose true class is y^* for some certain y^*, and 0 otherwise. Then we obtain a (potentially stronger) sufficient condition. Since the probability $p(y|y^*, x) = p(y|y^*)$ is class-conditional, we have

$$\inf_{\Omega_x \subset \mathcal{X}} \frac{\frac{1}{p(\Omega_x)} - 1}{\mathbb{E}_{y \sim p(y|\Omega_x)} \left[\frac{p(y|\Omega_x)}{p(y)} - 1 \right]}$$

$$= \inf_{y^*} \frac{\frac{1}{p(y^*)} - 1}{\mathbb{E}_{y \sim p(y|y^*)} \left[\frac{p(y|y^*)}{p(y)} - 1 \right]}$$

By requiring $\beta > \inf_{y^*} \frac{\frac{1}{p(y^*)} - 1}{\mathbb{E}_{y \sim p(y|y^*)} \left[\frac{p(y|y^*)}{p(y)} - 1 \right]}$, we obtain a sufficient condition for IB_β learnability. □

Appendix A.10.2. Proof of Corollary 2

Proof. We again use Theorem 4. Since Y is a deterministic function of X, let $Y = f(X)$. By the assumption that Y contains at least one value y such that its probability $p(y) > 0$, we let Ω_x contain only x such that $f(x) = y$. Substituting into Equation (5), we have

$$\frac{\frac{1}{p(\Omega_x)} - 1}{\mathbb{E}_{y \sim p(y|\Omega_x)}\left[\frac{p(y|\Omega_x)}{p(y)} - 1\right]}$$

$$= \frac{\frac{1}{p(y)} - 1}{\mathbb{E}_{y \sim p(y|\Omega_x)}\left[\frac{1}{p(y)} - 1\right]}$$

$$= \frac{\frac{1}{p(y)} - 1}{\frac{1}{p(y)} - 1}$$

$$= 1$$

□

Therefore, the sufficient condition becomes $\beta > 1$.

Appendix A.11. β_0, Hypercontractivity Coefficient, Contraction Coefficient, $\beta_0[h(x)]$, and Maximum Correlation

In this section, we prove the relations between the IB-Learnability threshold β_0, the hypercontractivity coefficient $\xi(X;Y)$, the contraction coefficient $\eta_{KL}(p(y|x), p(x))$, $\beta_0[h(x)]$ in Equation (2), and maximum correlation $\rho_m(X,Y)$, as follows:

$$\frac{1}{\beta_0} = \xi(X;Y) = \eta_{KL}(p(y|x), p(x)) \geq \sup_{h(x)} \frac{1}{\beta_0[h(x)]} = \rho_m^2(X;Y) \quad (A16)$$

Proof. The hypercontractivity coefficient ξ is defined as [16]:

$$\xi(X;Y) \equiv \sup_{Z-X-Y} \frac{I(Y;Z)}{I(X;Z)}$$

By our definition of IB-learnability, (X, Y) is IB-Learnable iff there exists Z obeying the Markov chain $Z - X - Y$, such that

$$I(X;Z) - \beta \cdot I(Y;Z) < 0 = IB_\beta(X,Y;Z)|_{p(z|x)=p(z)}$$

Or equivalently there exists Z obeying the Markov chain $Z - X - Y$ such that

$$0 < \frac{1}{\beta} < \frac{I(Y;Z)}{I(X;Z)} \quad (A17)$$

By Theorem 1, the IB-Learnability region for β is $(\beta_0, +\infty)$, or equivalently the IB-Learnability region for $1/\beta$ is

$$0 < \frac{1}{\beta} < \frac{1}{\beta_0} \quad (A18)$$

Comparing Equations (A17) and (A18), we have that

$$\frac{1}{\beta_0} = \sup_{Z-X-Y} \frac{I(Y;Z)}{I(X;Z)} = \xi(X;Y) \quad (A19)$$

In Anantharam et al. [16], the authors prove that

$$\xi(X;Y) = \eta_{\text{KL}}(p(y|x), p(x)) \tag{A20}$$

where the contraction coefficient $\eta_{\text{KL}}(p(y|x), p(x))$ is defined as

$$\eta_{\text{KL}}(p(y|x), p(x)) = \sup_{r(x) \neq p(x)} \frac{\mathbb{D}_{\text{KL}}(r(y)||p(y))}{\mathbb{D}_{\text{KL}}(r(x)||p(x))}$$

where $p(y) = \mathbb{E}_{x \sim p(x)}[p(y|x)]$ and $r(y) = \mathbb{E}_{x \sim r(x)}[p(y|x)]$. Treating $p(y|x)$ as a channel, the contraction coefficient measures how much the two distributions $r(x)$ and $p(x)$ becomes "nearer" (as measured by the KL-divergence) after passing through the channel.

In Anantharam et al. [16], the authors also provide a counterexample to an earlier result by Erkip and Cover [31] that incorrectly proved $\xi(X;Y) = \rho_m^2(X;Y)$. In the specific counterexample Anantharam et al. [16] design, $\xi(X;Y) > \rho_m^2(X;Y)$.

The maximum correlation is defined as $\rho_m(X;Y) \equiv \max_{f,g} \mathbb{E}[f(X)g(Y)]$ where $f(X)$ and $g(Y)$ are real-valued random variables such that $\mathbb{E}[f(X)] = \mathbb{E}[g(Y)] = 0$ and $\mathbb{E}[f^2(X)] = \mathbb{E}[g^2(Y)] = 1$ [20,21].

Now we prove $\xi(X;Y) \geq \rho_m^2(X;Y)$, based on Theorem 3. To see this, we use the alternate characterization of $\rho_m(X;Y)$ by Rényi [32]:

$$\rho_m^2(X;Y) = \max_{f(X): \mathbb{E}[f(X)]=0, \mathbb{E}[f^2(X)]=1} \mathbb{E}[(\mathbb{E}[f(X)|Y])^2] \tag{A21}$$

Denoting $\overline{h} = \mathbb{E}_{p(x)}[h(x)]$, we can transform $\beta_0[h(x)]$ in Equation (2) as follows:

$$\beta_0[h(x)] = \frac{\mathbb{E}_{x \sim p(x)}[h(x)^2] - \left(\mathbb{E}_{x \sim p(x)}[h(x)]\right)^2}{\mathbb{E}_{y \sim p(y)}\left[\left(\mathbb{E}_{x \sim p(x|y)}[h(x)]\right)^2\right] - \left(\mathbb{E}_{x \sim p(x)}[h(x)]\right)^2}$$

$$= \frac{\mathbb{E}_{x \sim p(x)}[h(x)^2] - \overline{h}^2}{\mathbb{E}_{y \sim p(y)}\left[\left(\mathbb{E}_{x \sim p(x|y)}[h(x)]\right)^2\right] - \overline{h}^2}$$

$$= \frac{\mathbb{E}_{x \sim p(x)}[(h(x) - \overline{h})^2]}{\mathbb{E}_{y \sim p(y)}\left[\left(\mathbb{E}_{x \sim p(x|y)}[h(x) - \overline{h}]\right)^2\right]}$$

$$= \frac{1}{\mathbb{E}_{y \sim p(y)}\left[\left(\mathbb{E}_{x \sim p(x|y)}[f(x)]\right)^2\right]}$$

$$= \frac{1}{\mathbb{E}[(\mathbb{E}[f(X)|Y])^2]}$$

where we denote $f(x) = \frac{h(x) - \overline{h}}{\left(\mathbb{E}_{x \sim p(x)}[(h(x) - \overline{h})^2]\right)^{1/2}}$, so that $\mathbb{E}[f(X)] = 0$ and $\mathbb{E}[f^2(X)] = 1$.

Combined with Equation (A21), we have

$$\sup_{h(x)} \frac{1}{\beta_0[h(x)]} = \rho_m^2(X;Y) \tag{A22}$$

Our Theorem 3 states that

$$\sup_{h(x)} \frac{1}{\beta_0[h(x)]} \leq \frac{1}{\beta_0} \tag{A23}$$

49

Combining Equations (A18), (A22) and Equation (A23), we have

$$\rho_m^2(X;Y) \leq \xi(X;Y) \tag{A24}$$

In summary, the relations among the quantities are:

$$\frac{1}{\beta_0} = \xi(X;Y) = \eta_{\mathrm{KL}}(p(y|x), p(x)) \geq \sup_{h(x)} \frac{1}{\beta_0[h(x)]} = \rho_m^2(X;Y) \tag{A25}$$

□

Appendix A.12. Experiment Details

We use the Variational Information Bottleneck (VIB) objective from [5]. For the synthetic experiment, the latent Z has dimension of 2. The encoder is a neural net with 2 hidden layers, each of which has 128 neurons with ReLU activation. The last layer has linear activation and 4 output neurons; the first two parameterize the mean of a Gaussian and the last two parameterize the log variance. The decoder is a neural net with 1 hidden layer with 128 neurons and ReLU activation. Its last layer has linear activation and outputs the logit for the class labels. It uses a mixture of Gaussian prior with 500 components (for the experiment with class overlap, 256 components), each of which is a 2D Gaussian with learnable mean and log variance, and the weights for the components are also learnable. For the MNIST experiment, the architecture is mostly the same, except the following: (1) for Z, we let it have dimension of 256. (2) For the prior, we use standard Gaussian with diagonal covariance matrix.

For all experiments, we use Adam [33] optimizer with default parameters. We do not add any explicit regularization. We use learning rate of 10^{-4} and have a learning rate decay of $\frac{1}{1+0.01 \times \mathrm{epoch}}$. We train in total 2000 epochs with mini-batch size of 500.

For estimation of the observed β_0 in Figure 3, in the $I(X;Z)$ vs. β_i curve (β_i denotes the i-th β), we take the mean and standard deviation of $I(X;Z)$ for the lowest 5 β_i values, denoting as μ_β, σ_β ($I(Y;Z)$ has similar behavior, but since we are minimizing $I(X;Z) - \beta \cdot I(Y;Z)$, the onset of nonzero $I(X;Z)$ is less prone to noise). When $I(X;Z)$ is greater than $\mu_\beta + 3\sigma_\beta$, we regard it as learning a non-trivial representation, and take the average of β_i and β_{i-1} as the experimentally estimated onset of learning. We also inspect manually and confirm that it is consistent with human intuition.

For estimating β_0 using Algorithm 1, at step 6 we use the following discrete search algorithm. We fix $i_{\mathrm{left}} = 1$ and gradually narrow down the range $[a, b]$ of i_{right}, starting from $[1, N]$. At each iteration, we set a tentative new range $[a', b']$, where $a' = 0.8a + 0.2b$, $b' = 0.2a + 0.8b$, and calculate $\tilde{\beta}_{0,a'} = \mathbf{Get\beta}(P_{y|x}, p_y, \Omega_{a'})$, $\tilde{\beta}_{0,b'} = \mathbf{Get\beta}(P_{y|x}, p_y, \Omega_{b'})$ where $\Omega_{a'} = \{1, 2, ...a'\}$ and $\Omega_{b'} = \{1, 2, ...b'\}$. If $\tilde{\beta}_{0,a'} < \tilde{\beta}_{0,a}$, let $a \leftarrow a'$. If $\tilde{\beta}_{0,b'} < \tilde{\beta}_{0,b}$, let $b \leftarrow b'$. In other words, we narrow down the range of i_{right} if we find that the Ω given by the left or right boundary gives a lower $\tilde{\beta}_0$ value. The process stops when both $\tilde{\beta}_{0,a'}$ and $\tilde{\beta}_{0,b'}$ stop improving (which we find always happens when $b' = a' + 1$), and we return the smaller of the final $\tilde{\beta}_{0,a'}$ and $\tilde{\beta}_{0,b'}$ as $\tilde{\beta}_0$.

For estimation of $p(y|x)$ for (2') Algorithm 1 and (3') $\hat{\eta}_{\mathrm{KL}}$ for both synthetic and MNIST experiments, we use a 3-layer neuron net where each hidden layer has 128 neurons and ReLU activation. The last layer has linear activation. The objective is cross-entropy loss. We use Adam [33] optimizer with a learning rate of 10^{-4}, and train for 100 epochs (after which the validation loss does not go down).

For estimating β_0 via (3') $\hat{\eta}_{\mathrm{KL}}$ by the algorithm in [18], we use the code from the GitHub repository provided by the paper (At https://github.com/wgao9/hypercontractivity), using the same $p(y|x)$ employed for (2') Algorithm 1. Since our datasets are classification tasks, we use $A_{ij} = p(y_j|x_i)/p(y_j)$ instead of the kernel density for estimating matrix A; we take the maximum of 10 runs as estimation of μ.

CIFAR10 Details

We trained a deterministic 28 × 10 wide resnet [34,35], using the open source implementation from Cubuk et al. [36]. However, we extended the final 10 dimensional logits of that model through another 3 layer MLP classifier, in order to keep the inference network architecture identical between this model and the VIB models we describe below. During training, we dynamically added label noise according to the class confusion matrix in Table A1. The mean label noise averaged across the 10 classes is 20%. After that model had converged, we used it to estimate β_0 with Algorithm 1. Even with 20% label noise, β_0 was estimated to be 1.0483.

Table A1. Class confusion matrix used in CIFAR10 experiments. The value in row i, column j means for class i, the probability of labeling it as class j. The mean confusion across the classes is 20%.

	Plane	Auto.	Bird	Cat	Deer	Dog	Frog	Horse	Ship	Truck
Plane	0.82232	0.00238	0.021	0.00069	0.00108	0	0.00017	0.00019	0.1473	0.00489
Auto.	0.00233	0.83419	0.00009	0.00011	0	0.00001	0.00002	0	0.00946	0.15379
Bird	0.03139	0.00026	0.76082	0.0095	0.07764	0.01389	0.1031	0.00309	0.00031	0
Cat	0.00096	0.0001	0.00273	0.69325	0.00557	0.28067	0.01471	0.00191	0.00002	0.0001
Deer	0.00199	0	0.03866	0.00542	0.83435	0.01273	0.02567	0.08066	0.00052	0.00001
Dog	0	0.00004	0.00391	0.2498	0.00531	0.73191	0.00477	0.00423	0.00001	0
Frog	0.00067	0.00008	0.06303	0.05025	0.0337	0.00842	0.8433	0	0.00054	0
Horse	0.00157	0.00006	0.00649	0.00295	0.13058	0.02287	0	0.83328	0.00023	0.00196
Ship	0.1288	0.01668	0.00029	0.00002	0.00164	0.00006	0.00027	0.00017	0.83385	0.01822
Truck	0.01007	0.15107	0	0.00015	0.00001	0.00001	0	0.00048	0.02549	0.81273

We then trained 73 different VIB models using the same 28 × 10 wide resnet architecture for the encoder, parameterizing the mean of a 10-dimensional unit variance Gaussian. Samples from the encoder distribution were fed to the same 3 layer MLP classifier architecture used in the deterministic model. The marginal distributions were mixtures of 500 fully covariate 10-dimensional Gaussians, all parameters of which are trained. The VIB models had β ranging from 1.02 to 2.0 by steps of 0.02, plus an extra set ranging from 1.04 to 1.06 by steps of 0.001 to ensure we captured the empirical β_0 with high precision.

However, this particular VIB architecture does not start learning until $\beta > 2.5$, so none of these models would train as described. (A given architecture trained using maximum likelihood and with no stochastic layers will tend to have higher effective capacity than the same architecture with a stochastic layer that has a fixed but non-trivial variance, even though those two architectures have exactly the same number of learnable parameters.) Instead, we started them all at $\beta = 100$, and annealed β down to the corresponding target over 10,000 training gradient steps. The models continued to train for another 200,000 gradient steps after that. In all cases, the models converged to essentially their final accuracy within 20,000 additional gradient steps after annealing was completed. They were stable over the remaining ~180,000 gradient steps.

References

1. Tishby, N.; Pereira, F.C.; Bialek, W. The information bottleneck method. *arXiv* **2000**, arXiv:physics/0004057.
2. Shannon, C.E. A Mathematical Theory of Communication. *Bell Syst. Tech. J.* **1948**, *27*, 379–423. [CrossRef]
3. Chechik, G.; Globerson, A.; Tishby, N.; Weiss, Y. Information bottleneck for Gaussian variables. *J. Mach. Learn. Res.* **2005**, *6*, 165–188.
4. Rey, M.; Roth, V. Meta-Gaussian information bottleneck. In *Advances in Neural Information Processing Systems*; lNIPS: San Diego, CA, USA, 2012; pp. 1916–1924.
5. Alemi, A.A.; Fischer, I.; Dillon, J.V.; Murphy, K. Deep variational information bottleneck. *arXiv* **2016**, arXiv:1612.00410.

6. Chalk, M.; Marre, O.; Tkacik, G. Relevant sparse codes with variational information bottleneck. In *Advances in Neural Information Processing Systems*; NIPS: San Diego, CA, USA, 2016; pp. 1957–1965.
7. Fischer, I. The Conditional Entropy Bottleneck. 2018. Available online: https://openreview.net/forum?id=rkVOXhAqY7 (accessed on 20 September 2019).
8. Strouse, D.; Schwab, D.J. The deterministic information bottleneck. *Neural Comput.* **2017**, *29*, 1611–1630. [CrossRef] [PubMed]
9. Kolchinsky, A.; Tracey, B.D.; Van Kuyk, S. Caveats for information bottleneck in deterministic scenarios. In Proceedings of the International Conference on Learning Representations (ICLR), New Orleans, LA, USA, 30 April 2019.
10. Strouse, D.; Schwab, D.J. The information bottleneck and geometric clustering. *arXiv* **2017**, arXiv:1712.09657.
11. Achille, A.; Soatto, S. Emergence of invariance and disentanglement in deep representations. *J. Mach. Learn. Res.* **2018**, *19*, 1947–1980.
12. Achille, A.; Soatto, S. Information dropout: Learning optimal representations through noisy computation. *IEEE Trans. Pattern Anal. Mach. Intell.* **2018**. [CrossRef]
13. LeCun, Y.; Bottou, L.; Bengio, Y.; Haffner, P. Gradient-based learning applied to document recognition. *Proc. IEEE* **1998**, *86*, 2278–2324. [CrossRef]
14. Krizhevsky, A.; Hinton, G. *Learning Multiple Layers of Features from Tiny Images*; Technical Report; University of Toronto: Toronto, ON, Canada, 2009.
15. Achille, A.; Mbeng, G.; Soatto, S. The Dynamics of Differential Learning I: Information-Dynamics and Task Reachability. *arXiv* **2018**, arXiv:1810.02440.
16. Anantharam, V.; Gohari, A.; Kamath, S.; Nair, C. On maximal correlation, hypercontractivity, and the data processing inequality studied by Erkip and Cover. *arXiv* **2013**, arXiv:1304.6133.
17. Polyanskiy, Y.; Wu, Y. Strong data-processing inequalities for channels and Bayesian networks. In *Convexity and Concentration*; Springer: Berlin/Heidelberg, Germany, 2017; pp. 211–249.
18. Kim, H.; Gao, W.; Kannan, S.; Oh, S.; Viswanath, P. Discovering potential correlations via hypercontractivity. In *Advances in Neural Information Processing Systems*; NIPS: San Diego, CA, USA, 2017; pp. 4577–4587.
19. Lin, H.W.; Tegmark, M. Criticality in formal languages and statistical physics. *arXiv* **2016**, arXiv:1606.06737.
20. Hirschfeld, H.O. A connection between correlation and contingency. In *Mathematical Proceedings of the Cambridge Philosophical Society*; Cambridge University Press: Cambridge, UK, 1935; Volume 31, pp. 520–524.
21. Gebelein, H. Das statistische Problem der Korrelation als Variations-und Eigenwertproblem und sein Zusammenhang mit der Ausgleichsrechnung. *ZAMM-J. Appl. Math. Mech. Für Angew. Math. Und Mech.* **1941**, *21*, 364–379. [CrossRef]
22. Angluin, D.; Laird, P. Learning from noisy examples. *Mach. Learn.* **1988**, *2*, 343–370. [CrossRef]
23. Natarajan, N.; Dhillon, I.S.; Ravikumar, P.K.; Tewari, A. Learning with noisy labels. In *Advances in Neural Information Processing Systems*; NIPS: San Diego, CA, USA, 2013; pp. 1196–1204.
24. Liu, T.; Tao, D. Classification with noisy labels by importance reweighting. *IEEE Trans. Pattern Anal. Mach. Intell.* **2016**, *38*, 447–461. [CrossRef] [PubMed]
25. Xiao, T.; Xia, T.; Yang, Y.; Huang, C.; Wang, X. Learning from massive noisy labeled data for image classification. In Proceedings of the IEEE Conference on Computer Vision and Pattern Recognition, Boston, MA, USA, 7–12 June 2015; pp. 2691–2699.
26. Northcutt, C.G.; Wu, T.; Chuang, I.L. Learning with confident examples: Rank pruning for robust classification with noisy labels. *arXiv* **2017**, arXiv:1705.01936.
27. van den Oord, A.; Kalchbrenner, N.; Espeholt, L.; Kavukcuoglu, K.; Vinyals, O.; Graves, A. Conditional Image Generation with PixelCNN Decoders. In *Advances in Neural Information Processing Systems 29*; Lee, D.D., Sugiyama, M., Luxburg, U.V., Guyon, I., Garnett, R., Eds.; Curran Associates, Inc.: Red Hook, NY, USA, 2016; pp. 4790–4798.
28. Salimans, T.; Karpathy, A.; Chen, X.; Kingma, D.P. PixelCNN++: A PixelCNN Implementation with Discretized Logistic Mixture Likelihood and Other Modifications. In Proceedings of the International Conference on Learning Representations (ICLR), Toulon, France, 24–26 April 2017.
29. Kraskov, A.; Stögbauer, H.; Grassberger, P. Estimating mutual information. *Phys. Rev. E* **2004**, *69*, 066138. [CrossRef]
30. Gelfand, I.M.; Silverman, R.A. *Calculus of Variations*; Courier Corporation: North Chelmsford, MA, USA, 2000.

31. Erkip, E.; Cover, T.M. The efficiency of investment information. *IEEE Trans. Inf. Theory* **1998**, *44*, 1026–1040. [CrossRef]
32. Rényi, A. On measures of dependence. *Acta Math. Hung.* **1959**, *10*, 441–451. [CrossRef]
33. Kingma, D.P.; Ba, J. Adam: A method for stochastic optimization. *arXiv* **2014**, arXiv:1412.6980.
34. He, K.; Zhang, X.; Ren, S.; Sun, J. Deep Residual Learning for Image Recognition. In Proceedings of the IEEE Conference on Computer Vision and Pattern Recognition (CVPR), Las Vegas, NV, USA, 27–30 June 2016.
35. Zagoruyko, S.; Komodakis, N. Wide Residual Networks. *arXiv* **2016**, arXiv: 1605.07146.
36. Cubuk, E.D.; Zoph, B.; Mane, D.; Vasudevan, V.; Le, Q.V. Autoaugment: Learning augmentation policies from data. *arXiv* **2018**, arXiv:1805.09501.

© 2019 by the authors. Licensee MDPI, Basel, Switzerland. This article is an open access article distributed under the terms and conditions of the Creative Commons Attribution (CC BY) license (http://creativecommons.org/licenses/by/4.0/).

Article

Markov Information Bottleneck to Improve Information Flow in Stochastic Neural Networks

Thanh Tang Nguyen [1,*,†] and Jaesik Choi [2,*,‡]

1 Applied Artificial Intelligence Institute, Deakin University, Geelong VIC 3220, Australia
2 Graduate School of Artificial Intelligence, Korea Advanced Institute of Science and Technology, Daejeon 34141, Korea
* Correspondence: thanhnt@deakin.edu.au (T.T.N.); jaesik.choi@kaist.ac.kr (J.C.)
† Part of this work was done at Ulsan National Institute of Science and Technology, Ulsan 44919, Korea.
‡ Part of this work was done at Ulsan National Institute of Science and Technology, Ulsan 44919, Korea; part of the work was done at KAIST.

Received: 8 September 2019; Accepted: 30 September 2019; Published: 6 October 2019

Abstract: While rate distortion theory compresses data under a distortion constraint, information bottleneck (IB) generalizes rate distortion theory to learning problems by replacing a distortion constraint with a constraint of relevant information. In this work, we further extend IB to multiple Markov bottlenecks (i.e., latent variables that form a Markov chain), namely Markov information bottleneck (MIB), which particularly fits better in the context of stochastic neural networks (SNNs) than the original IB. We show that Markov bottlenecks cannot simultaneously achieve their information optimality in a non-collapse MIB, and thus devise an optimality compromise. With MIB, we take the novel perspective that each layer of an SNN is a bottleneck whose learning goal is to encode relevant information in a compressed form from the data. The inference from a hidden layer to the output layer is then interpreted as a variational approximation to the layer's decoding of relevant information in the MIB. As a consequence of this perspective, the maximum likelihood estimate (MLE) principle in the context of SNNs becomes a special case of the variational MIB. We show that, compared to MLE, the variational MIB can encourage better information flow in SNNs in both principle and practice, and empirically improve performance in classification, adversarial robustness, and multi-modal learning in MNIST.

Keywords: information bottleneck; stochastic neural networks; variational inference; machine learning

---

1. Introduction

The information bottleneck (IB) principle [1] extracts relevant information about a target variable Y from an input variable X via a *single* bottleneck variable Z. In particular, it constructs a *bottleneck* variable $Z = Z(X)$ that is a *compressed* version of X but preserves as much *relevant* information in X about Y as possible. This principle of introducing relevant information under compression finds vast applications in clustering problems [2], neural network compression [3], disentanglement learning [4–7], and reinforcement learning [8,9]. In addition, there have been many variants of the original IB principle, such as multivariate IB [10], Gaussian IB [11], meta-Gaussian IB [12], deterministic IB [13], and variational IB [14]. Despite these vast applications and variants of IB, alongside the theoretical analysis of the IB principle in neural networks [15,16], the context of stochastic neural networks in which mutual information can be most naturally well-defined [17] has not been sufficiently studied from the IB insight. In this work, we are particularly interested in this context in which multiple stochastic variables are constructed for representation in the form of a Markov chain.

Stochastic neural networks (SNNs) are a general class of neural networks with stochastic neurons in the computation graph. There has been an active line of research in SNNs, including restricted Boltzmann machines (RBMs) [18], deep belief networks (DBNs) [19], sigmoid belief networks (SBNs) [20], and stochastic feed-forward neural networks (SFFNs) [21]. One of the advantages of SNNs is that they can induce rich multi-modal distributions in the output space [20] and enable exploration in reinforcement learning [22]. For learning SNNs (and deep neural networks in general), the maximum likelihood estimate (MLE) principle (in its various forms, such as maximum log-likelihood or Kullback–Leibler divergence) has generally been a de-facto standard. The MLE principle maximizes the likelihood of the model for observing the entire training data. However, this principle is generic and not specially tailored to the hierarchical structure of neural networks. Particularly, MLE treats the entire neural network as a whole without considering the explicit contribution of its hidden layers to model learning. As a result, the information contained within the hidden structure may not be adequately modified to capture the data regularities reflecting a target variable. Thus, it is reasonable to ask if the MLE principle effectively and sufficiently exploits a neural network's representative power, and whether there is a better alternative.

Contributions. In this paper, (i) we propose Markov information bottleneck (MIB), a variant of the IB principle for multiple Markov bottlenecks that directly offers an alternative learning principle for SNNs. In MIB, there are multiple bottleneck variables (as opposed to one single bottleneck variable in the original IB) that form a Markov chain. These multiple Markov bottlenecks sequentially extract relevant information for a learning task. From the perspective of MIB, each layer of an SNN is a bottleneck whose information is encoded from the data via the network parameters connecting the layer to the data layer. (ii) We show that in a non-collapse MIB, the information optimality is not simultaneously achievable for all bottlenecks; thus, an optimality compromise is devised. (iii) When applied to SNNs for a learning task, we interpret the inference from a hidden layer to the output layer in SNNs as a variational approximation to that layer's intractable decoding of relevant information. Consequently, the variational MIB in SNNs generalizes the MLE principle. We demonstrate via a simple analytical argument and synthetic experiment that MLE is unable to learn a good information representation, while the variational MIB can. (iv) We then empirically show that MIB improves the performance in classification, adversarial robust learning, and multi-modal learning in the standard hand-digit recognition data MNIST [23]. This work is an extended version of our preprint [24] and the first author's Master thesis [25].

2. Related Work

There have been many extensions of the original IB framework [1]. One natural consideration is to extend it to continuous variables, yet under special settings where the optimal information representation is analytic [11,12]. Another direction uses alternative measures for compression and/or relevance in IB [13]. Since the optimal information representation in IB is tractable only in limited settings such as discrete variables [1], Gaussian variables [11], and meta-Gaussian variables [12], scaling the IB solution using neural networks and variational inference is a very successful extension [14]. The closest extension to our MIB is multivariate IB [10], in which they define multi-information to capture the dependence among the elements of a multivariate variable. However, in MIB, we do not focus on capturing such multi-information but rather the optimal information sequentially processed by a Markov chain of (possibly multivariate) bottleneck variables.

The line of work applying the IB principle to learn information representation in neural networks is also relevant to our approach. For example, Reference [15] proposes the use of the mutual information of a hidden layer with the input layer and the output layer to quantify the performance of neural networks. However, it is not clear as to how the IB information optimality changes in multiple bottlenecks in a neural network and how we can approximate the IB solutions in this high-dimensional context. In addition, MLE is a standard learning principle for neural networks. It has been shown that the IB principle is mathematically equivalent to the MLE principle in the multinomial mixture model for the clustering problem when the input distribution X is uniform or has a large sample size [26]. However, it is also not clear how these two principles are related to each other in the context of neural networks. Moreover, regarding the feasibility of the IB principle for representation learning in neural networks, Reference [17] analyzes two critical issues of mutual information that representation learning might suffer from: indefinite in deterministic encoding, and invariant under bijective transformations. These are inherent properties of mutual information which are also studied in, for example, [7,27,28]. In MIB, we share with [17] the same insight in these caveats by considering only the scenario where mutual information is well defined. This also explains our rationale in applying MIB to stochastic neural networks.

Deep learning compression schemes [3,29] loosely bear some similarity with our work. Both of the directions aim for a more compressed and useful neural networks for given tasks. The critical distinction is that deep learning compression schemes attempt to produce a smaller-sized neural network with similar performance of a larger one so that the network can be efficiently deployed in small devices such as mobile phones. This task therefore involves size-reduction techniques such as neural network pruning, low-rank factorization, transferred convolution filters and knowledge distillation [29] . On the other hand, our work asks an important representation learning question that given a neural network, what learning principles are the best we can do to improve the information content learned from the data for a given task? In this work, we attempt to address this question via the perspective that a neural network is a set of stochastic variables that sequentially encode information into its layers. We then explicitly improve the information flow (in the sense of more compressed but relevant information) for each layer via our introduced Markov Information Bottleneck framework.

3. Preliminaries

3.1. Notations

We denote random variables (RVs) by capital letters (e.g., X), and their specific realization value by the corresponding lowercase letter (e.g., x). We write $X \perp Y$ (respectively, $X \not\perp Y$) to indicate that X and Y are independent (respectively, not independent). We denote a Markov chain by $Y \to X \to Z$, that is, Y and Z are conditionally independent given X, or $Y \perp Z | X$. We use the integral notation when taking expectation (e.g., $\int p(x)f(x)dx$) over the distribution of a random variable regardless of whether the variable is discrete or continuous. We also adopt the following conventions from [27] for defining entropy (denoted by H), mutual information (denoted by I), and Kullback–Leibler (KL) divergence (denoted by D_{KL}): $0 \log \frac{0}{0} = 0, 0 \log \frac{0}{q} = 0, p \log \frac{p}{0} = \infty$.

3.2. Information Bottleneck

Given a (possibly unknown) data joint distribution $p(X, Y)$, the IB framework constructs a *bottleneck* variable $Z = Z(X)$ that is a *compressed* version of X but preserves as much *relevant* information in X about Y as possible. The compression of the representation Z is quantized by $I(Z; X)$, the mutual information of Z and X. The relevance in Z, the amount of information Z contains about Y, is specified by $I(Z; Y)$. The optimal representation Z satisfying a certain compression–relevance trade-off constraint is then determined via minimization of the following Lagrangian $\mathcal{L}_{IB}[p(z|x)] = I(Z; X) - \beta I(Z; Y)$, where β is a positive Lagrangian multiplier that controls the trade-off. Due to the convexity of Lagrangian and constrained conditions with respect to the encoders $\{p(z|x)\}$, the Karush–Kuhn–Tucker (KKT) conditions for this constrained minimization problem become the sufficient and necessary conditions for finding the optimal encoders $\{p(z|x)\}$. By solving the KKT conditions, we can obtain the optimal encoders which can be expressed in an energy-based form as the following:

$$\arg\min_{p(z|x)} \mathcal{L}_{IB}[p(z|x)] \propto p(z) \exp\left(-\beta D_{KL}\left[p(Y|x) \| p(Y|z)\right]\right), \quad (1)$$

where $p(z) = \int p(z|x)p(x)dx$.

4. Markov Information Bottleneck

Given a data joint distribution $p(X, Y)$ which is possibly only observed via a set of i.i.d. samples $S = \{(x_i, y_i)_{i=1}^N\}$, an information representation Z for $p(X, Y)$ is said to be good if it encodes sufficient relevant information in X about Y in a compressed manner. Ideally, Z summarizes only the relevant information in X about Y and discards all the irrelevant information; more formally, Z is a minimal sufficient statistic for Y. Such information representation is desirable because it can capture the regularities in the data and is helpful for generalization in learning problems [30,31]. Our main interest is in solving the optimal information representation for a latent variable Z that has Markov structure, that is, $Z = (Z_1, Z_2, \ldots, Z_L)$, where $Z_1 \to Z_2 \to \cdots \to Z_L$. The Markov structure is common in deep neural networks whose advantage is the powerful modeling capacity coming from multiple layers. In MIB, each encoder $p(z_{l+1}|x)$ relates the encoders of the previous bottlenecks in the Markov chain via Bayes' rule:

$$p(z_{l+1}|x) = \int p(z_{l+1}, z_{1:l}|x) dz_{1:l} = \int \prod_{i=1}^{l+1} p(z_i|z_{i-1}) dz_{1:l}, \forall 1 \leq l \leq L-1, \quad (2)$$

where $z_{1:l} := (z_1, \ldots, z_l)$ and $z_0 := x$. In addition, each *encoder* $p(z_l|x)$ corresponds to a unique decoder, namely *relevance decoder*, that decodes the relevant information in x about y from representation z_l:

$$p(y|z_l) = \int p(x,y) \frac{p(z_l|x)}{p(z_l)} dx. \quad (3)$$

In MIB, we further introduce a surrogate target variable $\hat{Y}$ (for the target variable Y) into the Markov chain: $Y \to X \to Z_l \to Z_{l+1} \to \hat{Y}$ (Figure 1). The purpose of the surrogate target variable becomes clear in the section on variational MIB.

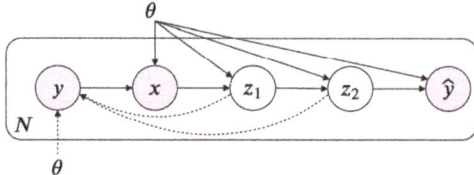

Figure 1. A directed graphical model for the Markov information bottleneck of two Markov bottlenecks. In a non-collapse Markov chain $Y \to X \to Z_1 \to Z_2$ with $0 < \beta_1, \beta_2 < \infty$, the information optimality in Z_1 prevents the information optimality in Z_2. Solid lines denote the encoders $p_\theta(z_i|x)$ (for $i \in \{1, 2\}$), dashed lines denote the variational approximations $p_\theta(\hat{y}|z_i)$ to the intractable *relevance* decoder $p_\theta(y|z_i)$. The variational relevance decoder $p_\theta(\hat{y}|z_i)$ encodes the information from z_i into a surrogate target variable $\hat{y}$. In the case of stochastic neural networks, z_i, θ, and the surrogate target variable represents the hidden layers, the network weights, and the output layer, respectively.

A trivial solution to the optimal information representation problem for Z is to apply the original IB principle for Z as a whole by computing the optimal IB solution in Equation (1). However, this solution ignores the Markov structure of Z. As a principled approach, leveraging the intrinsic structure of a problem can generally provide a new insight that goes beyond the limitation of the perspective that ignores such structure. Thus, in Markov information bottleneck (MIB), we explicitly leverage the Markov structure of Z to derive a principled and tractable approximate solution to the optimal information representation. We then empirically show that leveraging the intrinsic structure in the case of MIB is indeed beneficial for learning.

In MIB, we reframe the optimal information representation problem as multiple IB problems for each of the bottlenecks Z_l:

$$\min_{p(z_l|x)} \mathcal{L}_l[p(z_l|x)] := \min_{p(z_l|x)} \{I(Z_l;X) - \beta_l I(Z_l;Y)\}, \tag{4}$$

for all $1 \leq l \leq L$.

This extension is a natural approach for multiple bottlenecks because it aims for each bottleneck to achieve its own optimal information, and thus allows more relevant but compressed information to be encoded into Z. Another advantage is that we can leverage our understanding of the IB solution for each individual IB problem in Equation (4). Though this approach is promising and has good interpretation, there are two main challenges:

1. The Markov structure among Z_l prevents them from achieving their own information optimality simultaneously in non-trivial cases.
2. The intractability of $p(y|z_l)$ in Equation (3) and $p(z_l|x)$ in Equation (2) results in intractable mutual information in MIB.

In what follows, we formally establish and present the first challenge, the conflicting property of information optimality in Markov structure, in Theorem 1 followed by a simple compromise to overcome the information conflict. After that, we present variational MIB to address the second challenge.

Without loss of generality, we consider the case when $L = 2$ (the result trivially generalizes to $L > 2$). We first define the collapse mode of the representation Z to be the two extreme cases in which Z_2 either contains all the information in Z_1 about X or simply random noise:

Definition 1 (The *collapse* mode of MIB). *$Z = (Z_1, Z_2)$ is said to be in the collapse mode if it satisfies either of the following two conditions:*

1. *Z_2 is a sufficient statistic of Z_1 for X and Y (i.e., $Y \to X \to Z_2 \to Z_1$);*
2. *Z_2 is independent of Z_1.*

For example, if $Z_2 = f(Z_1)$ where f is a deterministic bijection, Z_2 is a sufficient statistic for X. We then establish the conflicting property of information optimality in the Markov representation Z via the following theorem:

Theorem 1 (*Conflicting Markov Information Optimality*). *Given X, Y, Z_1, and Z_2 such that $Y \to X \to Z_1 \to Z_2$ and $H(Y|X) > 0$, consider two constrained minimization problems:*

$$\arg\min_{p(z_l|x)} \mathcal{L}_l[p(z_l|x)] := \arg\min_{p(z_l|x)} \{I(Z_l;X) - \beta_l I(Z_l;Y)\}, l \in \{1,2\}, \tag{5}$$

where $0 < \beta_1 < \infty, 0 < \beta_2 < \infty$, and $p(z_2|x) = \int p(z_2|z_1)p(z_1|x)dz_1$. Then, the following two statements are equivalent:

1. *Z is not in the collapse mode;*
2. *The two optimal solutions to $\mathcal{L}_1$ and $\mathcal{L}_2$ in (5) are conflicting, that is, there is no single solution that minimizes $\mathcal{L}_1$ and $\mathcal{L}_2$ simultaneously.*

Theorem 1 suggests that the Markov information optimality conflicts for most cases of interest (e.g., stochastic neural networks, which we will present in detail in the next section). The values of β_1 and β_2 are important to control the ratio of the relevant information versus the irrelevant one presented in the bottlenecks. These values also determine the conflictability of multiple bottlenecks on the edge cases. Recall by the data processing inequality (DPI) [27] that for $Y \to X \to Z$, we have $0 \leq I(Z;X) \leq H(X)$ and $0 \leq I(Z;Y) \leq I(X;Y)$. If β_1 and β_2 go to infinity, the optimal bottlenecks

Z_1 and Z_2 are both deterministic functions of X and they do not conflict. When $\beta_1 = \beta_2 = 0$, the information about Y in X is maximally compressed in Z_1 and Z_2 (i.e., $Z_1 \perp X, Z_2 \perp X$), and they do not conflict. The optimal solutions conflict when $\beta_1 = 0$ and $\beta_2 > 0$, as the former leads to a maximally compressed Z_1 while the latter prefers an informative Z_2 (this contradicts the Markov structure $X \to Z_1 \to Z_2$, which indicates that maximal compression in Z_1 leads to maximal compression in Z_2).

We can also easily construct non-conflicting MIBs for $0 < \beta_1, \beta_2 < \infty$ that violate the condition. For example, if X and Y are jointly Gaussian, the optimal bottlenecks Z_1 and Z_2 are linear transforms of X and jointly Gaussian with X and Y [11]. In this case, Z_2 is a sufficient statistic of Z_1 for X. In the case of neural networks, we can also construct a simple but non-trivial neural network that can obtain a non-conflicting Markov information optimality. For example, consider a neural network of two hidden layers Z_1 and Z_2, where Z_1 is arbitrarily mapped from the input layer X but Z_2 is a sample mean of n samples i.i.d. drawn from the normal distribution $\mathcal{N}(Z_1; \Sigma)$. This construction guarantees that Z_2 is a sufficient statistic of Z_1 for X, and thus there is non-conflicting Markov information optimality.

Theorem 1 is a direct result of DPI if $\beta_i \in \{0, \infty\}$. In the case that $0 < \beta_i < \infty$, we trace down to the Lagrangian multiplier as in the original IB [1] to complete the proof. Formally, before proving Theorem 1, we first establish the two following lemmas. The first lemma expresses the uncertainty reduction in a Markov chain.

Lemma 1. *Given $Y \to X \to Z_1 \to Z_2$, we have*

$$I(Z_2; X) = I(Z_1; X) - I(Z_1; X|Z_2) \quad (6)$$

$$I(Z_2; Y) = I(Z_1; Y) - I(Z_1; Y|Z_2). \quad (7)$$

Proof. It follows from [27] that $I(X; Z_1; Z_2) = I(X; Z_2) + I(X; Z_1|Z_2) = I(X; Z_1) + I(X; Z_2|Z_1)$, but $I(X; Z_2|Z_1) = 0$ since $X \not\perp Z_2|Z_1$, hence Equation (6). The proof for Equation (7) is similar by replacing variable X with variable Y. (Q.E.D.) □

Lemma 2. *Given $Y \to X \to Z_1 \to Z_2, 0 < \beta_2 < \infty$ and $H(X|Y) > 0$, let us define the conditional information bottleneck objective:*

$$\mathcal{L}^c := \mathcal{L}^c[p(z_2|z_1), p(z_1|x)] := I(Z_1; X|Z_2) - \beta_2 I(Z_1; Y|Z_2). \quad (8)$$

If Z is not in the collapse mode, $\partial \mathcal{L}^c / \partial p(z_1|x)$ depends on $\{p(z_2|z_1)\}$.

Proof. Informally, if Z_2 in the conditional information bottleneck objective $\mathcal{L}^c$ is not a trivial transform of the bottleneck variable Z_1, Z_2 induces a non-trivial topology into the conditional information bottleneck objective. Formally, by the definition of the conditional mutual information

$$I(Z_1; X|Z_2) = \int \int \int p(x, z_1, z_2) \log \frac{p(z_1, x|z_2)}{p(z_1|z_2)p(x|z_2)} dz_2 dz_1 dx,$$

$I(Z_1; X|Z_2)$ depends on $p(x, z_1, z_2)$ as long as the presence of Z_2 in the conditional information bottleneck objective does not vanish (we will discuss the conditions for Z_2 to vanish in the final part of this proof). Note that due to the Markov chain $X \to Z_1 \to Z_2$, we have $p(x, z_1, z_2) = p(x)p(z_1|x)p(z_2|z_1)$.

Thus, $\partial I(Z_1; X|Z_2)/\partial p(z_1|x)$ depends on $p(z_2|z_1)$ as long as Z_2 does not vanish in the objective. Similarly, the same result also applies to $\partial I(Z_1; Y|Z_2)/\partial p(z_1|x)$. Hence, $\partial \mathcal{L}^c / \partial p(z_1|x)$ depends on $\{p(z_2|z_1)\}$ (note that $H(X|Y) > 0$ prevents the collapse of $\mathcal{L}^c$ when summing two mutual informations) if Z_2 does not vanish in the objective.

Now we discuss the vanishing condition for Z_2 in the objective. It follows from Lemma 1 that:

$$0 \leq I(Z_1; X|Z_2) \leq I(Z_1; X), \quad (9)$$
$$0 \leq I(Z_1; Y|Z_2) \leq I(Z_1; Y). \quad (10)$$

Note that Z_2 vanishes in $\mathcal{L}^c$ iff each of the mutual informations in $\mathcal{L}^c$ does not depend on Z_2 iff the equality in both (9) and (10) occur. If $I(Z_1; X|Z_2) = 0$, we have $Y \to X \to Z_2 \to Z_1$ (i.e., Z_2 is a sufficient statistic for X and Y), which also implies that $I(Z_1; Y|Z_2) = 0$. Similarly, $I(Z_1; X|Z_2) = I(Z_1; X)$ implies that Z_2 is independent of Z_1, which in turn implies that $I(Z_1; Y|Z_2) = I(Z_1; Y)$. (Q.E.D.) □

We now prove Theorem (1) by using Lemma (6) and Lemma (7).

Proof of Theorem 1. ($\Leftarrow$) This direction is obvious. When $I(Z_2; X) = I(Z_1; X)$ and $I(Z_2; Y) = I(Z_1; Y)$, or $I(Z_2; X) = 0$ and $I(Z_2; Y) = 0$, there is effectively only one optimization problem for $\mathcal{L}_1$, and this reduces into the original information bottleneck (with single bottleneck) [1].
($\Rightarrow$) First we prove that if Z is not in the collapse mode, the constrained minimization problems are conflicting. Assume, by contradiction, that there exists a solution that minimizes both $\mathcal{L}_1$ and $\mathcal{L}_2$ simultaneously, that is, $\exists p(z_1|x), p(z_2|z_1)$ s.t. $\mathcal{L}_1$ has a minimum at $\{p(z_1|x)\}$ and $\mathcal{L}_2$ has a minimum at $\{p(z_1|x), p(z_2|z_1)\}$. Note that $\{p(z_1|x)\}$ and $\{p(z_2|z_1)\}$ are independent variables for the optimization. By introducing Lagrangian multipliers $\lambda_1(x)$ and $\lambda_2(x)$ for the constraint $\int p(z_1|x)dz_1 = 1$ of $\mathcal{L}_1$ and $\mathcal{L}_2$, respectively, the stationarity in the Karush–Kuhn–Tucker (KKT) conditions becomes:

$$\frac{\partial L_1}{\partial p(z_1|x)} = 0, \quad (11)$$

$$\frac{\partial L_2}{\partial p(z_1|x)} = 0, \quad (12)$$

where L_1 and L_2 are the Lagrangians:

$$L_1[p(z_1|x), \lambda_1] := I(Z_1; X) - \beta_1 I(Z_1; Y) - \int \int \lambda_1(x) p(z_1|x) dz_1 dx \quad (13)$$

$$L_2[p(z_1|x), \lambda_2] := I(Z_2; X) - \beta_2 I(Z_2; Y) - \int \int \lambda_2(x) p(z_1|x) dz_1 dx. \quad (14)$$

It follows from Lemma 1 that:

$$L_2 - L_1 = (\beta_1 - \beta_2) I(Z_1; Y) - \mathcal{L}^c - \int \int (\lambda_2(x) - \lambda_1(x)) p(z_1|x) dz_1 dx, \quad (15)$$

where $\mathcal{L}^c = I(Z_1; X|Z_2) - \beta_2 I(Z_1; Y|Z_2)$ (defined in Lemma 2). We take the derivative w.r.t. $p(z_1|x)$ both sides of Equation (15) and use Equations (11)–(12):

$$\frac{\partial \mathcal{L}^c}{\partial p(z_1|x)} = (\beta_1 - \beta_2) \frac{\partial I(Z_1; Y)}{\partial p(z_1|x)} + \lambda_1(x) - \lambda_2(x). \quad (16)$$

Notice that the left hand side of Equation (16) strictly depends on $p(z_2|z_1)$ (Lemma 2) while the right hand side is independent of $\{p(z_2|z_1)\}$. This contradiction implies that the initial existence assumption is invalid, and thus implies the conclusion in Theorem 1. (Q.E.D.) □

4.1. Markov Information Optimality Compromise

Due to Theorem 1, we cannot simultaneously achieve the information optimality for all bottlenecks. Thus, we need some compromised approach to instead obtain a compromised optimality. We propose two simple compromise strategies, namely, JointMIB and GreedyMIB. JointMIB is a weighted sum of the IB objectives $\mathcal{L}^{joint} := \sum_{l=0}^{L} \gamma_l \mathcal{L}_l$ where $\gamma_l \geq 0$. The main idea of JointMIB is to simultaneously optimize all encoders. Even though each bottleneck might not achieve its individual optimality, their joint optimality encourages a joint compromise. On the other hand, `GreedyMIB` progressively solves the information optimality for each bottleneck given that the encoders for the previous bottlenecks are fixed. In other words, `GreedyMIB` tries to obtain the conditional optimality of a current bottleneck which is conditioned on the fixed greedy-optimal information of the previous bottlenecks.

4.2. Variational Markov Information Bottleneck

Due to the intractability of encoders in Equation (2) and relevance decoders in Equation (3), the resulting mutual information in Equation (4) is also intractable. In this section, we present variational methods to derive a lower bound on mutual information in MIB.

4.2.1. Approximate Relevance

Note that $I(Z_l; Y) = H(Y) - H(Y|Z_l)$, where $H(Y) = $ constant, which can be ignored in the minimization of $\mathcal{L}_l$. It follows from the non-negativity of KL divergence that:

$$H(Y|Z_l) = -\int p(y|z_l) p(z_l) \log p(y|z_l) dy dz_l \leq -\int p(y|z_l) p(z_l) \log p_v(y|z_l) dy dz_l$$
$$= -\mathbb{E}_{(X,Y)} \mathbb{E}_{Z_l|X} \log p_v(Y|Z_l) = -\mathbb{E}_{(X,Y)} \mathbb{E}_{Z_l|X} \log p(\hat{Y}|Z_l) =: \tilde{H}(Y|Z_l), \quad (17)$$

where we specifically use the relevance decoder for surrogate target variable $p_v(y|z_l) = p(\hat{y}|z_l)$ as a variational distribution to the intractable distribution $p(y|z_l)$:

$$p_v(y|z_l) := \mathbb{E}_{Z_L|z_l} [p(\hat{y}|Z_L)]. \quad (18)$$

The variational relevance $\tilde{I}(Z_l; Y) := H(Y) - \tilde{H}(Y|Z_l)$ is a lower bound on $I(Z_l; Y)$. This bound is tightest (i.e., zero gap) when the variational relevance decoder $p(\hat{y}|z_l)$ equals the relevance decoder $p(y|z_l)$. In what follows, we establish the relationship between the variational relevance and the log likelihood function, thus connecting MIB with the MLE principle:

Proposition 1 (Variational Relevance Inequalities). *Given the definition of variational relevance $\tilde{I}(Z_l; Y) = H(Y) - \tilde{H}(Y|Z_l)$ where $\tilde{H}(Y|Z_l)$ is defined in Equation (17), and $Z = (Z_1, ..., Z_L)$, we have:*

$$H(Y) + \mathbb{E}_{(X,Y)} \left[\log p(\hat{Y}|X)\right] = \tilde{I}(Z_0; Y) \geq \tilde{I}(Z_l; Y) \geq \tilde{I}(Z_{l+1}; Y) \geq \tilde{I}(Z_L; Y) = \tilde{I}(Z; Y), \quad (19)$$

for all $0 \leq l \leq L - 1$. where $Z = (Z_1, ..., Z_L)$.

Proposition 1 suggests that: (i) the log likelihood of $p(\hat{y}|x)$ (plus the constant output entropy $H(Y)$) is a special case of the variational relevance at bottleneck $Z_0 = X$; (ii) the log likelihood bound $H(Y) + \mathbb{E}_{(X,Y)} \left[\log p(\hat{Y}|X)\right]$ is an upper bound on the variational relevance for all the intermediate bottlenecks Z_l and for the composite bottleneck $Z = (Z_1, ..., Z_L)$. Therefore, maximizing the log likelihood, as in MLE, does not guarantee to increase the variational relevance for all the the intermediate bottlenecks and the composite bottleneck.

Proof. It follows from Jensen's inequality and the Markov chain that:

$$\int p(z_l|x) \log p(\hat{y}|z_l) dz_l = \int p(z_l|x) \log \left(\int p(\hat{y}|z_{l+1}) p(z_{l+1}|z_l) dz_{l+1}\right) dz_l$$
$$\geq \int p(z_l|x) \int p(z_{l+1}|z_l) \log p(\hat{y}|z_{l+1}) dz_{l+1} dz_l$$
$$= \int \int p(z_l|x) p(z_{l+1}|z_l) \log p(\hat{y}|z_{l+1}) dz_l dz_{l+1}$$
$$= \int p(z_{l+1}|x) \log p(\hat{y}|z_{l+1}) dz_{l+1},$$

for all $0 \leq l \leq L-1$. Thus, we have:

$$\begin{aligned}
\tilde{I}(Z_l; Y) &= H(Y) - \tilde{H}(Y|Z_l) \\
&= H(Y) + \mathbb{E}_{(X,Y)} \mathbb{E}_{Z_l|X} \log p(\hat{Y}|Z_l) \\
&\geq H(Y) + \mathbb{E}_{(X,Y)} \mathbb{E}_{Z_{l+1}|X} \log p(\hat{Y}|Z_{l+1}) \\
&= \tilde{I}(Z_{l+1}; Y).
\end{aligned}$$

It also follows from the Markov chain that:

$$p(\hat{y}|z) = p(\hat{y}|z_L, z_{L-1}, ..., z_1) = p(\hat{y}|z_L).$$

Therefore, we have:

$$\begin{aligned}
\tilde{I}(Z; Y) &= H(Y) + \mathbb{E}_{(X,Y)} \mathbb{E}_{Z_L|Z_{L-1},...,Z_1|Z_0} \log p(\hat{Y}|Z_L) \\
&= H(Y) + \mathbb{E}_{(X,Y)} \mathbb{E}_{Z_L|X} \log p(\hat{Y}|Z_L) \\
&= \tilde{I}(Z_L; Y).
\end{aligned}$$

Finally, by the definition in Equation (17), we have:

$$\begin{aligned}
\tilde{I}(Z_0; Y) &= H(Y) + \mathbb{E}_{(X,Y)} \mathbb{E}_{Z_0|X} \log p(\hat{Y}|Z_0) \\
&= H(Y) + \mathbb{E}_{(X,Y)} \log p(\hat{Y}|X).
\end{aligned}$$

(Q.E.D.) □

4.2.2. Approximate Compression

In practice (e.g., in SNN presented in the next section), we can model the encoding between consecutive layers $p(z_l|z_{l-1})$ with an analytical form. However, the encoding of non-consecutive layers $p(z_l|x)$ for $l > 1$ is generally not analytic as it is a mixture of $p(z_l|z_{l-1})$. We thus propose to avoid directly estimating $I(Z_l; X)$ by instead resorting to its upper bound $I(Z_l; Z_{l-1})$ as its surrogate in the optimization. However, $I(Z_l; Z_{l-1})$ is still intractable as it involves the intractable marginal distribution $p(z_l) = \int p(z_l|x)p(x)dx$. We then approximate $I(Z_l; Z_{l-1})$ using a mean-field (factorized) variational distribution $q(z_l) = \prod_{i=1}^{n_l} q(z_{l,i})$ where $z_l = (z_{l,1}, \ldots, z_{l,n_l})$:

$$\begin{aligned}
I(Z_l; X) \leq I(Z_l; Z_{l-1}) &= \int p(z_l|z_{l-1}) p(z_{l-1}) \log \frac{p(z_l|z_{l-1})}{p(z_l)} dz_l dz_{l-1} \\
&\leq \int p(z_l|z_{l-1}) p(z_{l-1}) \log \frac{p(z_l|z_{l-1})}{q(z_l)} dz_l dz_{l-1} = \mathbb{E}_{Z_{l-1}} D_{KL}\left[p(Z_l|Z_{l-1}) || q(Z_l) \right] \\
&= \mathbb{E}_{Z_{l-1}} \sum_{i=1}^{n_l} D_{KL}\left[p(Z_{l,i}|Z_{l-1}) || q(Z_{l,i}) \right] =: \tilde{I}(Z_l; Z_{l-1}).
\end{aligned} \qquad (20)$$

The mean-field variational inference not only helps derive a tractable approximation but also encourages distributed representation by constraining each neuron to capture an independent factor of variation for the data [32]; thus, it can potentially represent an exponential number of concepts using independent factors.

5. Case Study: Learning Binary Stochastic Neural Networks

In this section, we officially connect the variational MIB in Section 4 to stochastic neural networks (SNNs). We consider an SNN with L hidden layers (without any feedback or skip connection) where the input layer X, the hidden layers Z_l for $1 \leq l \leq L$, and the output layer $\hat{Y}$ are considered as random variables. We use the convention that $Z_0 := X$, $Z_{L+1} := \hat{Y}$, and $Z_l = \emptyset$ for all $l \notin \{0, 1, \ldots, L, L+1\}$. Without any feedback or skip connection, Y, X, Z_l, Z_{l+1}, and $\hat{Y}$ form a Markov chain in that order. The output layer $\hat{Y}$ is the surrogate target variable presented in Section 4. The role of SNNs is therefore reduced to transforming from one random variable to another via the Markov chain $X \to Z_l \to Z_{l+1} \to \hat{Y}$ such that it achieves the good information representation (i.e., the compression–relevance tradeoff) for each layer. With the MLE principle, the learning in SNNs is performed by maximizing the log likelihood $\mathbb{E}_{(X,Y)}\left[\log p(\hat{Y}|X)\right]$. However, maximizing the log likelihood does not guarantee to improve the variational relevance for all the intermediate bottlenecks and the composite bottleneck (Proposition 1).

Algorithm 1: JointMIB

Input: data $S_0 \leftarrow (x_i, y_i)_{i=1}^N \sim p_D(x, y)$, layer IB weights γ_l, information tradeoff β_l, number of particles for Monte Carlo simulation M.
Output: θ
Initialization: θ

1 **while** *not converged* **do**
2   **for** $i = 1$ *to* L **do**
3        $S_i \leftarrow \emptyset$
4   **for** $z_{i-1} \in S_{i-1}$ **do**
           /* Monte Carlo simulates M particles $z_i^{(k)}$ given each z_{i-1} */
5            $S_i \leftarrow S_i \cup \{z_i^{(k)} : 1 \le k \le M\}$ where $z_i^{(k)} \sim p(z_i|z_{i-1})$
6   **end**
7   **end**
     /* Estimate the variational relevance and compression using Monte Carlo samples */
8     $\tilde{I}(Z_l; Y) \leftarrow$ Equation (17) and $\{S_i\}_{i=0}^L$
9     $\tilde{I}(Z_l; Z_{l-1}) \leftarrow$ Equation (20) and $\{S_i\}_{i=0}^L$
10   $\tilde{\mathcal{L}}^{joint}(\theta) \leftarrow \sum_{l=0}^L \gamma_l \left(-\tilde{I}(Z_l; Y) + \beta_l \tilde{I}(Z_l; Z_{l-1})\right)$
11   $g \leftarrow \frac{\partial}{\partial \theta} \tilde{\mathcal{L}}^{joint}(\theta)$ /* Using the Raiko estimator in Binary SNNs */
12
13   $\theta \leftarrow \theta - \nu g$ /* Update using SGD */
14 **end**

We here instead combine the variational MIB and the MIB compromise to derive a practical learning principle that encourages compression and relevance for each layer, improving the information flow in SNNs. To make it concrete and simple, we consider a simple network architecture: binary stochastic feed-forward (fully-connected) neural networks (SFNNs). In binary SFNNs, we use a sigmoid as the activation function: $p(z_l = 1|z_{l-1}) = \sigma(W_{l-1}z_{l-1} + b_{l-1})$, where $\sigma(.)$ is the (element-wise) sigmoid function, W_{l-1} is the network weights connecting layer $l-1$ to layer l, b_{l-1} is a bias vector, and $Z_l \in \{0,1\}^{n_l}$. Let us define $\tilde{\mathcal{L}}_l := -\tilde{I}(Z_l; Y) + \beta_l \tilde{I}(Z_l; Z_{l-1})$, where $\tilde{I}(Z_l; Y)$ and $\tilde{I}(Z_l; Z_{l-1})$ are the approximate relevance and compression defined in Equation (17) and (20), respectively. Note that the position of β_l here is slightly different from its position in Equation (4). In Equation (4), β_l is associated with the relevance term to respect the convention of the original IB, while here it is associated with the compression term for practical reasons. In practice, the contribution of $\tilde{I}(Z_l; Y)$ is higher than $\tilde{I}(Z_l; Z_{l-1})$. In computing $\tilde{I}(Z_l; Y)$ and $\tilde{I}(Z_l; Z_{l-1})$, any expectation with respect to $p(z_l|z_{l-1})$ is approximated by Monte Carlo simulation in which we sample M particles

$z_l \sim p(z_l | z_{l-1})$. Regarding the information optimality compromise, we combine the variational MIB objectives into a weighted sum in `JointMIB`:

$$\tilde{\mathcal{L}}^{joint} := \sum_{l=0}^{L} \gamma_l \tilde{\mathcal{L}}_l, \tag{21}$$

where $\gamma_l \geq 0$. In `GreedyMIB`, we greedily minimize $\tilde{\mathcal{L}}_l$ for each $0 \leq l \leq L$. We also make each $q(Z_{l,i})$ a learnable Bernoulli distribution. The `JointMIB` is presented in Algorithm 1. The Monte Carlo sampling operation of Algorithm 1 in stochastic neural networks precludes the backpropagation in a computation graph. It becomes even more challenging with binary stochastic neural networks, as it is not well-defined to compute gradients w.r.t. discrete-valued variables. Fortunately, we can find approximate gradients, which have been proved to be efficient in practice: the REINFORCE estimator [33,34], the straight-through estimator [35], the generalized EM algorithm [20], and the Raiko (biased) estimator [21]. Especially, we found that the Raiko gradient estimator works best in our specific setting and thus deployed it in this application. In the Raiko estimator, the gradient of a bottleneck particle $z_{l,i} \sim p(z_{l,i} = 1 | z_{l-1}) = \sigma(a_i^{(l)})$ is propagated only through the deterministic term $\sigma(a_i^{(l)})$: $\frac{\partial z_{l,i}}{\partial \theta} \approx \frac{\partial \sigma(a_i^{(l)})}{\partial \theta}$.

6. Experimental Evaluation

We evaluated the effectiveness of the MIB framework on binary SNNs in synthetic data and MNIST hand-digit recognition data [23]. Each data sample in MNIST is a 28 × 28 gray-scale image representing a handwritten digit from 0 to 9. The dataset is split into 60000 training samples and 1000 test samples. In the synthetic data, we visualized the learning dynamics of the SNNs trained with the variational MIB variants (i.e., `JointMIB` and `GreedyMIB`), and those trained with MLE. In MNIST, we evaluate the effectiveness of the variational MIB variants by comparing them against the baselines MLE and VIB [14] in classification, adversarial robustness and multi-modal learning problems. We make the code for our framework publicly available at https://github.com/thanhnguyentang/pib.

6.1. Synthetic Data: Learning Dynamics of Variational MIB

To better understand how MIB modified the information within the layers during the learning process, we visualized the compression and relevance of each layer over the course of training of stochastic feed-forward neural networks (SFNNs) [21], `JointMIB`, and `GreedyMIB` in synthetic data. *SFNN* is different from *MIB* only in the objective functions: *SFNN* is trained with the negative log likelihood while *MIB* is trained with the variational MIB objective. To simplify our analysis, we considered a binary decision problem where X is 12 binary inputs making up $2^{12} = 4096$ equally likely input patterns and Y is a binary variable equally distributed among 4096 input patterns [16]. The base neural network architecture had 4 hidden layers with widths: 10–8–6–4 neurons. Since the network architecture was small, we could precisely compute the true compression $I_x := I(Z_i; X)$ and true relevance $I_y := I(Z_i; Y)$ over training epochs. We fixed $\beta_l = \beta = 10^{-4}$ for both `JointMIB`, trained five different randomly initialized neural networks for each comparative model with stochastic gradient descent (SGD) up to 20,000 epochs on 80% of the data, and averaged the mutual information. In `JointMIB`, we set $\gamma_l = \gamma = 1, \forall l$.

Figure 2 provides a visualization of the learning dynamics of SFNN versus `JointMIB` on the information plane (I_x, I_y). Firstly, we observed a common trend in the learning dynamics of MLE (in the SFNN model) and `JointMIB` frameworks. Both principles allow the network to gradually encode more information about X and the relevant information about Y into the hidden layers at the beginning as $I(Z_i; X)$ and $I(Z_i; Y)$ both increase. Intuitively, in order for the representations Z_l to make sense of the task, the representations should encode enough information about X; thus, $I(Z_l; X)$ should increase. This is especially true for shallow layers because, due to the Markov chain property, the shallower a layer, the greater its burden of carrying enough information to make sense of a task. Especially, we can observe that the increase of $I(Z_l; X)$ slowed down at some point for the deeper layers for both $SFNN$ and MIB. This slowing effect was especially stronger in MIB where the compression is explicitly encouraged during the learning. Secondly, MIB was different from MLE in the maximum level of relevance at each layer and the number of epochs to encode the same level of relevance. JointMIB at $l = 1$ needed only about 4.68% of the training epochs to achieve at least the same level of relevance in all layers of SFNN at the final epoch. In addition, MLE was unable to encode the network layers to reach the maximum level of relevance enabled by MIB (we also trained SFNN up to 100,000 epochs and observed that the level of relevance of each layer never reached the value of 0.8 bits).

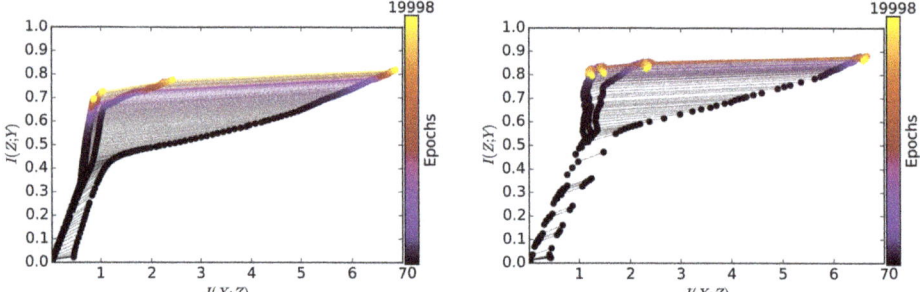

Figure 2. The learning dynamics of the stochastic feed-forward (fully-connected) neural network (SFNN) (**left**) and JointMIB (**right**). The color indicates the training epochs while each node in a color in the graph represents $(I(Z_l; X), I(Z_l; Y))$ at the corresponding epoch. Note that at each epoch, $I(Z_l; X) \geq I(Z_{l+1}; X), \forall l$ (data processing inequality—DPI). JointMIB jointly encodes relevant information into every layer of stochastic neural networks (SNNs), while keeping each layer informatively concise. Compared to maximum likelihood estimation (MLE), the level of relevant information encoded by JointMIB increased more quickly over training epochs and reached a higher value. MIB: Markov information bottleneck.

There is also a subtle observation in Figure 2 that the relevance for MIB increased until some point before decreasing, while the relevance for SFNN increased until some point where the value almost stayed the same without a noticeable decrease. This could be explained by the fact that that the MIB objective can eventually allow the encoding of relevant information into each layer to its optimal information trade-off at some point. After this point, if training is continued, due to the mismatch between the exact MIB objective and its variational bound, the further minimization of the variational bound would decrease $I(Z_l; Y)$. Consequently, in order for $\beta_l I(Z_l; X) - I(Z_l; Y)$ to be small, $I(Z_l; X)$ also needs to decrease after this point to compensate for the decrease in $I(Z_l; Y)$. In the case of SFNN (trained with MLE), the MLE objective reaches its local minimum before the information of each layer can even reach its optimal information trade-off (if ever). This also suggests that MIB is better than MLE in terms of exploiting information for each layer during the learning.

GreedyMIB also obtained the representation of higher relevance as compared to MLE (Figure 3). GreedyMIB at $l = 1$ needed only about 17.95% of the training epochs to achieve at least the same level of relevance in all layers of the SFNN at the final epoch. Recall that in GreedyMIB at $l = 1$ the MIB principle is applied only to the first hidden layer. The layer representation at the final epoch gradually shifts to the left (i.e., more compressed) while not degrading the relevance over the greedy training from layer 1 to layer 4 in Figure 3.

We also see the compression effect that the compression constraints within the MIB framework prevented the layer representation from shifting to the right (in the information plane) during the encoding of relevant information (e.g., it slowed down the increase of $I(Z_l; X)$ during the information encoding, keeping the representation more concise). As compared with JointMIB, GreedyMIB also obtained a comparable information representation.

To conclude, two main advantages of MIB as compared to MLE are: (i) MIB can improve the information representation in SNNs in terms of higher relevance while keeping the information in each layer concise during encoding; (ii) MIB uses much fewer training epochs to obtain such information representation.

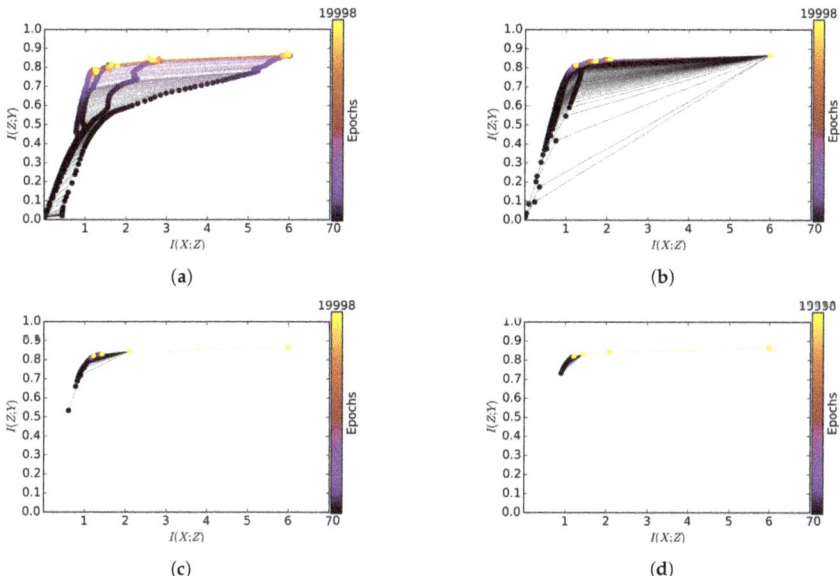

Figure 3. Subfigures (**a**), (**b**), (**c**), and (**d**) represent GreedyMIB's encoding of relevant information into layers $1 \leq l \leq 4$, respectively. GreedyMIB greedily encodes relevant information into each layer given the encoded information of the previous layers. GreedyMIB also achieved a significantly higher level of relevant information at each layer compared to MLE.

6.2. Image Classification

In this experiment, we compared JointMIB and GreedyMIB with three other comparative models which used the same network architecture without any explicit regularizer: (1) a standard deterministic neural network (DET) which simply treated each hidden layer as deterministic; (2) a stochastic feed-forward neural network (SFNN) [21] which is a binary stochastic neural network as in MIB but is trained with the MLE principle; and (3) variational information bottleneck (VIB) [14], which uses the entire deterministic network as an encoder, adds an extra stochastic layer as a out-of-network bottleneck variable, and is then trained with the IB principle on that single bottleneck layer. The base network architecture in this experiment had two hidden layers with 512 sigmoid-activated neurons per layer. These models were trained in MNIST [23].

Adopted from the common practice, we used the last 10,000 images of the training set as a validation (holdout) set for tuning hyperparameters. We then retrained the models from scratch in the full training set with the best validated configuration. We trained each of the five models with the same set of five different initializations and reported the average results over the set. For the stochastic models (all except DET), we drew $M = 32$ samples per stochastic layer during both training and inference, and performed inference 10 times at test time to report the mean classification errors for MNIST. The value of $M = 32$ is empirically reasonable in this experiment, as illustrated in Figure 4.

For JointMIB and GreedyMIB, we set $\gamma_l = 1$ (in JointMIB only) and $\beta_l = \beta, \forall 1 \leq l \leq L$, tuned β on a linear log scale $\beta \in \{10^{-i} : 1 \leq i \leq 10\}$. We found $\beta = 10^{-4}$ worked best for both models (Figure 5). For VIB, we found that $\beta = 10^{-3}$ worked best on MNIST. We trained all the models on MNIST with Adadelta optimization [36], except for VIB for which we used Adam optimization [37], as we found that they worked best in the validation set.

The results are shown in Table 1. It shows that JointMIB substantially outperformed DET, MLE, and VIB on MNIST while GreedyMIB outperformed only DET and underperformed SFNN. Though JointMIB and GreedyMIB could have comparable information representation, as illustrated in the synthetic experiment in Section 6.1, in practice, it can be harder to obtain a comparable information representation for GreedyMIB. In GreedyMIB, it is necessary to train each layer greedily in order to obtain its information representation. The greedy nature makes it difficult to determine when would be a good time to stop the training and conclude the information representation for each layer. In addition, training greedily is expensive. JointMIB makes it more efficient by jointly obtaining a compromised information representation in each layer. Thus, it allows the compromised information representations of all the layers to jointly interact with each other during the learning. In principle, it is also harder to obtain good information representation in GreedyMIB. Due to the conflicting information optimality in MIB (Theorem 1), the good encoder for the first layer does not guarantee a good information trade-off in the the deeper layers. Though JointMIB also suffers from the conflicting information optimality, jointly and explicitly inducing relevant but compressed information into each layer of a neural network via MIBs as in JointMIB can make it easier for the training.

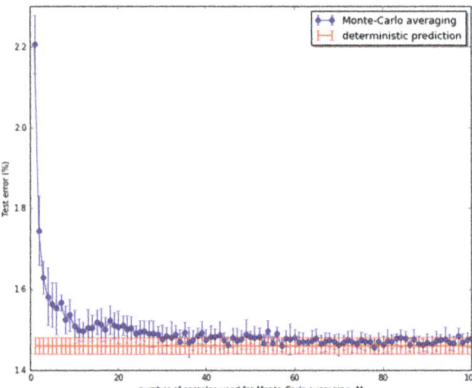

Figure 4. The value of M versus validation error. $M = 32$ gave a reasonably good performance as compared to other larger values.

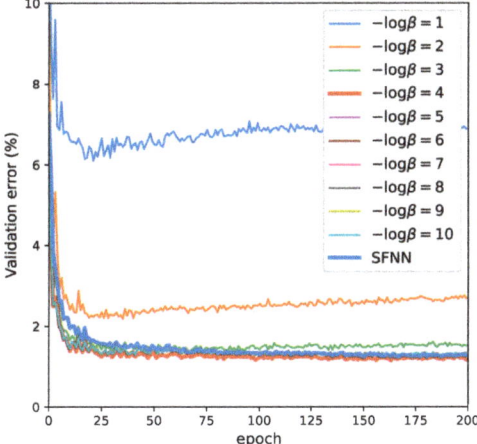

Figure 5. The learning curve of JointMIB and SFNN in the MNIST validation set. Either a too-large or too-small value of β could hurt the generalization of learning. While a large value of β introduces aggressive compression, a smaller value allows more irrelevant information into the representation. In this experiment, we found that $\beta = 10^{-4}$ was the best trade-off hyperparameter in JointMIB for this experiment.

Table 1. The performance of the variational MIB variants (i.e., JointMIB and GreedyMIB) for classification and adversarial robustness on MNIST in comparison with MLE and variational information bottleneck (VIB). MIB explicitly induces compression–relevance trade-offs in each layer during the training, which outperforms and is more adversarially robust than the other models of the same architecture. DET: deterministic neural network.

Model	Classification MNIST (Error %)	Adv. Robustness (%)	
		Targeted	Untargeted
DET	1.73	00.00	00.00
VIB [14]	1.45	83.70	93.10
SFNN [21]	1.44	83.00	95.20
GreedyMIB	1.54	83.21	94.30
JointMIB	**1.36**	**84.16**	**96.00**

6.3. Robustness against Adversarial Attacks

We consider here the adversarial robustness of neural networks trained with MIBs. Neural networks are prone to adversarial attacks which disturb the input pixels by small amounts that are imperceptible to humans [38,39]. Adversarial attacks generally fall into two categories: untargeted and targeted attacks. An untargeted adversarial attack $\mathcal{A}$ maps the target model M and an input image x into an adversarially perturbed image x': $\mathcal{A} : (M, x) \to x'$, and is considered successful if it can fool the model $M(x) \neq M(x')$. A targeted attack, on the other hand, has an additional target label l: $\mathcal{A} : (M, x, l) \to x'$, and is considered successful if $M(x') = l \neq M(x)$.

We performed adversarial attacks on the neural networks trained with MLE and MIB, and used the accuracy on adversarially perturbed versions of the test set to rank a model's robustness. In addition, we used the L_2 attack method for both targeted and untargeted attacks [40], which has shown to be the most effective attack algorithm with smaller perturbations. Specifically, we attacked the same four comparative models described from the previous experiment on the first 1000 samples of the MNIST test set. For the targeted attacks, we targeted each image into the other 9 labels other than the true label of the image. We used the same hyperparameters as in the classification experiment. The value of $\beta = \beta_l = 10^{-4}$ was also reasonable for this adversarial robustness task (Figure 6).

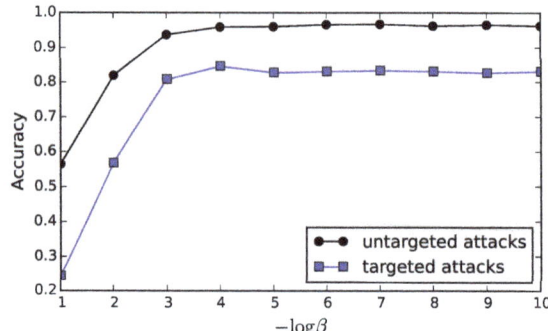

Figure 6. Adversarial robustness of JointMIB for various values of β. Introducing aggressive compression (i.e., large values of β) reduced adversarial robustness while smaller values of β introduced comparable robustness. The best information trade-off for targeted attacks was at $\beta = 10^{-4}$ in this experiment.

The results are shown in Table 1. Firstly, it was expected that the adversarial robustness accuracy in the targeted attacks would be smaller than that in the untargeted attacks because the targeted attacks are more challenging for the neural networks to overcome than untargeted attacks. This result is consistent in our experiment. Secondly, the deterministic model DET was totally fooled by all the attacks. It is known that stochasticity in neural networks improves adversarial robustness, which is consistent with our experiment as SFNN was significantly more adversarially robust than DET. Thirdly, VIB had comparable adversarial robustness to SFNN even if VIB had "less stochasticity" than SFNN (VIB had one stochastic layer while all hidden layers of the SFNN were stochastic). We hypothesize that this is because VIB performance was compensated with the IB principle for its stochastic layer. Finally, JointMIB was more adversarially robust than the other models. Again, GreedyMIB was not very effective in adversarial robustness (it was worse than VIB in the targeted attack and SFNN in the untargeted attack). We hypothesize that this relates to the difficulty for GreedyMIB to have a good information representation for all layers. In conclusion, this experiment suggests that explicitly and jointly inducing compression and relevance into each layer has a good potential of being more adversarially robust for neural networks.

6.4. Multi-Modal Learning

One of the main advantages of stochastic neural networks is their ability to model structured output space in which a one-to-many mapping is required. A binary stochastic variable z_l of dimensionality n_l can take on 2^{n_l} different states, each of which would give a different $\hat{y}$. Thus, the conditional distribution $p(\hat{y}|x)$ in stochastic neural networks is multi-modal. Hence in this experiment, we evaluated how MIB affected the multi-modal learning capability of SNNs.

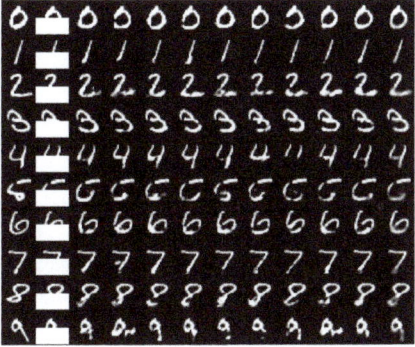 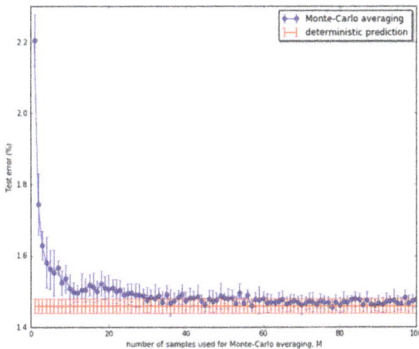

Figure 7. Samples drawn from the prediction of the lower half of the MNIST test data digits based on the upper half for JointMIB (**right**, after 60 epochs) and SFNN (**left**, after 200 epochs). The leftmost column is the original MNIST test digit followed by the masked out digits and nine samples. The rightmost column was obtained by averaging over all generated samples of bottlenecks drawn from the prediction. The figures illustrate the capability of modeling structured output space using JointMIB and SFNN. JointMIB generated more recognizable digits within much fewer training epochs.

In this experiment, we followed [21] and predicted the lower half of the MNIST digits using the upper half as inputs. We used the same neural network architecture of 392–512–512–392 for JointMIB and SFNN and trained them with SGD with a constant learning rate of 0.01 (due to the under-performance of GreedyMIB from the previous experiments and its expensive training, we compared only JointMIB with SFNN in this experiment). We trained the models on the full training set of 60,000 images and tested with the test set. For JointMIB, we also used $\beta_l = \beta = 10^{-4}$. The results of JointMIB at epoch 60 and MLE at epoch 200 are shown in Figure 7. Firstly, JointMIB could generate digit variations which were more recognizable than those generated by MLE. In particular, some samples of digits 2, 4, 5, and 7 generated by MLE were distorted, while all digit samples generated by JointMIB were recognizable. Secondly, JointMIB used much fewer epochs to achieve good samples. In JointMIB, we trained only up to 60 epochs while in MLE, we trained up to 200 epochs but did not observe as good samples in between. This further highlights the advantage of MIB in obtaining good information representation in much fewer training epochs. Furthermore, we expect that the advantage of inducing compression and relevance into each layer by JointMIB is particularly helpful for multi-modal learning because in multi-modal learning, the modes generated in each hidden layer are critical for representing multiple modes. While MLE ignores the explicit contribution of each layer to the information representation of the neural network, JointMIB explicitly takes into account the compression and relevance of each layer.

7. Discussion and Future Work

In this work, we introduce Markov Information Bottleneck, an extension of the original Information Bottleneck to the context where a representation is multiple stochastic variables that form a Markov chain. In this context, we show that one cannot simply directly apply the original IB principle to each variable as their information optimality is conflicting for most of the interesting cases. We suggest a simple but efficient fix via a joint compromise. In this scheme, we jointly combine the information trade-offs of each variable into a weighted sum, encouraging the information trade-offs for all the variables better off during the learning. In particular in the context of Stochastic Neural Networks, we present the variational inference to estimate the compression and relevance for each bottleneck. As a result, the variational MIB turns the intractable decoding of each bottleneck approximately into an efficient inference for that bottleneck. This variational approximation turns out to generalize the MLE principle in the context of Stochastic Neural Networks. We empirically demonstrate the effectiveness of MIB by comparing it with the baselines using MLE principle and Variational Information Bottleneck in classification, adversarial robustness and multi-modal learning. The empirical performance supports the potential benefit of explicitly inducing compression and relevance into each layer (e.g., in a jointly manner), presenting a special link between information representation and the performance in classification, adversarial robustness and multi-modal learning.

One limitation of our current approach is the number of samples generated via $z_l \sim p(z_l|z_{l-1})$ used to estimate the variational compression and relevance scales exponentially with the number of layers. This is however a common drawback for performing inference in fully stochastic neural networks. This difficulty can be overcome by using partially stochastic neural networks. In addition, the Monte Carlo sampling to estimate the variational mutual information, though unbiased, is of high variance and sample inefficiency. This sample inefficiency limitation can be overcome by resorting to more advanced methods of estimating mutual information such as [41,42]. The MIB framework also admits several possible future extensions including scaling the framework to bigger networks and real-valued stochastic neural networks. The extension to real-valued stochastic neural networks are straightforward by, e.g., constructing a Gaussian layer for modeling $p(z_l|z_{l-1})$ and using reparameterization tricks [43] to perform back-propagation via sampling. Another dimension of improvement is to study hyperparameter effect of MIB. This current work only considers equal $\gamma_l = \gamma$ for JointMIB and equal $\beta_l = \beta$, and tuned β via grid search. We can use, e.g., Bayesian optimization [44] to efficiently tune γ_l and β_l with expandable bounds. In addition, we believe that the challenge of applying our methods to more advanced datasets such as Imagenet [45] is partly associated with that of scaling the stochastic neural network as we tend to need more expressive models for more challenging datasets. Given this perspective, the challenge to scale to large datasets can be partially addressed with the solutions from scaling stochastic neural networks some of which we suggest above. Furthermore, we believe that, as one of the main messages from our work, explicitly inducing compressed and relevant information (e.g., via mutual information as in MIB) into many intermediate layers can be more beneficial to large-scale tasks than simply resorting to the MLE principle. An intuition is to think of this as a way for *information-theoretic regularization for intermediate layers*. Finally, a followup important question to ask is whether there is any theoretical and stronger empirical link between an improved information representation (e.g., in the MIB sense) and the generalization of neural networks. This connection might be intuitively correct but a systematically empirical study or a theoretical suggestion are an important future research direction.

Author Contributions: Conceptualization by J.C. and T.T.N.; writing and conducting experiments supervised by J.C.; methodology, software, validation, formal analysis, investigation, visualization, and writing of the original draft by T.T.N.

Funding: This work was supported by the Institute for Information and Communications Technology Planning and Evaluation (IITP) grant (No.2017-0-01779, A machine learning and statistical inference framework for explainable artificial intelligence).

Conflicts of Interest: The authors declare no conflicts of interest.

Abbreviations

The following abbreviations are used in this manuscript:

IB	Information Bottleneck
MIB	Markov Information Bottleneck
SNN	Stochastic Neural Network
DPI	Data Processing Inequality
MLE	Maximum Likelihood Estimation
SGD	Stochastic Gradient Descent
SFNN	Stochastic Feed-forward Neural Network
VIB	Variational Information Bottleneck

References

1. Tishby, N.; Pereira, F.C.; Bialek, W. The information bottleneck method. In Proceedings of the Annual Allerton Conference on Communication, Control and Computing, Monticello, IL, USA, 22–24 September 1999.
2. Strouse, D.; Schwab, D.J. The Information Bottleneck and Geometric Clustering. *Neural Comput.* **2019**, *31*, doi:10.1162/neco_a_01136. [CrossRef] [PubMed]
3. Dai, B.; Zhu, C.; Guo, B.; Wipf, D.P. Compressing Neural Networks using the Variational Information Bottleneck. In Proceedings of the 35th International Conference on Machine Learning (ICML 2018) Stockholmsmässan, Stockholm, Sweden, 10–15 July 2018; pp. 1143–1152.
4. Achille, A.; Soatto, S. Emergence of Invariance and Disentanglement in Deep Representations. *J. Mach. Learn. Res.* **2018**, *19*, 50:1–50:34.
5. Yamada, M.; Heecheol, K.; Miyoshi, K.; Yamakawa, H. FAVAE: Sequence Disentanglement using Information Bottleneck Principle. *arXiv* **2019**, arXiv:1902.08341.
6. Jeon, I.; Lee, W.; Kim, G. IB-GAN: Disentangled Representation Learning with Information Bottleneck GAN. 2019. Available online: https://openreview.net/forum?id=ryljV2A5KX (accessed on 25 September 2019)
7. Tschannen, M.; Djolonga, J.; Rubenstein, P.K.; Gelly, S.; Lucic, M. On Mutual Information Maximization for Representation Learning. *arXiv* **2019**, arXiv:1907.13625.
8. Tishby, N.; Polani, D. Information Theory of Decisions and Actions. In *Perception-Action Cycle*; Springer: New York, NY, USA, 2011; pp. 601–636.
9. Goyal, A.; Islam, R.; Strouse, D.; Ahmed, Z.; Larochelle, H.; Botvinick, M.; Levine, S.; Bengio, Y. Transfer and Exploration via the Information Bottleneck. In Proceedings of the International Conference on Learning Representations, New Orleans, LA, USA, 6–9 May 2019.
10. Friedman, N.; Mosenzon, O.; Slonim, N.; Tishby, N. Multivariate Information Bottleneck. In Proceedings of the Seventeenth conference on Uncertainty in artificial intelligence, Seattle, WA, USA, 2–5 August, 2001.
11. Chechik, G.; Globerson, A.; Tishby, N.; Weiss, Y. Information Bottleneck for Gaussian Variables. *J. Mach. Learn. Res.* **2005**, *6*, 165–188.
12. Rey, M.; Roth, V. Meta-Gaussian Information Bottleneck. In Proceedings of the Annual Conference on Neural Information Processing Systems, NIPS, Lake Tahoe, NV, USA, 3–6 December 2012; pp. 1925–1933.
13. Strouse, D.; Schwab, D.J. The Deterministic Information Bottleneck. In Proceedings of the Thirty-Second Conference on Uncertainty in Artificial Intelligence (UAI 2016), New York, NY, USA, 25–29 June 2016; Ihler, A.T., Janzing, D., Eds.; AUAI Press: Corvallis, OR, USA, 2016.
14. Alemi, A.A.; Fischer, I.; Dillon, J.V.; Murphy, K. Deep Variational Information Bottleneck. In Proceedings of the International Conference on Learning Representations (ICLR), Toulon, France, 24–26 April 2017.
15. Tishby, N.; Zaslavsky, N. Deep learning and the information bottleneck principle. In Proceedings of the IEEE Information Theory Workshop (ITW), Jerusalem, Israel, 26 April–1 May 2015; pp. 1–5.
16. Shwartz-Ziv, R.; Tishby, N. Opening the Black Box of Deep Neural Networks via Information. *arXiv* **2017**, arXiv:1703.00810.
17. Amjad, R.A.; Geiger, B.C. Learning Representations for Neural Network-Based Classification Using the Information Bottleneck Principle. *IEEE Trans. Pattern Anal. Mach. Intell.* **2019**. [CrossRef] [PubMed]
18. Hinton, G.E. Training Products of Experts by Minimizing Contrastive Divergence. *Neural Comput.* **2002**, *14*, 1771–1800, doi:10.1162/089976602760128018. [CrossRef] [PubMed]
19. Hinton, G.E.; Osindero, S.; Teh, Y.W. A Fast Learning Algorithm for Deep Belief Nets. *Neural Comput.* **2006**, *18*, 1527–1554, doi:10.1162/neco.2006.18.7.1527. [CrossRef] [PubMed]
20. Tang, Y.; Salakhutdinov, R. Learning Stochastic Feedforward Neural Networks. In *Advances in Neural Information Processing Systems 26: Proceedings of the 27th Annual Conference on Neural Information Processing Systems 2013, Lake Tahoe, NV, USA, 5–10 December 2013*; Burges, C.J.C., Bottou, L., Ghahramani, Z., Weinberger, K.Q., Eds.; Curran: Norwich, UK, 2013; pp. 530–538.
21. Raiko, T.; Berglund, M.; Alain, G.; Dinh, L. Techniques for Learning Binary Stochastic Feedforward Neural Networks. In Proceedings of the International Conference on Learning Representations (ICLR), San Diego, CA, USA, 7–9 May 2015.
22. Florensa, C.; Duan, Y.; Abbeel, P. Stochastic Neural Networks for Hierarchical Reinforcement Learning. In Proceedings of the 5th International Conference on Learning Representations (ICLR 2017), Toulon, France, 24–26 April 2017.

23. LeCun, Y.; Bottou, L.; Bengio, Y.; Haffner, P. Gradient-based learning applied to document recognition. *Proc. IEEE* **1998**, *86*, 2278–2324. [CrossRef]
24. Nguyen, T.T.; Choi, J. Layer-wise Learning of Stochastic Neural Networks with Information Bottleneck. *arXiv* **2017**, arXiv:1712.01272.
25. Nguyen, T.T. Parametric Information Bottleneck to Optimize Stochastic Neural Networks. Master's Thesis, Ulsan National Institute of Science and Technology, Ulsan, Korea, 2018.
26. Slonim, N. Information Bottleneck Theory and Applications. Ph.D. Thesis, Hebrew University of Jerusalem, Jerusalem, Israel, 2003.
27. Cover, T.M.; Thomas, J.A. *Elements of Information Theory*; Wiley Series in Telecommunications and Signal Processing; Wiley: New York, NY, USA, 2006.
28. Saxe, A.M.; Bansal, Y.; Dapello, J.; Advani, M.; Kolchinsky, A.; Tracey, B.D.; Cox, D.D. On the Information Bottleneck Theory of Deep Learning. In Proceedings of the 6th International Conference on Learning Representations (ICLR 2018), Vancouver, BC, Canada, 30 April–3 May 2018.
29. Cheng, Y.; Wang, D.; Zhou, P.; Zhang, T. A Survey of Model Compression and Acceleration for Deep Neural Networks. *arXiv* **2017**, arXiv:1710.09282.
30. Rasmussen, C.E.; Ghahramani, Z. Occam's Razor. In *Advances in Neural Information Processing Systems*; MIT Press: Cambridge, MA, USA, 2001, pp. 294–300.
31. Arora, S.; Ge, R.; Neyshabur, B.; Zhang, Y. Stronger generalization bounds for deep nets via a compression approach. In Proceedings of the 35th International Conference on Machine Learning, ICML 2018, Stockholmsmässan, Stockholm, Sweden, 10–15 July 2018; pp. 254–263.
32. Bengio, Y. Learning Deep Architectures for AI. *Found. Trends Mach. Learn.* **2009**, *2*, 1–127. [CrossRef]
33. Williams, R.J. Simple Statistical Gradient-Following Algorithms for Connectionist Reinforcement Learning. *Mach. Learn.* **1992**, *8*, 229–256, doi:10.1007/BF00992696. [CrossRef]
34. Bengio, Y.; Léonard, N.; Courville, A.C. Estimating or Propagating Gradients Through Stochastic Neurons for Conditional Computation. *arXiv* **2013**, arXiv:1308.3432.
35. Hinton, G. Lecture 9.3—Using Noise as a Regularizer. In *Neural Networks for Machine Learning*; University of Toronto: Toronto, ON, USA, 2016.
36. Zeiler, M.D. ADADELTA: An Adaptive Learning Rate Method. *arXiv* **2012**, arXiv:1212.5701,
37. Kingma, D.P.; Ba, J. Adam: A Method for Stochastic Optimization. In Proceedings of the 3rd International Conference on Learning Representations, ICLR, 2015, San Diego, CA, USA, 7–9 May 2015.
38. Szegedy, C.; Zaremba, W.; Sutskever, I.; Bruna, J.; Erhan, D.; Goodfellow, I.J.; Fergus, R. Intriguing properties of neural networks. In Proceedings of the 2nd International Conference on Learning Representations, ICLR 2014, Banff, AB, Canada, 14–16 April 2014.
39. Nguyen, A.M.; Yosinski, J.; Clune, J. Deep neural networks are easily fooled: High confidence predictions for unrecognizable images. In Proceedings of the IEEE Conference on Computer Vision and Pattern Recognition (CVPR 2015), Boston, MA, USA, 7–12 June 2015; IEEE Computer Society: Washington, DC, USA, 2015; pp. 427–436, doi:10.1109/CVPR.2015.7298640. [CrossRef]
40. Carlini, N.; Wagner, D.A. Towards Evaluating the Robustness of Neural Networks. In Proceedings of the 2017 IEEE Symposium on Security and Privacy (SP 2017), San Jose, CA, USA, 22–26 May 2017; IEEE Computer Society: Washington, DC, USA, 2017; pp. 39–57, doi:10.1109/SP.2017.49. [CrossRef]
41. Lin, X.; Sur, I.; Nastase, S.A.; Divakaran, A.; Hasson, U.; Amer, M.R. Data-Efficient Mutual Information Neural Estimator. *arXiv* **2019**, arXiv:1905.03319.
42. Belghazi, M. I.; Baratin, A.; Rajeshwar, S.; Baratin, A.; Ozair, S.; Bengio, Y.; Hjelm, R.D.; Courville, A.C. Mutual Information Neural Estimation. In Proceedings of the 35th International Conference on Machine Learning, ICML 2018, Stockholmsässan, Stockholm, Sweden, 10–15 July 2018; pp. 530–539.
43. Kingma, D.P.; Welling, M. Auto-Encoding Variational Bayes. In Proceedings of the 2nd International Conference on Learning Representations, ICLR 2014, Banff, AB, Canada, 14–16 April 2014.

44. Ha, H.; Rana, S.; Gupta, S.; Nguyen, T.; Tran-The, H.; Venkatesh, S. Bayesian Optimization with Unknown Search Space. In Proceedings of the Advances in Neural Information Processing Systems 33: Annual Conference on Neural Information Processing Systems 2019, Vancouver, BC, Canada, 8–14 December 2019.
45. Deng, J.; Dong, W.; Socher, R.; Li, L.J.; Li, K.; Fei-Fei, L. ImageNet: A Large-Scale Hierarchical Image Database. In Proceedings of the 2009 IEEE Conference on Computer Vision and Pattern Recognition, Miami, Florida, 20–25 June 2009.

© 2019 by the authors. Licensee MDPI, Basel, Switzerland. This article is an open access article distributed under the terms and conditions of the Creative Commons Attribution (CC BY) license (http://creativecommons.org/licenses/by/4.0/).

Article
Nonlinear Information Bottleneck

Artemy Kolchinsky [1,*], Brendan D. Tracey [1,2] and David H. Wolpert [1,3,4]

[1] Santa Fe Institute, 1399 Hyde Park Road, Santa Fe, NM 87501, USA; tracey.brendan@gmail.com (B.D.T.); david.h.wolpert@gmail.com (D.H.W.)
[2] Department of Aeronautics & Astronautics, Massachusetts Institute of Technology, Cambridge, MA 02139, USA
[3] Complexity Science Hub, 1080 Vienna, Austria
[4] Center for Bio-Social Complex Systems, Arizona State University, Tempe, AZ 85281, USA
* Correspondence: artemyk@gmail.com

Received: 16 October 2019; Accepted: 28 November 2019; Published: 30 November 2019

Abstract: Information bottleneck (IB) is a technique for extracting information in one random variable X that is relevant for predicting another random variable Y. IB works by encoding X in a compressed "bottleneck" random variable M from which Y can be accurately decoded. However, finding the optimal bottleneck variable involves a difficult optimization problem, which until recently has been considered for only two limited cases: discrete X and Y with small state spaces, and continuous X and Y with a Gaussian joint distribution (in which case optimal encoding and decoding maps are linear). We propose a method for performing IB on arbitrarily-distributed discrete and/or continuous X and Y, while allowing for nonlinear encoding and decoding maps. Our approach relies on a novel non-parametric upper bound for mutual information. We describe how to implement our method using neural networks. We then show that it achieves better performance than the recently-proposed "variational IB" method on several real-world datasets.

Keywords: information bottleneck; mutual information; representation learning; neural networks

1. Introduction

Imagine that one has two random variables, an "input" random variable X and an "output" random variable Y, and that one wishes to use X to predict Y. In some situations, it is useful to extract a compressed representation of X that is relevant for predicting Y. This problem is formally considered by the *information bottleneck* (IB) method [1–3]. IB proposes to find a "bottleneck" variable M which maximizes prediction, formulated in terms of the mutual information $I(Y; M)$, given a constraint on compression, formulated in terms of the mutual information $I(X; M)$. Formally, this can be stated in terms of the constrained optimization problem

$$\arg\max_{M \in \Delta} I(Y; M) \quad \text{s.t.} \quad I(X; M) \leq R, \qquad (1)$$

where Δ is the set of random variables M that obey the Markov condition $Y - X - M$ [4–6]. This Markov condition states that M is conditionally independent of Y given X, and it guarantees that any information that M has about Y is extracted from X. The maximal value of $I(Y; M)$ for each possible compression value R forms what is called the *IB curve* [1].

The following example illustrates how IB might be used. Suppose that a remote weather station makes detailed recordings of meteorological data (X), which are then encoded and sent to a central server (M) and used to predict weather conditions for the next day (Y). If the channel between the weather station and server has low capacity, then the information transmitted from the weather station to the server must be compressed. Minimizing the IB objective amounts to finding a compressed representation of meteorological data which can be transmitted across a low capacity channel (have low $I(X; M)$) and used to optimally predict future weather (have high $I(Y; M)$). The IB curve specifies the trade-off between channel capacity and accurate prediction.

Numerous applications of IB exist in domains such as clustering [7,8], coding theory and quantization [9–12], speech and image recognition [13–17], and cognitive science [18]. Several recent papers have also drawn connections between IB and supervised learning, in particular, classification using neural networks [19,20]. In this context, X typically represents input vectors, Y the output classes, and M the intermediate representations used by the network, such as the activity of hidden layer(s) [21]. Existing research has considered whether intermediate representations that are optimal in the IB sense (i.e., close to the IB curve) may be better in terms of generalization error [21–23], robustness to adversarial inputs [24], detection of out-of-distribution data [25], or provide more "interesting" or "useful" intermediate representations of inputs [26]. Other related research has investigated whether stochastic gradient descent (SGD) training dynamics may drive hidden layer representations towards IB optimality [27,28].

In practice, optimal bottleneck variables are usually not found by solving the constrained optimization problem of Equation (1), but rather by finding M that maximize the so-called *IB Lagrangian* [1,6,22],

$$\mathcal{L}_{\text{IB}}(M) := I(Y; M) - \beta I(X; M). \tag{2}$$

$\mathcal{L}_{\text{IB}}$ is the Lagrangian relaxation [29] of the constrained optimization problem of Equation (1), and β is a Lagrange multiplier that enforces the constraint $I(X; M) \leq R$. In practice, $\beta \in [0, 1]$ serves as a parameter that controls the trade-off between compression and prediction. As $\beta \to 1$, IB will favor maximal compression of X; for $\beta = 1$ (or any $\beta \geq 1$) the optimal M will satisfy $I(X; M) = I(Y; M) = 0$. As $\beta \to 0$, IB will favor prediction of Y; for $\beta = 0$ (or any $\beta \leq 0$), there is no penalty on $I(X; M)$ and the optimal M will satisfy $I(Y; M) = I(X; Y)$, the maximum possible. It is typically easier to optimize $\mathcal{L}_{\text{IB}}$ than Equation (1), since the latter involves a complicated non-linear constraint. For this reason, optimizing $\mathcal{L}_{\text{IB}}$ has become standard in the IB literature [1,6,19,20,22,24,30,31].

However, in recent work [32] we showed that whenever Y is a deterministic function of X (or close to being one), optimizing $\mathcal{L}_{\text{IB}}$ is not longer equivalent to optimizing Equation (1). In fact, when Y is a deterministic function of X, the same M will optimize $\mathcal{L}_{\text{IB}}$ for all values of β, meaning that the IB curve cannot be explored by optimizing $\mathcal{L}_{\text{IB}}$ while sweeping β. This is a serious issue in supervised learning scenarios (as well as some other domains), where it is very common for the output Y to be a deterministic function of the input X. Nonetheless, the IB curve can still be explored by optimizing the following simple modification of the IB Lagrangian, which we called the *squared-IB Lagrangian* [32],

$$\mathcal{L}_{\text{sqIB}}(M) := I(Y; M) - \beta I(X; M)^2 \tag{3}$$

where $\beta \geq 0$ is again a parameter that controls the trade-off between compression and prediction. Unlike the case for $\mathcal{L}_{\text{IB}}$, there is always a one-to-one correspondence between M that optimize $\mathcal{L}_{\text{sqIB}}$ and solutions to Equation (1), regardless of the relationship between X and Y. In the language of optimization theory, the squared-IB Lagrangian is a "scalarization" of the multi-objective problem $\{\min I(X; M), \max I(Y; M)\}$ [33]. Importantly, unlike $\mathcal{L}_{\text{IB}}$, there can be non-trivial optimizers of $\mathcal{L}_{\text{sqIB}}$ even for $\beta \geq 1$; the relationship between β and corresponding solutions on the IB curve has been analyzed in [34]. In that work, it was also shown that the objective function of Equation (3) is part of a general family of objectives $I(Y; M) - \beta F(I(X; M))$, where F is any monotonically-increasing and strictly convex function, all of which can be used to explore the IB curve.

Unfortunately, optimizing the IB Lagrangian and squared-IB Lagrangian remains a difficult problem. First, both objectives are non-convex, so there is no guarantee that a global optimum can be found. Second, finding even a local optimum requires evaluating the mutual information terms $I(X; M)$ and $I(Y; M)$, which can involve intractable integrals. For this reason, until recently IB has been mainly developed for two limited cases. The first case is where X and Y are discrete-valued and have a small number of possible outcomes [1]. There, one can explicitly represent the full *encoding map* (the condition probability distribution of M given X) during optimization, and the relevant integrals become tractable finite sums. The second case is when X and Y are continuous-valued and jointly Gaussian. Here, the IB optimization problem can be solved analytically, and the resulting encoding and decoding maps are linear [31].

In this work, we propose a method for performing IB in much more general settings, which we call *nonlinear information bottleneck*, or *nonlinear IB* for short. Our method assumes that M is a continuous-valued random variable, but X and Y can be either discrete-valued (possibly with many states) or continuous-valued, and with any desired joint distribution. Furthermore, as suggested by the term nonlinear IB, the encoding and decoding maps can be nonlinear.

To carry out nonlinear IB, we derive a lower bound on $\mathcal{L}_{\text{IB}}$ (or, where appropriate, $\mathcal{L}_{\text{sqIB}}$) which can be maximized using gradient-based methods. As we describe in the next section, our approach makes use of the following techniques:

- We represent the distribution over X and Y using a finite number of data samples.
- We represent the encoding map $p(m|x)$ and the *decoding map* $p(y|m)$ as parameterized conditional distributions.
- We use a variational lower bound for the prediction term $I(Y; M)$, and non-parametric upper bound for the compression term $I(X; M)$, which we developed in earlier work [35].

Note that three recent papers have suggested other ways of optimizing the IB Lagrangian in general settings [24,36,37]. These papers use variational upper bounds on the compression term $I(X; M)$, which is different from our non-parametric upper bound. A detailed comparison is provided in Section 3. In that section, we also relate our approach to other work in machine learning.

In Section 4, we explain how to implement our approach using standard neural network techniques. We demonstrate its performance on several real-world datasets, and compare it to the recently-proposed *variational IB* method [24].

2. Proposed Approach

In the following, we use $H(\cdot)$ for Shannon entropy, $I(\cdot; \cdot)$ for mutual information [MI], $D_{\text{KL}}(\cdot \| \cdot)$ for Kullback–Leibler [KL] divergence. All information-theoretic quantities are in units of bits, and all logs are base-2. We use $\mathcal{N}(\mu, \Sigma)$ to indicate the probability density function of a multivariate Gaussian with mean μ and covariance matrix Σ. We use notation like $\mathbb{E}_{P(X)}[f(X)] = \int P(x) f(x)\, dx$ to indicate expectations, where $f(x)$ is some function and $P(x)$ some probability distribution. We use $\delta(\cdot, \cdot)$ for the Kronecker delta.

Let the input random variable X and the output random variable Y be distributed according to some joint distribution $Q(x, y)$, with marginals indicated by $Q(y)$ and $Q(x)$. We assume that we are provided with a "training dataset" $\mathcal{D} = \{(x_1, y_1), \ldots, (x_N, y_N)\}$, which contains N input–output pairs sampled IID from $Q(x, y)$. Let M indicate the bottleneck random variable, with outcomes in $\mathbb{R}^d$. In the derivations in this section, we assume that X and Y are continuous-valued, but our approach extends immediately to the discrete case (with some integrals replaced by sums).

Let the conditional probability $P_\theta(m|x)$ indicate a parameterized *encoding map* from input X to the bottleneck variable M, where θ is a vector of parameters. Given an encoding map, one can compute the MI between X and M, $I_\theta(X; M)$, using the joint distribution $Q_\theta(x, m) := P_\theta(m|x) Q(x)$. Similarly, one can compute the MI between Y and M, $I_\theta(Y; M)$, using the joint distribution

$$Q_\theta(y, m) := \int P_\theta(m|x) Q(x, y)\, dx. \tag{4}$$

We now consider the IB Lagrangian, Equation (2), as a function of the encoding map parameters,

$$\mathcal{L}_{\text{IB}}(\theta) := I_\theta(Y; M) - \beta I_\theta(X; M). \tag{5}$$

In this parametric setting, we seek parameter values that maximize $\mathcal{L}_{IB}(\theta)$. Unfortunately, this optimization problem is usually intractable due to the difficulty of computing the integrals in Equation (4) and in the MI terms of Equation (5). Nonetheless, it is possible to carry out an approximate form of IB by maximizing a tractable lower bound on $\mathcal{L}_{IB}$, which we now derive.

First, consider any conditional probability $P_\phi(y|m)$ of outputs given bottleneck variable, where ϕ is a vector of parameters, which we call the *(variational) decoding map*. Given $P_\phi(y|m)$, the non-negativity of KL divergence leads to the following variational lower bound on the first MI term in Equation (5),

$$\begin{aligned} I_\theta(Y;M) &= H(Q(Y)) - H(Q_\theta(Y|M)) \\ &\geq H(Q(Y)) - H(Q_\theta(Y|M)) - D_{KL}(Q_\theta(Y|M)\|P_\phi(Y|M)) \\ &= H(Q(Y)) + \mathbb{E}_{Q_\theta(Y,M)}\left[\log P_\phi(Y|M)\right], \end{aligned} \qquad (6)$$

where in the last line we've used the following identity,

$$-\mathbb{E}_{Q_\theta(Y,M)}\left[\log P_\phi(Y|M)\right] = D_{KL}(Q_\theta(Y|M)\|P_\phi(Y|M)) + H(Q_\theta(Y|M)). \qquad (7)$$

Note that the inequality of Equation (6) holds for any choice of $P_\phi(y|m)$, and becomes an equality when $P_\phi(y|m)$ is equal to the "optimal" decoding map $Q_\theta(y|m)$ (as would be computed from Equation (4)). Moreover, the bound becomes tighter as the KL divergence between $P_\phi(y|m)$ and $Q_\theta(y|m)$ gets smaller. Below, we will maximize the RHS of Equation (6) with respect to ϕ, thereby bringing $P_\phi(y|m)$ closer to $Q_\theta(y|m)$.

It remains to upper bound the $I_\theta(X;M)$ term in Equation (5). To proceed, we first approximate the joint distribution of X and Y with the empirical distribution in the training dataset,

$$Q(x,y) \approx \frac{1}{N}\sum_i \delta(x_i,x)\delta(y_i,y). \qquad (8)$$

We then assume that the encoding map is the sum of a deterministic function $f_\theta(x)$ plus Gaussian noise,

$$M = f_\theta(X) + Z, \qquad (9)$$

where $(Z|X=x) \sim \mathcal{N}(f_\theta(x), \Sigma_\theta(x))$. Note that the noise covariance $\Sigma_\theta(x)$ can depend both on the parameters θ and the outcome of X (i.e., the noise can be heteroscedastic). Combining Equation (8) and Equation (9) implies that the bottleneck variable M will be distributed as a mixture of N equally-weighted Gaussian components, with component i having distribution $\mathcal{N}(f_\theta(x_i), \Sigma_\theta(x_i))$. We can then employ the following non-parametric upper bound on MI, which was derived in a recent paper [35]:

$$I_\theta(X;M) \leq \hat{I}_\theta(X;M) := -\frac{1}{N}\sum_i \log \frac{1}{N}\sum_j e^{-D_{KL}\left[\mathcal{N}(f_\theta(x_i),\Sigma_\theta(x_i))\|\mathcal{N}(f_\theta(x_j),\Sigma_\theta(x_j))\right]}. \qquad (10)$$

(Note that the published version of [35] contains some typos which are corrected in the latest arXiv version at arxiv.org/abs/1706.02419.)

Equation (10) bounds the MI in terms of the pairwise KL divergences between the Gaussian components of the mixture distribution of M. It is useful because the KL divergence between two d-dimensional Gaussians has a closed-form expression,

$$D_{KL}\left[\mathcal{N}(\mu',\Sigma')\|\mathcal{N}(\mu,\Sigma)\right] = \frac{1}{2}\left[\ln\frac{\det\Sigma}{\det\Sigma'} + (\mu'-\mu)\Sigma^{-1}(\mu'-\mu) + \mathrm{tr}(\Sigma^{-1}\Sigma') - d\right]. \qquad (11)$$

Furthermore, in the special case when all components have the same covariance and can be grouped into well-separated clusters, the upper bound of Equation (10) becomes tight [35]. As we will see below, this special case is a commonly encountered solution to the optimization problem considered here.

Combining Equation (6) and Equation (10) provides the following tractable lower bound for the IB Lagrangian,

$$\mathcal{L}_{\text{IB}}(\theta) \geq \hat{\mathcal{L}}_{\text{IB}}(\theta, \phi) := \mathbb{E}_{Q_\theta(Y,M)}\left[\log P_\phi(Y|M)\right] - \beta \hat{I}_\theta(X; M) \quad (12)$$

where we dropped the additive constant $H(Q(Y))$ (which does not depend on the parameter values and is therefore irrelevant for optimization). We refer to Equation (12) as the *nonlinear IB objective*.

As mentioned in the introduction, in cases where Y is a deterministic function of X (or close to being one), it is no longer possible to explore the IB curve by optimizing the IB Lagrangian for different values of β [19,32,34]. Nonetheless, it is always possible to explore the IB curve by instead optimizing the squared-IB Lagrangian, Equation (3). The above derivations also lead to the following tractable lower bound for the squared-IB Lagrangian,

$$\mathcal{L}_{\text{sqIB}}(\theta) \geq \hat{\mathcal{L}}_{\text{sqIB}}(\theta, \phi) := \mathbb{E}_{Q_\theta(Y,M)}\left[\log P_\phi(Y|M)\right] - \beta \left[\hat{I}_\theta(X; M)\right]^2. \quad (13)$$

Note that maximizing the expectation term $\mathbb{E}_{Q_\theta(Y,M)}\left[\log P_\phi(Y|M)\right]$ is equivalent to minimizing the usual cross-entropy loss in supervised learning. (Note that mean squared error, the typical loss function used for training regression models, can also be interpreted as a cross-entropy term [38] (pp. 132–134).) From this point of view, Equation (12) and Equation (13) can be interpreted as adding an information-theoretic regularization term to the regular objective of supervised learning.

For optimization purposes, the compression term $\hat{I}_\theta(X; M)$ can be computed from data using Equations (10) and (11), while the expectation term $\mathbb{E}_{Q_\theta(Y,M)}\left[\log P_\phi(Y|M)\right]$ can be estimated as $\mathbb{E}_{Q_\theta(Y,M)}\left[\log P_\phi(Y|M)\right] \approx \frac{1}{N}\sum_i \log P_\phi(y_i|m_i)$, where m_i indicates samples from $P_\theta(m|x_i)$. Assuming that f_θ is differentiable with respect to θ and P_ϕ is differentiable with respect to ϕ, the optimal θ and ϕ can be selected by using gradient-based methods to maximize Equation (12) or Equation (13), as desired. In practice, this optimization will typically be done using stochastic gradient descent (SGD), i.e., by computing the gradient using randomly sampled mini-batches rather than the whole training dataset. In fact, mini-batching becomes necessary for large datasets, since evaluating $\hat{I}_\theta(X; M)$ involves $O(n^2)$ operations, where n is the number of data points in the batch used to compute the gradient, which becomes prohibitively slow for very large n. At the same time, $\hat{I}_\theta(X; M)$ is closely related to kernel-density estimators [35], and it is known that the number of samples required for accurate kernel-density estimates grows rapidly as dimensionality increases [39]. Thus, mini-batches should not be too small when d (the dimensionality of the bottleneck variable) is large. In some cases, it may be useful to estimate the gradient of $\mathbb{E}_{Q_\theta(Y,M)}\left[\log P_\phi(Y|M)\right]$ and the gradient of $\hat{I}_\theta(X; M)$ using mini-batches of different sizes. More implementation details are discussed below in Section 4.1.

Note that the approach described here is somewhat different (and simpler) than in previous versions of this manuscript [40,41]. In previous versions, we represented the marginal distribution $Q(x)$ with a mixture of Gaussians, rather than with the empirical distribution in the training data. However, we found that this increased complexity but was not necessary for good performance. Furthermore, we previously focused only on optimizing a bound on the IB Lagrangian, Equation (12). In subsequent work [32], we showed that the IB Lagrangian is inadequate for many supervised learning scenarios, including some of those explored in Section 4.2, and that the squared-IB Lagrangian should be used instead. In this work, we report performance when optimizing Equation (13), a bound on the squared-IB Lagrangian.

3. Relation to Prior Work

In this section, we relate our proposed method to prior work in machine learning.

3.1. Variational IB

Recently, there have been three other proposals for performing IB for continuous and possibly non-Gaussian random variables using neural networks [24,36,37], the most popular of which is called *variational IB* (VIB) [24]. As in our approach, these methods propose tractable lower bounds on the $\mathcal{L}_{\text{IB}}$ objective. They employ the same variational bound for the prediction MI term $I(Y; M)$ as our Equation (6). These methods differ from ours, however, in how they bound the compression term, $I_\theta(X; M)$. In particular, they all use some form of the following variational upper bound,

$$I_\theta(X; M) = D_{\text{KL}}(P_\theta(M|X) \| R(M)) - D_{\text{KL}}(P_\theta(M) \| R(M)) \leq D_{\text{KL}}(P_\theta(M|X) \| R(M)), \quad (14)$$

where R is some surrogate marginal distribution over the bottleneck variable M. Combining with Equation (6) leads to the following variational lower bound for $\mathcal{L}_{\text{IB}}$,

$$\mathcal{L}_{\text{IB}}(M) \geq \mathbb{E}_{Q_\theta(Y,M)} \left[\log P_\phi(Y|M) \right] - \beta D_{\text{KL}}(P_\theta(M|X) \| R(M)) + \text{const}. \quad (15)$$

The three aforementioned papers differ in how they define the surrogate marginal distribution R. In [24], R is a standard multivariate normal distribution, $\mathcal{N}(0, \mathbf{I})$. In [36], R is a product of Student's t-distributions. The scale and shape parameters of each t-distribution are optimized during training, in this way tightening the bound in Equation (14). In [37], two surrogate distributions are considered, the improper log-uniform and the log-normal, with the appropriate choice depending on the particular activation function (non-linearity) used in the neural network.

In addition, the encoding map $P_\theta(m|x)$ in [36] and [24] is a deterministic function plus Gaussian noise, same as in Equation (9). In [37], the encoding map consists of a deterministic function with multiplicative, rather than additive, noise.

These alternative methods have potential advantages and disadvantages compared to our approach. On one hand, they are more computationally efficient: Our non-parametric estimator of $\hat{I}_\theta(X; M)$ requires $O(n^2)$ operations per mini-batch (where n is the size of the mini-batch), while the variational bound of Equation (14) requires $O(n)$ operations. On the other hand, our non-parametric estimator is expected to give a better estimate of the true MI $I(X; M)$ [35]. We provide a comparison between our approach and variational IB [25] in Section 4.2.

3.2. Neural Networks and Kernel Density Entropy Estimates

A key component of our approach is using a differentiable upper bound on MI, $\hat{I}_\theta(X; M)$. As discussed in [35], this bound is related to non-parametric kernel-density estimators of MI. See [42–46] for related work on using neural networks to optimize non-parametric estimates of information-theoretic functions. This technique can also be related to kernel-based estimation of the likelihood of held-out data for neural networks (e.g., [47]). In these later approaches, however, the likelihood of held-out data is estimated only once, as a diagnostic measure once learning is complete. We instead propose to train the network by directly incorporating our non-parametric estimator $\hat{I}_\theta(X; M)$ in the objective function.

3.3. Auto-Encoders

Auto-encoders are unsupervised learning architectures that learn to reconstruct a copy of the input X, while using some intermediate representations (such as a hidden layer in a neural network). Auto-encoders have some conceptual relationships to IB, in that the intermediate representations are sometimes restricted in terms of dimensionality, or with information-theoretic penalties on hidden layer coding length [48,49]. Similar penalties have also been explored in a supervised learning scenario in [50]. In that work, however, hidden layer states were treated as discrete-valued, limiting the flexibility and information capacity of hidden representations.

More recently, *denoising auto-encoders* [51] have attracted attention. Denoising auto-encoders constrain the amount of information passing from input to hidden layers by injecting noise into the hidden layer activity, similarly to our noisy mapping from the input to the bottleneck layer. Previous work on auto-encoders has considered either penalizing hidden layer coding length *or* injecting noise into the map, rather than combing the two as we do here. Moreover, denoising auto-encoders do not have a notion of an "optimal" noise level, since less noise will always improve prediction error on the training data. Thus, they cannot directly adapt the noise level (as done in our method).

Finally, *variational auto-encoders* [52] [VAEs] are recently-proposed architectures which learn generative models from unsupervised data (i.e., after training, they can be used to generate new samples that resemble training data). Interestingly, the objective optimized in VAE training, called "ELBO", contains both a prediction term and a compression term and can be seen as a special case of the variational IB objective [24,37,53,54]. In principle, it may be fruitful to replace the compression term in the ELBO with our MI estimator $\hat{I}_\theta(X; M)$. Given our reported performance below, this may result in better compression, though it might also complicate sampling from the latent variable space. We leave this line of research for future work.

4. Experiments

In this section, we first explain how to implement nonlinear IB using neural network techniques. We then evaluate its on several datasets, and compare it to the variational IB (VIB) method. We demonstrate that, compared to VIB, nonlinear IB achieves better performance and uncovers different kinds of representations.

4.1. Implementation

Any implementation of nonlinear IB requires a way to compute the encoding map $P_\theta(m|x)$ and decoding map $P_\phi(y|m)$, as well as a way to choose the parameters of these maps so as to maximize the nonlinear IB objective. Here we explain how this can be done using standard neural network methods.

The encoding map $P_\theta(m|x)$, Equation (9), is implemented in the following way: First, several neural network layers with parameters θ implement the (possibly nonlinear) deterministic function $f_\theta(x)$. The output of these layers is then added to zero-centered Gaussian noise with covariance $\Sigma_\theta(x)$, which becomes the state of the *bottleneck layer*. This is typically done via the "reparameterization trick" [52], in which samples of Gaussian noise are passed through several deterministic layers (whose parameters are also indicated by θ) and then added to $f_\theta(x)$. Note that due to the presence of noise, the neural network is stochastic: even with parameters held constant, different states of the bottleneck layer are sampled during different NN evaluations. This stochasticity guarantees that the mutual information $I(X; M)$ is finite [26,28].

In all of the experiments described below, the encoding map consists of two layers with 128 ReLU neurons each, following by a layer of 5 linear neurons. In addition, for simplicity we use a simple homoscedastic noise model: $\Sigma_\theta(x) = \sigma^2 \mathbf{I}$, where σ^2 is a parameter the sets the scale of the noise variance. This noise model permits us to rewrite the MI bound of Equation (10) in terms of the following simple expression,

$$\hat{I}_\theta(X; M) = -\frac{1}{N} \sum_i \log \frac{1}{N} \sum_j e^{-\frac{1}{2\sigma^2}\|f_\theta(x_i) - f_\theta(x_j)\|_2^2}. \tag{16}$$

For purposes of comparison, we use this same homoscedastic noise model for both nonlinear IB and for VIB (note that the original VIB paper [24] used a heteroscedastic noise model; investigating the performance of nonlinear IB with heteroscedastic noise remains for future work).

In our runs, the noise parameter σ^2 was one of the trainable parameters in θ. The initial value of σ^2 should be chosen with some care. If the initial σ^2 is too small, the Gaussian components that make up the mixture distribution of M will be many standard deviations away from each other and $\hat{I}_\theta(X; M)$ (as well as $I(X; M)$) will be exponentially close to the constant $\log N$ [35]. In this case, the gradient of the compression term $\hat{I}_\theta(X; M)$ with respect to θ will also be exponentially small, and the optimizer will not be able to learn to compress. On the other hand, when σ^2 is too large, the resulting noise can swamp gradient information arising from the accuracy (cross-entropy) term, cause the optimizer to collapse to a "trivial" maximally-compressed model in which $I(X;M) \approx I(Y;M) \approx 0$. Nonetheless, the optimization is robust to several orders of magnitude of variation of the initial value of σ^2. In the experiments below, we uses the initial value $\sigma^2 = 1$, which works sufficiently well in practice. (Note that the scale of the noise can also be trained by changing the parameters of the 5-neuron linear layer; thus, in our neural network architecture, having a trainable σ^2 is not strictly necessary.)

To implement the decoding map $P_\phi(y|m)$, the bottleneck layer states are passed through several deterministic neural network layers with parameters ϕ. In the experiments described below, the decoding map is implemented with a single layer with 128 ReLU neurons, followed by a linear output layer. The log decoding probability ($\log P_\phi(y|m)$) is then evaluated using the network output and an appropriately-chosen cost function: cross-entropy loss of the softmax of the output for classification, and mean squared error (MSE) of the output for regression.

In the experiments below, we use nonlinear IB to optimize the bound on the "squared-IB Lagrangian", Equation (13), rather than the bound on the IB Lagrangian, Equation (12). For comparison purposes, we also optimize the following "squared" version of the VIB objective, Equation (15),

$$\mathcal{L}_{\text{sq-VIB}} := \mathbb{E}_{Q_\theta(Y,M)}\big[\log P_\phi(Y|M)\big] - \beta\big[D_{\text{KL}}(P_\theta(M|X)\|R(M))\big]^2. \tag{17}$$

As in the original VIB paper, we take $R(m)$ to be the standard Gaussian $\mathcal{N}(0,\mathbf{I})$. We found that optimizing the squared-IB bounds, Equation (13) and Equation (17), produced quantitatively similar results to optimizing Equation (12) and Equation (15), but was more numerically robust when exploring the full range of the IB curve. For an explanation of why this occurs, see the discussion and analysis in [32]. We report performance of nonlinear IB and VIB when optimizing bounds on the IB Lagrangian, Equation (12) and Equation (15), in the Supplementary Material.

We use the Adam [55] optimizer with standard TensorFlow settings and mini-batches of size 256. To avoid over-fitting, we use early stopping: we split the training data into 80% actual training data and 20% validation data; training is stopped once the objective on the validation dataset did not improve for 50 epochs.

A TensorFlow implementation of our approach is provided at https://github.com/artemyk/nonlinearIB. An independent PyTorch implementation is available at https://github.com/burklight/nonlinear-IB-PyTorch.

4.2. Results

We report the performance of nonlinear IB on two different classification datasets (MNIST and FashionMNIST) and one regression dataset (California housing prices). We also compare it with the recently-proposed variational IB (VIB) method [24]. Here we focus purely on the ability of these methods to optimize the IB objective on training and testing data. We leave for future work comparisons of these methods in terms of adversarial robustness [24], detection of out-of-distribution data [25], and other desirable characteristics that may emerge from IB training.

We optimize both the nonlinear IB (Equation (13)) and the VIB (Equation (17)) objectives for different values of β, producing a series of models that explore the trade-off between compression and prediction. We vary $\beta \in [10^{-3}, 2]$ for classification tasks and $\beta \in [10^{-5}, 2]$ for the regression task. These ranges were chosen empirically so that the resulting models fully explore the IB curve.

To report our results, we use *information plane* (info-plane) diagrams [27], which visualize the performance of different models in terms of the compression term ($I(X;M)$, the x-axis) and the prediction term ($I(Y;M)$, the y-axis) both on training and testing data. For the info-plane diagrams, we use Monte Carlo sampling to get an accurate estimate of $I(X;M)$ terms. To estimate the $I(Y;M) = H(Y) - H(Y|M)$ term, we use two different approaches. For classification datasets, we approximate $H(Y)$ using the empirical entropy of the class labels in the dataset, and approximate the conditional entropy with the cross-entropy loss, $H(Y|M) \approx -\mathbb{E}_{Q_\theta(Y,M)}\bigl[\log P_\phi(Y|M)\bigr]$. Note that the resulting MI estimate is an additive constant away from the cross-entropy loss. For the regression dataset, we approximate $H(Y)$ via the entropy of a Gaussian with variance Var(Y), and approximate $H(Y|M)$ via the entropy of a Gaussian with variance equal to the mean-squared-error. This results in the estimate $I(Y;M) \approx \frac{1}{2}\log(\text{Var}(Y)/\text{MSE})$. Finally, we also use scatter plots to visualize the activity of the hidden layer for models trained with different objectives.

We first consider the *MNIST* dataset of hand-drawn digits, which contains 60,000 training images and 10,000 testing images. Each image is 28-by-28 pixels (784 total pixels, so $X \in \mathbb{R}^{784}$), and is classified into 1 of 10 classes corresponding to the digit identity ($Y \in \{1,\ldots,10\}$).

The top row of Figure 1 shows $I(Y;M)$ and $I(X;M)$ values achieved by nonlinear IB and VIB on the MNIST dataset. As can be seen, nonlinear IB achieves better prediction values at the same level of compression than VIB, both on training and testing data. The difference is especially marked near the "corner point" $I(X;M) = I(Y;M) \approx \log 10$ (which corresponds to maximal compression, given perfect prediction), where nonlinear IB achieved ≈ 0.1 bits better prediction at the same compression level (see also Table 1).

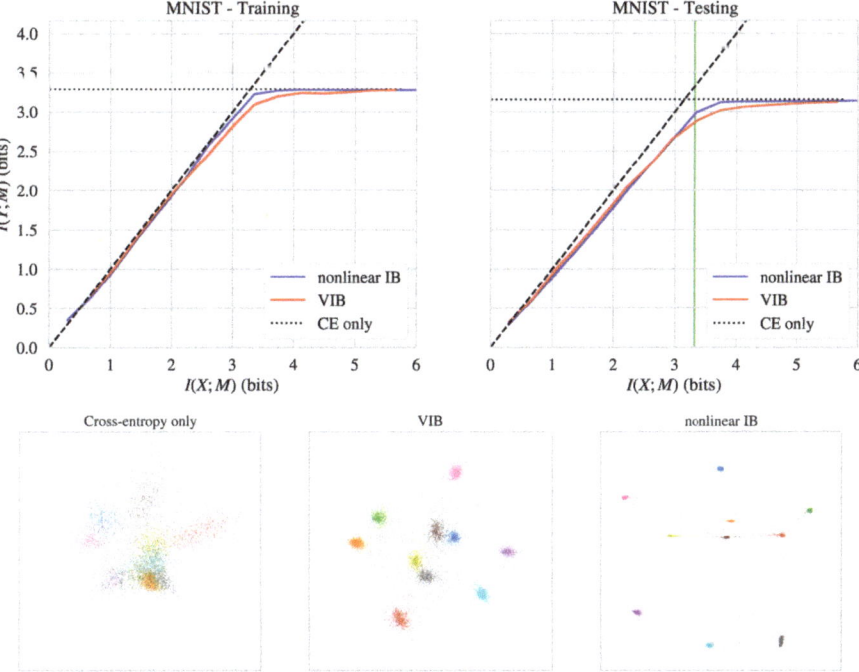

Figure 1. Top row: Info-plane diagrams for nonlinear IB and variational IB (VIB) on the MNIST training (**left**) and testing (**right**) data. The solid lines indicate means across five runs, shaded region indicates the standard error of the mean. The black dashed line is the data-processing inequality bound $I(Y;M) \leq I(X;M)$, the black dotted line indicates the value of $I(Y;M)$ achieved by a baseline model trained only to optimize cross-entropy. **Bottom row**: Principal component analysis (PCA) projection of bottleneck layer activity (on testing data, no noise) for models trained with regular cross-entropy loss (**left**), VIB (**middle**), and nonlinear IB (**right**) objectives. The location of the nonlinear IB and VIB models shown in the bottom row is indicated with the green vertical line in the top right panel.

Further insight is provided by considering the bottleneck representations found when training with nonlinear IB versus VIB versus regular cross-entropy loss. To visualize these bottleneck representations, we selected three models: a baseline model trained only to optimize cross-entropy loss, a model trained with nonlinear IB, and a model trained with VIB (the latter two models were chosen to both have $I(X;M) \approx \log 10$). We then measured the activity of their 5-neuron bottleneck hidden layer on the testing dataset, projected down to two dimensions using principal component analysis (PCA). Figure 1 visualizes these two-dimensional projections for these three models, with colors indicating class label (digit identity). Training with VIB and nonlinear IB objectives causes inputs corresponding to different digits to fall into well-separated clusters, unlike training with cross-entropy loss. Moreover, the clustering is particularly tight for nonlinear IB, meaning that the bottleneck states carry almost no information about input vectors beyond class identity. Note that in this regime, where Gaussian components are grouped into tightly separate clusters, our MI upper bound $\hat{I}_\theta(X;M)$ becomes exact [35].

Table 1. Amount of prediction $I(Y;M)$ achieved at compression level $I(X;M) = \log 10$ for both nonlinear IB and VIB.

Dataset		Nonlinear IB	VIB
MNIST	Training	3.22	3.09
	Testing	2.99	2.88
FashionMNIST	Training	2.85	2.67
	Testing	2.58	2.46
California housing	Training	1.37	1.26
	Testing	1.13	1.07

In the next experiment, we considered the recently-proposed *FashionMNIST* dataset. FashionMNIST has the same structure as the MNIST dataset (28 × 28 images grouped into 10 classes, with 60,000 training and 10,000 testing images). Instead of hand-written digits, however, FashionMNIST includes images of clothes labeled with classes such as "Dress", "Coat", and "Sneaker". This dataset was designed as a drop-in replacement for MNIST which addresses the problem that MNIST is too easy for modern machine learning (e.g., it is fairly straightforward to achieve ≈99% test accuracy on MNIST) [56]. FashionMNIST is a more difficult dataset, with typical test accuracies of ≈90%–95%.

The top row Figure 2 shows $I(Y;M)$ and $I(X;M)$ values achieved by nonlinear IB and VIB on the FashionMNIST dataset. Compared to VIB, nonlinear IB again achieves better prediction values at the same level of compression, both on training and testing data. The difficulty of FashionMNIST is evident in the fact that neither method gets very close to the corner point $I(X;M) = I(Y;M) \approx \log 10$. Nonetheless, nonlinear IB performed better than VIB at a range of compression values, often extracting ≈ 0.15 additional bits of prediction at the same compression level (see also Table 1).

As for MNIST, we consider the bottleneck representations uncovered when training on FashionMNIST with cross-entropy loss only versus nonlinear IB versus VIB (the latter two models were chosen to have $I(X;M) \approx \log 10$). We measured the activity of the 5-neuron bottleneck layer on the testing dataset, projected down to two dimensions using PCA. The bottom row of Figure 2 visualizes these two-dimensional projections for these three models, with colors indicating class label (digit identity). It can again be seen that models trained with VIB and nonlinear IB map inputs into separated clusters, but that the clusters are significantly tighter for nonlinear IB.

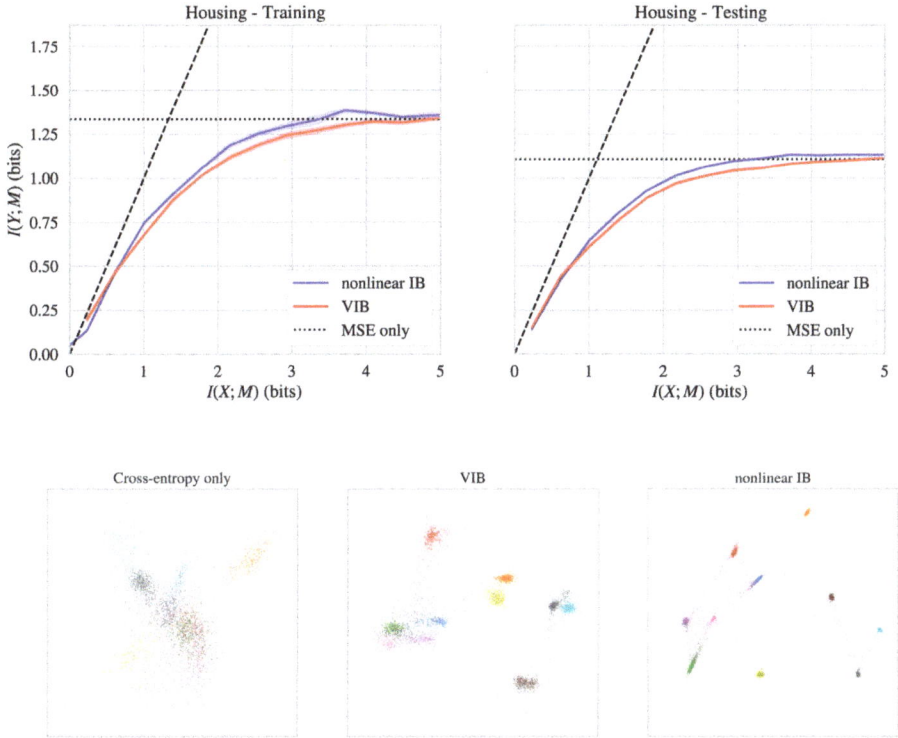

Figure 2. Top row: Info-plane diagrams for nonlinear IB and VIB on the FashionMNIST dataset. **Bottom row**: PCA projection of bottleneck layer activations for models trained only to optimize cross-entropy (**left**), VIB (**middle**), and nonlinear IB (**right**) objectives. See caption of Figure 1 for details.

In our final experiment, we considered the *California housing prices* dataset. This is a regression dataset based on the 1990 California census, originally published in [57] (we use the version distributed with the scikit-learn package [58]). It consists of $N = 20,640$ total samples, with one dependent variable (the house price) and 8 independent variables (such as "longitude", "latitude", and "number of rooms"). We used the log-transformed house price as the dependent variable Y (this made the distribution of Y closer to a Gaussian). To prepare the training and testing data, we first dropped 992 samples in which the house price was equal to or greater than \$500,000 (prices were clipped at this upper value in the dataset, which distorted the distribution of the dependent variable). We then randomly split the remaining samples into an 80% training and 20% testing dataset (the training dataset was then further split into the actual training dataset and a validation dataset, see above).

The top row of Figure 3 shows $I(Y; M)$ and $I(X; M)$ values achieved by nonlinear IB and VIB on the California housing prices dataset. Nonlinear IB achieves better prediction values at the same level of compression than VIB, both on training and testing data (see also Table 1). As for the other datasets, we also show the bottleneck representations uncovered when training on California housing prices dataset with MSE loss only versus nonlinear IB versus VIB (the latter two models were chosen to have $I(X; M) \approx \log 10$). The bottom row of Figure 3 visualizes the two-dimensional PCA projections of bottleneck layer activity for these three models, with colors indicating the dependent variable (log housing price). The bottleneck representations uncovered when training with MSE loss only and when training with VIB were somewhat similar. Nonlinear IB, however, finds a different and almost perfectly one-dimensional bottleneck representation. In fact, for the nonlinear IB model, the first principal component explains 99.8% of the variance in bottleneck layer activity on testing data. For the models trained with MSE loss and VIB, the first principal component explains only 76.6% and 69% of the variance, respectively. The one-dimensional representation uncovered by nonlinear IB compresses away all information about the input vectors which is not relevant for predicting the dependent variable.

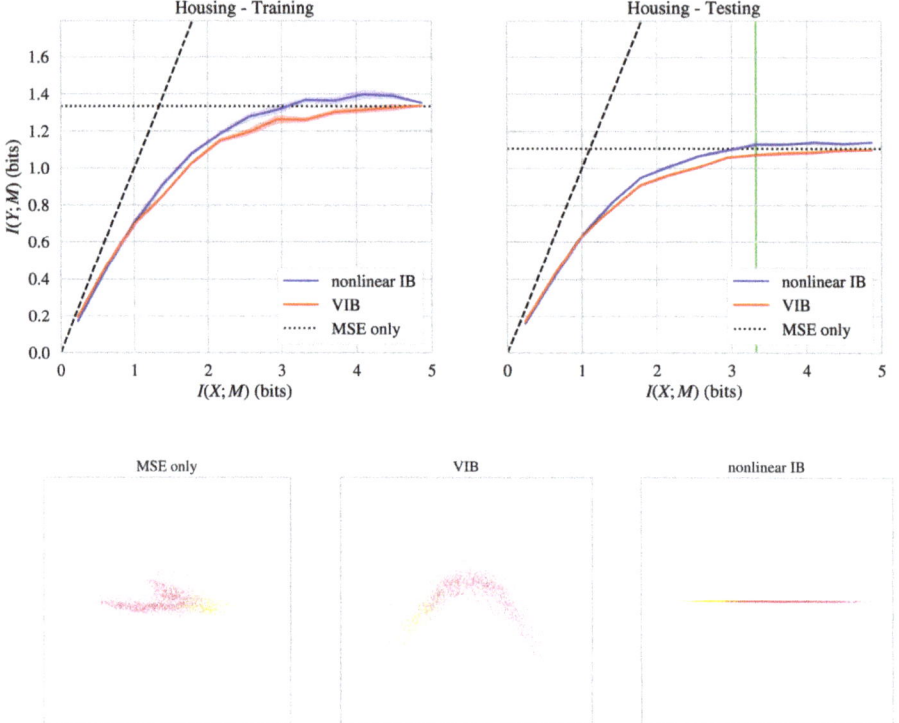

Figure 3. Top row: Information plane diagrams for nonlinear IB and VIB on the California housing prices dataset. **Bottom row**: PCA projection of bottleneck layer activations for models trained only to optimize mean squared error (MSE) (**left**), VIB (**middle**), and nonlinear IB (**right**) objectives. See caption of Figure 1 for details.

We finish by presenting some of our numerical results in Table 1. In particular, we quantify the amount of prediction, $I(Y; M)$, achieved when training with nonlinear IB and VIB at the compression level $I(X; M) = \log 10$, for training and testing datasets of the three datasets considered above. Nonlinear IB consistently achieves better prediction at a fixed level of compression.

5. Conclusions

We propose "nonlinear IB", a method for exploring the information bottleneck [IB] trade-off curve in a general setting. We allow the input and output variables to be discrete or continuous (though we assume a continuous bottleneck variable). We also allow for arbitrary (e.g., non-Gaussian) joint distributions over inputs and outputs and for non-linear encoding and decoding maps. We gain this generality by exploiting a new tractable and differentiable bound on the IB objective.

We describe how to implement our method using off-the-shelf neural network software, and apply it to several standard classification and regression problems. We find that nonlinear IB is able to effectively discover the tradeoff curve, and find solutions that are superior compared with competing methods. We also find that the intermediate representations discovered by nonlinear IB have visibly tighter clusters in the classification problems. In the regression problem, nonlinear IB discovers a one-dimensional intermediate representation.

We have successfully demonstrated the ability of nonlinear IB to explore the IB curve. It is possible that increased compression may lead to other benefits in supervised learning, such as improved generalization performance or increased robustness to adversarial inputs. Exploring its efficacy in these domains remains for future work.

Supplementary Materials: The following are available online at http://www.mdpi.com/1099-4300/21/12/1181/s1: Figure S1: Performance of nonlinear IB and VIB when optimizing bounds on regular IB objective.

Author Contributions: Conceptualization, A.K.; Funding acquisition, D.H.W.; Software, A.K. and B.D.T.; Visualization, A.K.; Writing—original draft, A.K.; Writing—review & editing, A.K., B.D.T. and D.H.W.

Funding: This research was funded by National Science Foundation: CHE-1648973; Foundational Questions Institute: FQXi-RFP-1622; Air Force Office of Scientific Research: A9550-15-1-0038.

Acknowledgments: We thank Steven Van Kuyk and Borja Rodríguez Gálvez for helpful comments. We would also like to thank the Santa Fe Institute for helping to support this research.

Conflicts of Interest: The authors declare no conflict of interest.

References

1. Tishby, N.; Pereira, F.; Bialek, W. The information bottleneck method. In Proceedings of the 37th Annual Allerton Conference on Communication, Control, and Computing, Monticello, IL, USA, 22–24 September 1999.
2. Dimitrov, A.G.; Miller, J.P. Neural coding and decoding: Communication channels and quantization. *Netw. Comput. Neural Syst.* **2001**, *12*, 441–472. [CrossRef]
3. Samengo, I. Information loss in an optimal maximum likelihood decoding. *Neural Comput.* **2002**, *14*, 771–779. [CrossRef]
4. Witsenhausen, H.; Wyner, A. A conditional entropy bound for a pair of discrete random variables. *IEEE Trans. Inf. Theory* **1975**, *21*, 493–501. [CrossRef]
5. Ahlswede, R.; Körner, J. Source Coding with Side Information and a Converse for Degraded Broadcast Channels. *IEEE Trans. Inf. Theory* **1975**, *21*, 629–637. [CrossRef]
6. Gilad-Bachrach, R.; Navot, A.; Tishby, N. An Information Theoretic Tradeoff between Complexity and Accuracy. In *Learning Theory and Kernel Machines*; Goos, G., Hartmanis, J., van Leeuwen, J., Schölkopf, B., Warmuth, M.K., Eds.; Springer: Berlin/Heidelberg, Germany, 2003; Volume 2777, pp. 595–609.
7. Slonim, N.; Tishby, N. Document clustering using word clusters via the information bottleneck method. In Proceedings of the 23rd Annual International ACM SIGIR Conference on Research and Development in Information Retrieval, Athens, Greece, 24–28 July 2000; pp. 208–215.
8. Tishby, N.; Slonim, N. Data clustering by markovian relaxation and the information bottleneck method. In *Advances in Neural Information Processing Systems 13 (NIPS 2000)*; MIT Press: Cambridge, MA, USA, 2001; pp. 640–646.
9. Cardinal, J. Compression of side information. In Proceedings of the 2003 International Conference on Multimedia and Expo, Baltimore, MD, USA, 6–9 July 2003; pp. 569–572.
10. Zeitler, G.; Koetter, R.; Bauch, G.; Widmer, J. Design of network coding functions in multihop relay networks. In Proceedings of the 2008 5th International Symposium on Turbo Codes and Related Topics, Lausanne, Switzerland, 1–5 September 2008; pp. 249–254.
11. Courtade, T.A.; Wesel, R.D. Multiterminal source coding with an entropy-based distortion measure. In Proceedings of the 2011 IEEE International Symposium on Information Theory, St. Petersburg, Russia, 31 July–5 August 2011; pp. 2040–2044.
12. Lazebnik, S.; Raginsky, M. Supervised learning of quantizer codebooks by information loss minimization. *IEEE Trans. Pattern Anal. Mach. Intell.* **2008**, *31*, 1294–1309. [CrossRef] [PubMed]
13. Winn, J.; Criminisi, A.; Minka, T. Object categorization by learned universal visual dictionary. In Proceedings of the Tenth IEEE International Conference on Computer Vision (ICCV'05) Volume 1, Beijing, China, 17–21 October 2005; Volume 2, pp. 1800–1807.

14. Hecht, R.M.; Noor, E.; Tishby, N. Speaker recognition by Gaussian information bottleneck. In Proceedings of the 10th Annual Conference of the International Speech Communication Association, Brighton, UK, 6–10 September 2009.
15. Yaman, S.; Pelecanos, J.; Sarikaya, R. Bottleneck features for speaker recognition. In Proceedings of the Speaker and Language Recognition Workshop, Singapore, 25–28 June 2012.
16. Van Kuyk, S.; Kleijn, W.B.; Hendriks, R.C. On the information rate of speech communication. In Proceedings of the 2017 IEEE International Conference on Acoustics, Speech and Signal Processing (ICASSP), New Orleans, LA, USA, 5–9 March 2017; pp. 5625–5629.
17. Van Kuyk, S. Speech Communication from an Information Theoretical Perspective. Ph.D. Thesis, Victoria University of Wellington, Wellington, New Zealand, 2019.
18. Zaslavsky, N.; Kemp, C.; Regier, T.; Tishby, N. Efficient compression in color naming and its evolution. *Proc. Natl. Acad. Sci. USA* **2018**, *115*, 7937–7942. [CrossRef] [PubMed]
19. Rodríguez Gálvez, B. The Information Bottleneck: Connections to Other Problems, Learning and Exploration of the IB Curve. Master's Thesis, KTH Royal Institute of Technology, Stockholm, Sweden, June 2019.
20. Hafez-Kolahi, H.; Kasaei, S. Information Bottleneck and its Applications in Deep Learning. *arXiv* **2019**, arXiv:1904.03743.
21. Tishby, N.; Zaslavsky, N. Deep learning and the information bottleneck principle. In Proceedings of the 2015 IEEE Information Theory Workshop (ITW), Jerusalem, Israel, 26 April–1 May 2015; pp. 1–5.
22. Shamir, O.; Sabato, S.; Tishby, N. Learning and generalization with the information bottleneck. *Theor. Comput. Sci.* **2010**, *411*, 2696–2711. [CrossRef]
23. Vera, M.; Piantanida, P.; Vega, L.R. The Role of the Information Bottleneck in Representation Learning. In Proceedings of the 2018 IEEE International Symposium on Information Theory (ISIT), Vail, CO, USA, 17–22 June 2018; pp. 1580–1584.
24. Alemi, A.A.; Fischer, I.; Dillon, J.V.; Murphy, K. Deep Variational Information Bottleneck. In Proceedings of the International Conference on Learning Representations (ICLR), Toulon, France, 24–26 April 2017.
25. Alemi, A.A.; Fischer, I.; Dillon, J.V. Uncertainty in the variational information bottleneck. *arXiv* **2018**, arXiv:1807.00906.
26. Amjad, R.A.; Geiger, B.C. Learning Representations for Neural Network-Based Classification Using the Information Bottleneck Principle. *arXiv* **2018**, arXiv:1802.09766.
27. Shwartz-Ziv, R.; Tishby, N. Opening the Black Box of Deep Neural Networks via Information. *arXiv* **2017**, arXiv:1703.00810.
28. Saxe, A.; Bansal, Y.; Dapello, J.; Advani, M.; Kolchinsky, A.; Tracey, B.; Cox, D. On the information bottleneck theory of deep learning. In Proceedings of the 6th International Conference on Learning Representations, Vancouver, BC, Canada, 30 April–3 May 2018.
29. Lemaréchal, C. Lagrangian relaxation. In *Computational Combinatorial Optimization*; Springer: Berlin/Heidelberg, Germany, 2001; pp. 112–156.
30. Cover, T.M.; Thomas, J.A. *Elements of Information Theory*; John Wiley & Sons: Hoboken, NJ, USA, 2012.
31. Chechik, G.; Globerson, A.; Tishby, N.; Weiss, Y. Information bottleneck for Gaussian variables. *J. Mach. Learn. Res.* **2005**, *6*, 165–188.
32. Kolchinsky, A.; Tracey, B.D.; Van Kuyk, S. Caveats for information bottleneck in deterministic scenarios. In Proceedings of the 6th International Conference on Learning Representations, Vancouver, BC, Canada, 30 April–3 May 2018.
33. Miettinen, K. *Nonlinear Multiobjective Optimization*; Springer: Boston, MA, USA, 1998. [CrossRef]
34. Rodríguez Gálvez, B.; Thobaben, R.; Skoglund, M. The Convex Information Bottleneck Lagrangian. *arXiv* **2019**, arXiv:1911.11000.
35. Kolchinsky, A.; Tracey, B.D. Estimating Mixture Entropy with Pairwise Distances. *Entropy* **2017**, *19*, 361. [CrossRef]
36. Chalk, M.; Marre, O.; Tkacik, G. Relevant sparse codes with variational information bottleneck. In Proceedings of the 2016 Conference on Neural Information Processing Systems (NIPS 2016), Barcelona, Spain, 5–10 December 2016; pp. 1957–1965.
37. Achille, A.; Soatto, S. Information Dropout: Learning optimal representations through noise. *arXiv* **2016**, arXiv:1611.01353.
38. Goodfellow, I.; Bengio, Y.; Courville, A. *Deep Learning*; MIT Press: Cambridge, MA, USA, 2016.

39. Silverman, B.W. *Density Estimation for Statistics and Data Analysis*; Routledge: New York, NY, USA, 2018.
40. Kolchinsky, A.; Wolpert, D.H. Supervised learning with information penalties. In Proceedings of the 2016 Conference on Neural Information Processing Systems (NIPS 2016), Barcelona, Spain, 5–10 December 2016.
41. Kolchinsky, A.; Tracey, B.D.; Wolpert, D.H. Nonlinear Information Bottleneck. *Entropy* **2019**, *21*, 1181. [CrossRef]
42. Schraudolph, N.N. Optimization of entropy with neural networks. Ph.D. Thesis, University of California, San Diego, CA, USA, 1995.
43. Schraudolph, N.N. Gradient-based manipulation of nonparametric entropy estimates. *IEEE Trans. Neural Netw.* **2004**, *15*, 828–837. [CrossRef]
44. Shwartz, S.; Zibulevsky, M.; Schechner, Y.Y. Fast kernel entropy estimation and optimization. *Signal Process.* **2005**, *85*, 1045–1058. [CrossRef]
45. Torkkola, K. Feature extraction by non-parametric mutual information maximization. *J. Mach. Learn. Res.* **2003**, *3*, 1415–1438.
46. Hlavávcková-Schindler, K.; Palus, M.; Vejmelka, M.; Bhattacharya, J. Causality detection based on information-theoretic approaches in time series analysis. *Phys. Rep.* **2007**, *441*, 1–46. [CrossRef]
47. Goodfellow, I.; Pouget-Abadie, J.; Mirza, M.; Xu, B.; Warde-Farley, D.; Ozair, S.; Courville, A.; Bengio, Y. Generative adversarial nets. In Proceedings of the 2014 Conference on Neural Information Processing Systems (NIPS 2014), Montreal, QC, Canada, 8–13 December 2014; pp. 2672–2680.
48. Hinton, G.E.; Zemel, R.S. Autoencoders, minimum description length, and Helmholtz free energy. In *Advances in Neural Information Processing Systems 7 (NIPS 1994)*; MIT Press: Cambridge, MA, USA, 1994; p 3.
49. Hinton, G.E.; Zemel, R.S. Minimizing Description Length in an Unsupervised Neural Network. Available online: https://www.cs.toronto.edu/~fritz/absps/mdlnn.pdf (accessed on 30 November 2019).
50. Deco, G.; Finnoff, W.; Zimmermann, H.G. Elimination of Overtraining by a Mutual Information Network. In *ICANN'93*; Gielen, S., Kappen, B., Eds.; Springer: London, UK, 1993; pp. 744–749. [CrossRef]
51. Vincent, P.; Larochelle, H.; Bengio, Y.; Manzagol, P.A. Extracting and composing robust features with denoising autoencoders. In Proceedings of the 25th International Conference on Machine Learning, Helsinki, Finland, 5–9 July 2008; pp. 1096–1103.
52. Kingma, D.P.; Welling, M. Auto-encoding variational bayes. In Proceedings of the International Conference on Learning Representations (ICLR 2014), Banff, AB, Canada, 14–16 April 2014.
53. Higgins, I.; Matthey, L.; Pal, A.; Burgess, C.; Glorot, X.; Botvinick, M.; Mohamed, S.; Lerchner, A. beta-VAE: Learning Basic Visual Concepts with a Constrained Variational Framework. In Proceedings of the International Conference on Learning Representations (ICLR 2017), Toulon, France, 24–26 April 2017.
54. Alemi, A.; Poole, B.; Fischer, I.; Dillon, J.; Saurous, R.A.; Murphy, K. Fixing a Broken ELBO. In Proceedings of the 35th International Conference on Machine Learning, Stockholm, Sweden, 10–15 July 2018; pp. 159–168.
55. Kingma, D.; Ba, J. Adam: A method for stochastic optimization. In Proceedings of the 3rd International Conference for Learning Representations (ICLR 2015), San Diego, CA, USA, 7–9 May 2015.
56. Xiao, H.; Rasul, K.; Vollgraf, R. Fashion-MNIST: A novel image dataset for benchmarking machine learning algorithms. *arXiv* **2017**, arXiv:1708.07747.
57. Pace, R.K.; Barry, R. Sparse spatial autoregressions. *Stat. Probab. Lett.* **1997**, *33*, 291–297. [CrossRef]
58. Pedregosa, F.; Varoquaux, G.; Gramfort, A.; Michel, V.; Thirion, B.; Grisel, O.; Blondel, M.; Prettenhofer, P.; Weiss, R.; Dubourg, V.; et al. Scikit-learn: Machine Learning in Python. *J. Mach. Learn. Res.* **2011**, *12*, 2825–2830.

© 2019 by the authors. Licensee MDPI, Basel, Switzerland. This article is an open access article distributed under the terms and conditions of the Creative Commons Attribution (CC BY) license (http://creativecommons.org/licenses/by/4.0/).

Article

Pareto-Optimal Data Compression for Binary Classification Tasks

Max Tegmark * and Tailin Wu

Department of Physics, MIT Kavli Institute & Center for Brains, Minds & Machines, Massachusetts Institute of Technology, Cambridge, MA 02139 USA; tailin@mit.edu
* Correspondence: Tegmark@mit.edu

Received: 18 October 2019; Accepted: 12 December 2019; Published: 19 December 2019

Abstract: The goal of lossy data compression is to reduce the storage cost of a data set X while retaining as much information as possible about something (Y) that you care about. For example, what aspects of an image X contain the most information about whether it depicts a cat? Mathematically, this corresponds to finding a mapping $X \to Z \equiv f(X)$ that maximizes the mutual information $I(Z,Y)$ while the entropy $H(Z)$ is kept below some fixed threshold. We present a new method for mapping out the Pareto frontier for classification tasks, reflecting the tradeoff between retained entropy and class information. We first show how a random variable X (an image, say) drawn from a class $Y \in \{1, ..., n\}$ can be distilled into a vector $W = f(X) \in \mathbb{R}^{n-1}$ losslessly, so that $I(W,Y) = I(X,Y)$; for example, for a binary classification task of cats and dogs, each image X is mapped into a single real number W retaining all information that helps distinguish cats from dogs. For the $n = 2$ case of binary classification, we then show how W can be further compressed into a discrete variable $Z = g_\beta(W) \in \{1, ..., m_\beta\}$ by binning W into m_β bins, in such a way that varying the parameter β sweeps out the full Pareto frontier, solving a generalization of the discrete information bottleneck (DIB) problem. We argue that the most interesting points on this frontier are "corners" maximizing $I(Z,Y)$ for a fixed number of bins $m = 2, 3, ...$ which can conveniently be found without multiobjective optimization. We apply this method to the CIFAR-10, MNIST and Fashion-MNIST datasets, illustrating how it can be interpreted as an information-theoretically optimal image clustering algorithm. We find that these Pareto frontiers are not concave, and that recently reported DIB phase transitions correspond to transitions between these corners, changing the number of clusters.

Keywords: information; bottleneck; compression; classification

1. Introduction

A core challenge in science, and in life quite generally, is data distillation: Keeping only a manageably small fraction of our available data X while retaining as much information as possible about something (Y) that we care about. For example, what aspects of an image contain the most information about whether it depicts a cat ($Y = 1$) rather than a dog ($Y = 2$)? Mathematically, this corresponds to finding a deterministic mapping $X \to Z \equiv g(X)$ that maximizes the mutual information $I(Z, Y)$ while the entropy $H(Z)$ is kept below some fixed threshold. The tradeoff between $H_* = H(Z)$ (bits stored) and $I_* = I(Z, Y)$ (useful bits) is described by a Pareto frontier, defined as

$$I_*(H_*) \equiv \sup_{\{g: H[g(X)] \leq H_*\}} I[g(X), Y], \qquad (1)$$

and illustrated in Figure 1 (this is for a toy example described below; we compute the Pareto frontier for our cat/dog example in Section 3). The shaded region is impossible because $I(Z, Y) \leq I(X, Y)$ and

$I(Z, Y) \leq H(Z)$. The colored dots correspond to random likelihood binnings into various numbers of bins, as described in the next section, and the upper envelope of all attainable points define the Pareto frontier. Its "corners", which are marked by black dots and maximize $I(Z, Y)$ for M bins ($M = 1, 2, ...$), are seen to lie close to the vertical dashed lines $H(Z) = \log M$, corresponding to all bins having equal size. We plot the H-axis flipped to conform with the tradition that up and to the right are more desirable. The core goal of this paper is to present a method for computing such Pareto frontiers.

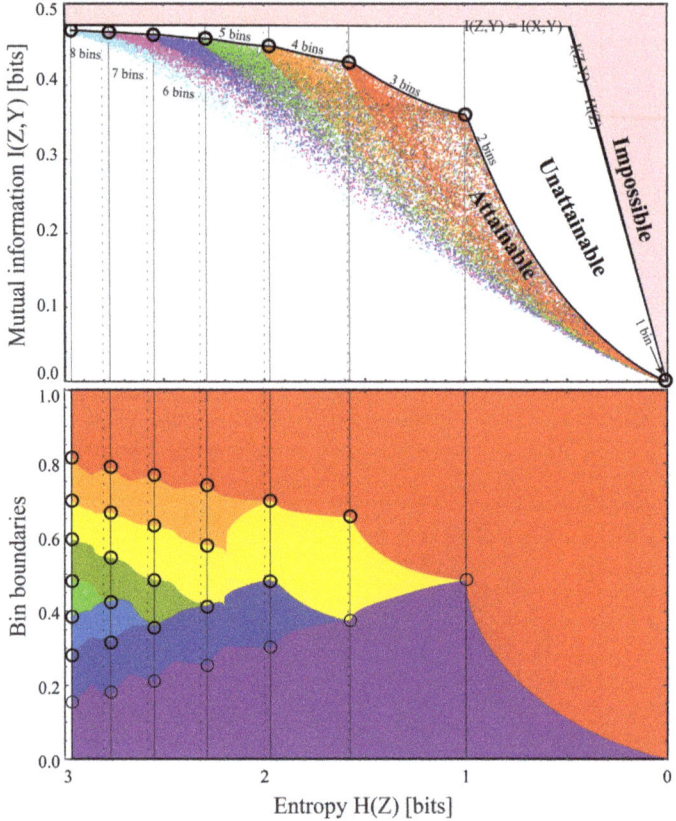

Figure 1. The Pareto frontier (top panel) for compressed versions $Z = g(X)$ of our warmup dataset $X \in [0, 1]^2$ with classes $Y \in \{1, 2\}$, showing the maximum attainable class information $I(Z, Y)$ for a given entropy $H(Z)$, mapped using the method described in this paper using the likelihood binning in the bottom panel.

Objectives and Relation to Prior Work

In other words, the goal of this paper is to analyze soft rather than hard classifiers: not to make the most accurate classifier, but rather to compute the Pareto frontier that reveals the most accurate (in an information-theoretic sense) classifier Z given a constraint on its bit content $H(Z)$. These optimal soft classifiers that we will derive (corresponding to points on the Pareto frontier) are useful for the same reason that other methods for lossy data compression methods are useful: Overfitting less and therefore generalizing better, among other things.

This Pareto frontier challenge is thus part of the broader quest for data distillation: Lossy data compression that retains as much as possible of the information that is useful to us. Ideally, the information can be partitioned into a set of independent chunks and sorted from most to least useful, enabling us to select the number of chunks to retain so as to optimize our tradeoff between utility and data size. Consider two random variables X and Y which may each be vectors or scalars. For simplicity, consider them to be discrete with finite entropy. (this discreteness restriction loses us no generality in practice, since since we can always discretize real numbers by rounding them to some very large number of significant digits.) For prediction tasks, we might interpret Y as the future state of a dynamical system that we wish to predict from the present state X. For classification tasks, we might interpret Y as a class label that we wish to predict from an image, sound, video or text string X. Let us now consider various forms of ideal data distillation, as summarized in Table 1.

Table 1. Data distillation: The relationship between principal component analysis (PCA), canonical correlation analysis (CCA), nonlinear autoencoders and nonlinear latent representations.

Random Vectors	What Is Distilled?	Probability Distribution	
		Gaussian	Non-Gaussian
1	Entropy $H(X) = \sum_i H(Z_i)$	PCA $\mathbf{z} = \mathbf{F}\mathbf{x}$	Autoencoder $Z = f(X)$
2	Mutual information $I(X,Y) = \sum_i I(Z_i, Z'_i)$	CCA $\mathbf{z} = \mathbf{F}\mathbf{x}$ $\mathbf{z}' = \mathbf{G}\mathbf{y}$	Latent reps $Z = f(X)$ $Z' = g(Y)$

If we distill X as a whole, then we would ideally like to find a function f such that the so-called latent representation $Z = f(X)$ retains the full entropy $H(X) = H(Z) = \sum H(Z_i)$, decomposed into independent parts with vanishing mutual infomation: $I(Z_i, Z_j) = \delta_{ij} H(Z_i)$. (Note that when implementing any distillation algorithm in practice, there is always a one-parameter tradeoff between compression and information retention which defines a Pareto frontier. A key advantage of the latent variables (or variable pairs) being statistically independent is that this allows the Pareto frontier to be trivially computed, by simply sorting them by decreasing information content and varying the number retained.)

For the special case where $X = \mathbf{x}$ is a vector with a multivariate Gaussian distribution, the optimal solution is Principal Component Analysis (PCA) [1], which has long been a workhorse of statistical physics and many other disciplines: Here f is simply a linear function mapping into the eigenbasis of the covariance matrix of $\mathbf{x}$. The general case remains unsolved, and it is easy to see that it is hard: If $X = c(Z)$ where c implements some state-of-the-art cryptographic code, then finding $f = c^{-1}$ (to recover the independent pieces of information and discard the useless parts) would generically require breaking the code. Great progress has nonetheless been made for many special cases, using techniques such as nonlinear autoencoders [2] and Generative Adversarial Networks (GANs) [3].

Now consider the case where we wish to distill X and Y separately, into $Z \equiv f(X)$ and $Z' = g(Y)$, retaining the mutual information between the two parts. Then we ideally have $I(X,Y) = \sum_i I(Z_i, Z'_i)$, $I(Z_i, Z_j) = \delta_{ij} H(Z_i)$, $I(Z'_i, Z'_j) = \delta_{ij} H(Z'_i)$, $I(Z_i, Z'_j) = \delta_{ij} I(Z_i, Z'_j)$. This problem has attracted great interest, especially for time series where $X = \mathbf{u}_i$ and $Y = \mathbf{u}_j$ for some sequence of states $\mathbf{u}_k$ ($k = 0, 1, 2, ...$) in physics or other fields, where one typically maps the state vectors $\mathbf{u}_i$ into some lower-dimensional vectors $f(\mathbf{u}_i)$, after which the prediction is carried out in this latent space. For the special case where X has a multivariate Gaussian distribution, the optimal solution is Canonical Correlation Analysis (CCA) [4]: Here both f and g are linear functions, computed via a singular-value decomposition (SVD) [5] of the cross-correlation matrix after prewhitening X and Y. The general case remains unsolved, and is obviously even harder than the above-mentioned 1-vector autoencoding problem. The recent work [6,7] review the state-of-the art as well as presenting Contrastive Predictive Coding and Dynamic Component Analysis, powerful new distillation techniques for time series, following the long tradition of setting $f = g$ even though this is provably not optimal for the Gaussian case as shown in [8].

The goal of this paper is to make progress in the lower right quadrant of Table 1. We will first show that if $Y \in \{1,2\}$ (as in binary classification tasks) and we can successfully train a classifier that correctly predicts the conditional probability distribution $p(Y|X)$, then it can be used to provide an exact solution to the distillation problem, losslessly distilling X into a single real variable $W = f(X)$. We will generalize this to an arbitrary classification problem $Y \in \{1,...,n\}$ by losslessly distilling X into a vector $W = f(X) \in \mathbb{R}^{n-1}$, although in this case, the components of the vector W may not be independent. We will then return to the binary classification case and provide a family of binnings that map W into an integer Z, allowing us to scan the full Pareto frontier reflecting the tradeoff between retained entropy and class information, illustrating the end-to-end procedure with the CIFAR-10, MNIST and Fashion-MNIST datasets. This is related to the work of [9] which maximizes $I(Z,Y)$ for a fixed number of bins (instead of for a fixed entropy), which corresponds to the "corners" seen in Figure 1.

This work is closely related to the Information Bottleneck (IB) method [10], which provides an insightful, principled approach for balancing compression against prediction [11]. Just as in our work, the IB method aims to find a random variable $Z = f(X)$ that loosely speaking retains as much information as possible about Y and as little other information as possible. The IB method implements this by maximizing the IB-objective

$$\mathcal{L}_{\text{IB}} = I(Z,Y) - \beta I(Z,X) \tag{2}$$

where the Lagrange multiplier β tunes the balance between knowing about Y and forgetting about X. Ref. [12] considered the alternative Deterministic Information Bottleneck (DIB) objective

$$\mathcal{L}_{\text{DIB}} = I(Z,Y) - \beta H(Z), \tag{3}$$

to close the loophole where Z retains random information that is independent of both X and Y. (which is possible if f is function that contains random components rather than fully deterministic. In contrast, if $Z = f(X)$ for some deterministic function f, which is typically not the case in the popular variational IB-implementation [13–15], then $H(Z|X) = 0$, so $I(Z,X) \equiv H(Z) - H(Z|X) = H(Z)$, which means the two objectives (2) and (3) are identical.)

However, there is a well-known problem with this DIB objective that occurs when $Z \in \mathbb{R}^n$ is continuous [16]: $H(Z)$ is strictly speaking infinite, since it requires an infinite amount of information to store the infinitely many decimals of a generic real number. While this infinity is normally regularized away by only defining $H(Z)$ up to an additive constant, which is irrelevant when minimizing Equation (3), the problem is that we can define a new rescaled random variable

$$Z' = aZ \tag{4}$$

for a constant $a \neq 0$ and obtain
$$I(Z',X) = I(Z,X) \tag{5}$$

and
$$H(Z') = H(Z) + n \log |a|. \tag{6}$$

(Throughout this paper, we take log to denote the logarithm in base 2, so that entropy and mutual information are measured in bits.) The last two equations imply that by choosing $|a| \ll 1$, we can make $H(Z')$ arbitrarily negative while keeping $I(Z',X)$ unchanged, thus making $\mathcal{L}_{\text{DIB}}$ arbitrarily negative. The objective $\mathcal{L}_{\text{DIB}}$ is therefore not bounded from below, and trying to minimize it will not produce an interesting result. We will eliminate this Z-rescaling problem by making Z discrete rather than continuous, so that $H(Z)$ is always well-defined and finite. Another challenge with the DIB objective of Equation (3), which we will also overcome, is that it maximizes a linear combination of the two axes in Figure 1, and can therefore only discover concave parts of the Pareto frontier, not convex ones (which are seen to dominate in Figure 1).

The rest of this paper is organized as follows: In Section 2.1, we will provide an exact solution for the binary classification problem where $Y \in \{1, 2\}$ by losslessly distilling X into a single real variable $Z = f(X)$. We also generalize this to an arbitrary classification problem $Y \in \{1, ..., n\}$ by losslessly distilling X into a vector $W = f(X) \in \mathbb{R}^{n-1}$, although the components of the vector W may not be independent. In Section 2.2, we return to the binary classification case and provide a family a binnings that map Z into an integer, allowing us to scan the full Pareto frontier reflecting the tradeoff between retained entropy and class information. We apply our method to various image datasets in Section 3 and discuss our conclusions in Section 4.

2. Method

Our algorithm for mapping the Pareto frontier transforms our original data set X in a series of steps which will be describe in turn below:

$$X \stackrel{w}{\mapsto} W \mapsto W_{\text{uniform}} \mapsto W_{\text{binned}} \mapsto W_{\text{sorted}} \stackrel{B}{\mapsto} Z. \tag{7}$$

As we will show, the first, second and fourth transformations retain all mutual information with the label Y, and the information loss about Y can be kept arbitrarily small in the third step. In contrast, the last step treats the information loss as a tuneable parameter that parameterizes the Pareto frontier.

2.1. Lossless Distillation for Classification Tasks

Our first step is to compress X (an image, say) into W, a set of $n - 1$ real numbers, in such a way that no class information is lost about $Y \in \{1, ..., n\}$.

Theorem 1 (Lossless Distillation Theorem). *For an arbitrary random variable X and a categorical random variable $Y \in \{1, ..., n\}$, we have*

$$P(Y|X) = P(Y|W), \tag{8}$$

where $W \equiv w(X) \in \mathbb{R}^{n-1}$ is defined by

$$w_i(X) \equiv P(Y = i|X). \tag{9}$$

Note that we ignore the nth component since it is redundant: $w_n(X) = 1 - \sum_i^{n-1} w_i(X)$.

Proof. Let S denote the domain of X, i.e., $X \in S$, and define the set-valued function

$$s(W) \equiv \{x \in S : w(x) = W\}.$$

These sets $s(W)$ form a partition of S parameterized by W, since they are disjoint and

$$\cup_{W \in \mathbb{R}^{n-1}} s(W) = S. \tag{10}$$

For example, if $S = \mathbb{R}^2$ and $n = 2$, then the sets $s(W)$ are simply contour curves of the conditional probability $W \equiv P(Y = 1|X) \in \mathbb{R}$. This partition enables us to uniquely specify X as the pair $\{W, X_W\}$ by first specifying which set $s[f(X)]$ it belongs to (determined by $W = f(X)$), and then specifying the particular element within that set, which we denote $X_W \in S(W)$. This implies that

$$P(Y|X) = P(Y|W, X_W) = P(Y|W), \tag{11}$$

completing the proof. The last equal sign follows from the fact that the conditional probability $P(Y|X)$ is independent of X_W, since it is by definition constant throughout the set $s(W)$. □

The following corollary implies that W is an optimal distillation of the information X has about Y, in the sense that it constitutes a lossless compression of said information: $I(W,Y) = I(X,Y)$ as shown, and the total information content (entropy) in W cannot exceed that of X since it is a deterministic function thereof.

Corollary 1. *With the same notation as above, we have*

$$I(X,Y) = I(W,Y). \tag{12}$$

Proof. For any two random variables, we have the identity $I(U,V) = H(V) - H(V|U)$, where $I(U,V)$ is their mutual information and $H(V|U)$ denotes conditional entropy. We thus obtain

$$\begin{aligned} I(X,Y) &= H(Y) - H(Y|X) = H(Y) + \langle \log P(Y|X) \rangle_{X,Y} \\ &= H(Y) + \langle \log P(Y|W) \rangle_{W,X_W,Y} \\ &= H(Y) + \langle \log P(Y|W) \rangle_{W,Y} \\ &= H(Y) - H(Y|W) = I(W,Y), \end{aligned} \tag{13}$$

which completes the proof. We obtain the second line by using $P(Y|X) = P(Y|W)$ from Theorem 1 and specifying X by W and X_W, and the third line since $P(Y|W)$ is independent of X_W, as above. □

In most situations of practical interest, the conditional probability distribution $P(Y|X)$ is not precisely known, but can be approximated by training a neural-network-based classifier that outputs the probability distribution for Y given any input X. We present such examples in Section 3. The better the classifier, the smaller the information loss $I(X,Y) - I(W,Y)$ will be, approaching zero in the limit of an optimal classifier.

2.2. Pareto-Optimal Compression for Binary Classification Tasks

Let us now focus on the special case where $n = 2$, i.e., binary classification tasks. For example, X may correspond to images of equal numbers of felines and canines to be classified despite challenges with variable lighting, occlusion, etc., as in Figure 2, and $Y \in \{1,2\}$ may correspond to the labels "cat" and "dog". In this case, Y contains $H(Y) = 1$ bit of information of which $I(X,Y) \leq 1$ bit is contained in X. Theorem 1 shows that for this case, all of this information about whether an image contains a cat or a dog can be compressed into a single number W which is not a bit like Y, but a real number between zero and one.

The goal of this section is find a class of functions g that perform Pareto-optimal lossy compression of W, mapping it into an integer $Z \equiv g(W)$ that maximizes $I(Z,Y)$ for a fixed entropy $H(Z)$. (Throughout this paper, we will use the term "Pareto-optimal" or "optimal" in this sense, i.e., maximizing $I(X,Y)$ for a fixed $H(Z)$.) The only input we need for our work in this section is the joint probability distribution $f_i(w) = P(Y = i, W = w)$, whose marginal distributions are the discrete probability distribution for P_i^Y for Y and the probability distribution f for W, which we will henceforth assume to be continuous:

$$f(w) \equiv \sum_{i=1}^{2} f_i(w), \tag{14}$$

$$P_i^Y \equiv P(Y = i) = \int_0^1 f_i(w)dw. \tag{15}$$

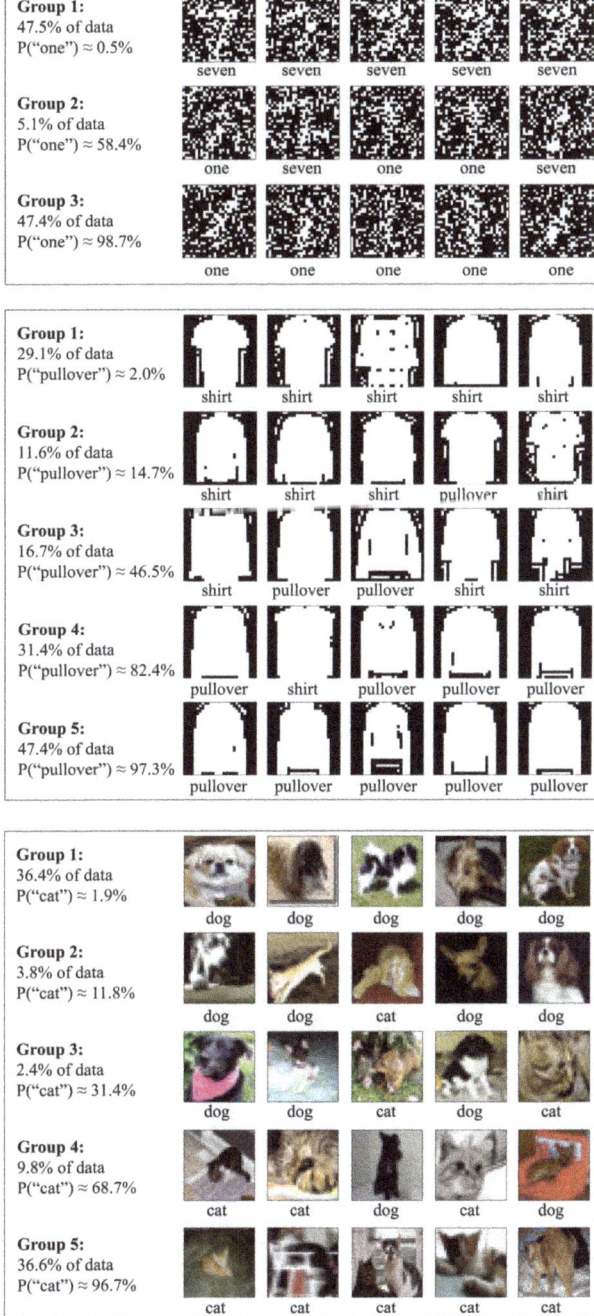

Figure 2. Sample data from Section 3. Images from MMNIST (**top**), Fashion-MNIST (**middle**) and CIFAR-10 are mapped into integers (group labels) $Z = f(X)$ retaining maximum mutual information with the class variable Y (ones/sevens, shirts/pullovers and cats/dogs, respectively) for 3, 5 and 5 groups, respectively. These mappings f correspond to Pareto frontier "corners".

2.2.1. Uniformization of W

For convenience and without loss of generality, we will henceforth assume that $f(w) = 1$, i.e., that W has a uniform distribution on the unit interval $[0, 1]$. We can do this because if W were not uniformly distributed, we could make it so by using the standard statistical technique of applying its cumulative probability distribution function to it

$$W \mapsto W' \equiv F(W), \quad F(w) \equiv \int_0^w f(w')dw', \qquad (16)$$

retaining all information — $I(W', Y) = I(W, Y)$ — since this procedure is invertible almost everywhere.

2.2.2. Binning W

Given a set of bin boundaries $b_1 < b_2 < ... < b_{n-1}$ grouped into a vector **b**, we define the integer-value contiguous binning function

$$B(x, \mathbf{b}) \equiv \begin{cases} 1 & \text{if } x < b_1 \\ k & \text{if } b_{k-1} < x \leq b_k \\ n & \text{if } x \geq b_{N-1} \end{cases} \qquad (17)$$

$B(x, \mathbf{b})$ can thus be interpreted as the ID of the bin into which x falls. Note that B is a monotonically increasing piecewise constant function of x that is shaped like an N-level staircase with $n - 1$ steps at $b_1, ..., b_{N-1}$.

Let us now bin W into N equispaced bins, by mapping it into an integer $W' \in \{1, ..., N\}$ (the bin ID) defined by

$$W' \equiv W_{\text{binned}} \equiv B(W, \mathbf{b}_N), \qquad (18)$$

where **b** is the vector with elements $b_j = j/N$, $j = 1, ..., N - 1$. As illustrated visually in Figure 3 and mathematically in Appendix A, binning $W \mapsto W'$ corresponds to creating a new random variable for which the conditional distribution $p_1(w) = P(Y = 1|W = w)$ is replaced by a piecewise constant function $\bar{p}_1(w)$, replacing the values in each bin by their average. The binned variable W' thus retains only information about which bin W falls into, discarding all information about the precise location within that bin. In the $N \to \infty$ limit of infinitesimal bins, $\bar{p}_1(w) \to p_1(w)$, and we expect the above-mentioned discarded information to become negligible. This intuition is formalized by Theorem A1 in Appendix A, which under mild smoothness assumptions ensuring that $p_1(w)$ is not pathological shows that

$$I(W', Y) \to I(W, Y) \quad \text{as} \quad N \to \infty, \qquad (19)$$

i.e., that we can make the binned data W' retain essentially all the class information from W as long as we use enough bins.

In practice, such as for the numerical experiments that we will present in Section 3, training data is never infinite and the conditional probability function $p_1(w)$ is never known to perfect accuracy. This means that the pedantic distinction between $I(W', Y) = I(W, Y)$ and $I(W', Y) \approx I(W, Y)$ for very large N is completely irrelevant in practice. In the rest of this paper, we will therefore work with the unbinned (W) and binned (W') data somewhat interchangeably below for convenience, occasionally dropping the apostrophy ' from W' when no confusion is caused.

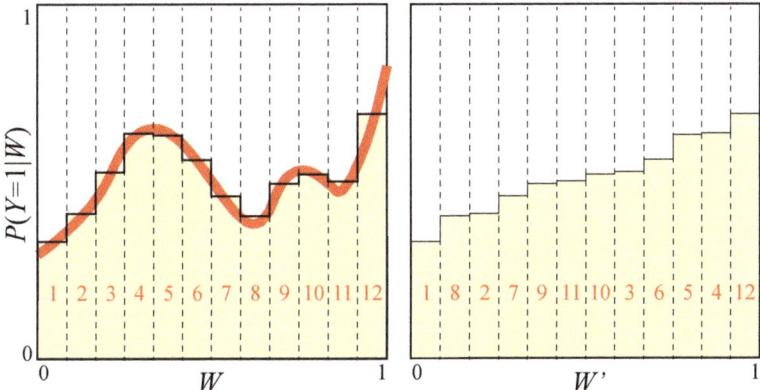

Figure 3. Essentially all information about Y is retained if W is binned into sufficiently narrow bins. Sorting the bins (**left**) to make the conditional probability monotonically increasing (**right**) changes neither this information nor the entropy.

2.2.3. Making the Conditional Probability Monotonic

For convenience and without loss of generality, we can assume that the conditional probability distribution $\tilde{p}_1(w)$ is a monotonically increasing function. We can do this because if this were not the case, we could make it so by sorting the bins by increasing conditional probability, as illustrated in Figure 3, because both the entropy $H(W')$ and the mutual information $I(W', Y)$ are left invariant by this renumbering/relabeling of the bins. The "cat" probability $P(Y=1)$ (the total shaded area in Figure 3) is of course also left unchanged by both this sorting and by the above-mentioned binning.

2.2.4. Proof that Pareto Frontier is Spanned by Contiguous Binnings

We are now finally ready to tackle the core goal of this paper: Mapping the Pareto frontier (H_*, I_*) of optimal data compression $X \mapsto Z$ that reflects the tradeoff between $H(Z)$ and $I(Z, Y)$. While fine-grained binning has no effect on the entropy $H(Y)$ and negligible effect on $I(W, Y)$, it dramatically reduces the entropy of our data, whereas $H(W) = \infty$ since W is continuous, $H(W') = \log N$ is finite, approaching infinity only in the limit of infinitely many infinitesimal bins. (Note that while this infinity, which reflects the infinite number of bits required to describe a single generic real number, is customarily eliminated by defining entropy only up to an overall additive constant, we will not follow that custom here, for the reason explained in the introduction.) Taken together, these scalings of I and H imply that the leftmost part of the Pareto frontier $I_*(H_*)$, defined by Equation (1) and illustrated in Figure 1, asymptotes to a horizontal line of height $I_* = I(X, Y)$ as $H_* \to \infty$.

To reach the interesting parts of the Pareto frontier further to the right, we must destroy some information about Y. We do this by defining

$$Z = g(W'), \qquad (20)$$

where the function g groups the tiny bins indexed by $W' \in \{1, ..., N\}$ into fewer ones indexed by $Z \in \{1, ..., M\}$, $M < N$. There are vast numbers of such possible groupings, since each group corresponds to one of the $2^N - 2$ nontrivial subsets of the tiny bins. Fortunately, as we will now prove, we need only consider the $\mathcal{O}(N^M)$ contiguous groupings, since non-contiguous ones are inferior and cannot lie on the Pareto frontier. Indeed, we will see that for the examples in Section 3, $M \lesssim 5$ suffices to capture the most interesting information.

Theorem 2 (Contiguous binning theorem). *If W has a uniform distribution and the conditional probability distribution $P(W|Y=1)$ is monotonically increasing, then all points (H_*, I_*) on the Pareto frontier correspond to binning W into contiguous intervals, i.e., if*

$$I(H_*) \equiv \sup_{\{g : H[g(W)] \leq H_*\}} I[g(W), Y], \tag{21}$$

then there exists a set of bin boundaries $b_1 < ... < b_{n-1}$ such that the binned variable $Z \equiv B(W, \mathbf{b}) \in \{1, ..., M\}$ satisfies $H(Z) = H_*$ and $I(Z, Y) = I_*$.

Proof. We prove this by contradiction: we will assume that there is a point (H_*, I_*) on the Pareto frontier to which we can come arbitrarily close with $(H(Z), I(Z, Y))$ for $Z \equiv g(X)$ for a compression function $g : \mathbb{R} \mapsto \{1, ..., M\}$ that is not a contiguous binning function, and obtain a contradiction by using g to construct another compression function $g'(W)$ lying above the Pareto frontier, with $H[g'(W)] = H_*$ and $I[g'(W), Y]) > I_*$. The joint probability distribution P_{ij} for the Z and Y is given by the Lebesgue integral

$$P_{ij} \equiv P(Z = i, Y = j) = \int f_j d\mu_i, \tag{22}$$

where $f_j(w)$ is the joint probability distribution for W and Y introduced earlier and μ_j is the set $\mu \equiv \{w \in [0,1] : g(w) = i\}$, i.e., the set of w-values that are grouped together into the ith large bin. We define the marginal and conditional probabilities

$$P_i \equiv P(Z = i) = P_{i1} + P_{i2}, \quad p_i \equiv P(Y = 1|Z = i) = \frac{P_{i1}}{P_i}. \tag{23}$$

Figure 4 illustrates the case where the binning function g corresponds to $M = 4$ large bins, the second of which consists of two non-contiguous regions that are grouped together; the shaded rectangles in the bottom panel have width P_i, height p_i and area $P_{ij} = P_i p_i$.

According to Theorem A2 in the Appendix B, we obtain the contradiction required to complete our proof (an alternative compression $Z' \equiv g'(W)$ above the Pareto frontier with $H(Z') = H_*$ and $I(Z', Y) > I_*$) if there are two different conditional probabilities $p_k \neq p_l$, and we can change g into g' so that the joint distribution P'_{ij} of Z' and Y changes in the following way:

1. Only P_{kj} and P_{lj} change,
2. both marginal distributions remain the same,
3. the new conditional probabilities p'_k and p'_l are further apart.

Figure 4 shows how this can be accomplished for non-contiguous binning: Let k be a bin with non-contiguous support set μ_k (bin 2 in the illustrated example), let l be a bin whose support μ_l (bin 4 in the example) contains a positive measure subset $\mu_l^{\text{mid}} \subset \mu_l$ within two parts μ_k^{left} and μ_k^{right} of μ_k, and define a new binning function $g'(w)$ that differs from $g(w)$ only by swapping a set $\mu^\epsilon \subset \mu_l^{\text{mid}}$ against a subset of either μ_k^{left} or μ_k^{right} of measure ϵ (in the illustrated example, the binning function change implementing this subset is shown with dotted lines). This swap leaves the total measure of both bins (and hence the marginal distribution P_i) unchanged, and also leaves $P(Y=1)$ unchanged. If $p_k < p_l$, we perform this swap between μ_l^{mid} an μ_k^{right} (as in the figure), and if $p_k > p_l$, we instead perform this swap between μ_l^{mid} an μ_k^{left}, in both cases guaranteeing that p_l and p_k move further apart (since $p(w)$ is monotonically increasing). This completes our proof by contradiction except for the case where $p_k = p_l$; in this case, we swap to entirely eliminate the discontiguity, and repeat our swapping procedure between other bins until we increase the entropy (again obtaining a contradiction) or end up with a fully contiguous binning (if needed, $g(w)'$ can be changed to eliminate any measure-zero subsets that ruin contiguity, since they leave the Lebesgue integral in Equation (22) unchanged.) □

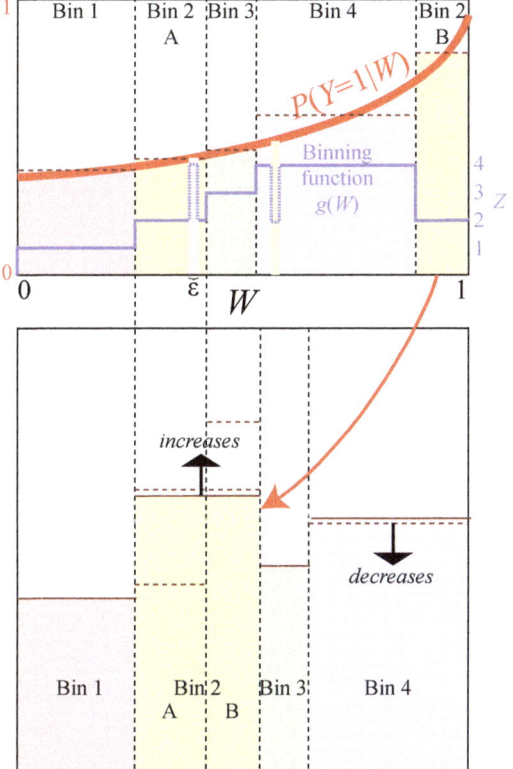

Figure 4. The reason that the Pareto frontier can never be reached using non-contiguous bins is that swapping parts of them against parts of an intermediate bin can increase $I(Z,X)$ while keeping $H(Z)$ constant. In this example, the binning function g assigns two separate W-intervals (top panel) to the same bin (bin 2) as seen in the bottom panel. The shaded rectangles have widths P_i, heights p_i and areas $P_{i1} = P_i p_1$. In the upper panel, the conditional probabilities p_i are monotonically increasing because they are averages of the monotonically increasing curve $p_1(w)$.

2.3. Mapping the Frontier

Theorem 2 implies that we can in practice find the Pareto frontier for any random variable X by searching the space of contiguous binnings of $W = w(X)$ after uniformization, binning and sorting. In practice, we can first try the two bin case by scanning the bin boundary $0 < b_1 < 1$, then trying the three bin case by trying bin boundaries $0 < b_1 < b_2 < 1$, then trying the four bin case, etc., as illustrated in Figure 1. Each of these cases corresponds to a standard multi-objective optimization problem aiming to maximize the two objectives $I(Z,Y)$ and $H(Z)$. We perform this optimization numerically with the AWS algorithm of [17] as described in the next section.

While the uniformization, binning and sorting procedures are helpful in practice as well as for for simplifying proofs, they are not necessary in practice. Since what we really care about is grouping into integrals containing similar conditional probabilities $p_1(w)$, not similar w-values, it is easy to see that binning horizontally after sorting is equivalent to binning vertically before sorting. In other words, we can eliminate the binning and sorting steps if we replace "horizontal" binning $g(W) = B(W, \mathbf{b})$ by "vertical" binning

$$g(W) = B[p_1(W), \mathbf{b}], \qquad (24)$$

where p_1 denotes the conditional probability as before.

3. Results

The purpose of this section is to examine how our method for Pareto-frontier mapping works in practice on various datasets, both to compare its performance with prior work and to gain insight into the shape and structure of the Pareto frontiers for well-known datasets such as the CIFAR-10 image database [18], the MNIST database of hand-written digits [19] and the Fashion-MNIST database of garment images [20]. Before doing this, however, let us build intuition for how our method works by testing on a much simpler toy model that is analytically solvable, where the accuracy of all approximations can be exactly determined.

3.1. Analytic Warmup Example

Let the random variables $X = (x_1, x_2) \in [0,1]^2$ and $Y \in \{1,2\}$ be defined by the bivariate probability distribution

$$f(X,Y) = \begin{cases} 2x_1 x_2 & \text{if } Y = 1, \\ 2(1-x_1)(1-x_2) & \text{if } Y = 2, \end{cases} \tag{25}$$

which corresponds to x_1 and x_2 being two independent and identically distributed random variables with triangle distribution $f(x_i) = x_i$ if $Y = 1$, but flipped $x_i \mapsto 1 - x_i$ if $Y = 2$. This gives $H(Y) = 1$ bit and mutual information

$$I(X,Y) = 1 - \frac{\pi^2 - 4}{16 \ln 2} \approx 0.4707 \text{ bits.} \tag{26}$$

The compressed random variable $W = w(X) \in \mathbb{R}$ defined by Equation (9) is thus

$$W = P(Y=1|X) = \frac{x_1 x_2}{x_1 x_2 + (1-x_1)(1-x_2)}. \tag{27}$$

After defining $Z \equiv B(W, \mathbf{b})$ for a vector $\mathbf{b}$ of bin boundaries, a straightforward calculation shows that the joint probability distribution of Y and the binned variable Z is given by

$$P_{ij} \equiv P(Z=i, Y=j) = F_j(b_{i+1}) - F_j(b_i), \tag{28}$$

where the cumulative distribution function $F_j(w) \equiv P(W<w, Y=j)$ is given by

$$\begin{aligned} F_1(w) &= \frac{w^2 \left[(2w-1)(5-4w) + 2(1-w^2)\log(w^{-1}-1)\right]}{2(2w-1)^4}, \\ F_2(w) &= \frac{1}{2} - F_1(1-w). \end{aligned} \tag{29}$$

Computing $I(W,Y)$ using this probability distribution recovers exactly the same mutual information $I \approx 0.4707$ bits as in Equation (26), as we proved in Theorem 1.

3.2. The Pareto Frontier

Given any binning vector $\mathbf{b}$, we can plot a corresponding point $(H[Z], I[Z,Y])$ in Figure 1 by computing $I(Z,Y) = H(Z) + H(Y) - H(Z,Y)$, $H(Z,Y) = -\sum P_{ij} \log P_{ij}$, etc., where P_{ij} is given by Equation (28).

The figure shows 6000 random binnings each for $M = 3, \dots, 8$ bins; as we have proven, the upper envelope of points corresponding to all possible (contiguos) binnings defines the Pareto frontier. The Pareto frontier begins with the black dot at $(0,0)$ (the lower right corner), since $M = 1$ bin obviously destroys all information. The $M = 2$ bin case corresponds to a 1-dimensional closed curve parametrized by the single parameter b_1 that specifies the boundary between the two bins: It runs from $(0,0)$ when $b_1 = 1$, moves to the left until $H(Z) = 1$ when $b_1 = 0.5$, and returns to $(0,0)$ when $b_1 = 1$. The $b_1 < 0.5$ and $b_1 > 0.5$ branches are indistinguishable in Figure 1 because of the symmetry of our warmup problem, but in generic cases, a closed loop can be seen where only the upper part defines the Pareto frontier.

More generally, we see that the set of all binnings into $M > 2$ bins maps the vector **b** of $M - 1$ bin boundaries into a contiguous region in Figure 1. The inferior white region region below can also be reached if we use non-contiguous binnings.

The Pareto Frontier is seen to resemble the top of a circus tent, with convex segments separated by "corners" where the derivative vanishes, corresponding to a change in the number of bins. We can understand the origin of these corners by considering what happens when adding a new bin of infinitesimal size ϵ. As long as $p_i(w)$ is continuous, this changes all probabilites P_{ij} by amounts $\delta P_{ij} = \mathcal{O}(\epsilon)$, and the probabilities corresponding to the new bin (which used to vanish) will now be $\mathcal{O}(\epsilon)$. The function $\epsilon \log \epsilon$ has infinite derivative at $\epsilon = 0$, blowing up as $\mathcal{O}(\log \epsilon)$, which implies that the entropy increase $\delta H(Z) = \mathcal{O}(-\log \epsilon)$. In contrast, a straightforward calculation shows that all $\log \epsilon$-terms cancel when computing the mutual information, which changes only by $\delta I(Z, Y) = \mathcal{O}(\epsilon)$. As we birth a new bin and move leftward from one of the black dots in Figure 1, the initial slope of the Pareto frontier is thus

$$\lim_{\epsilon \to 0} \frac{\delta I(Z, Y)}{\delta H(Z)} = 0. \tag{30}$$

In other words, the Pareto frontier starts out *horizontally* to the left of each of its corners in Figure 1. Indeed, the corners are "soft" in the sense that the derivative of the Pareto Frontier is continuous and vanishes at the corners: For a given number of bins, $I(X, Z)$ by definition takes its global maximum at the corresponding corner, so the derivative $\partial I(Z, Y)/\partial H(Z)$ vanishes also as we approach the corner from the right. The first corner (the transition from 2 to 3 bins) can nonetheless look fairly sharp because the 2-bin curve turns around rather abruptly, and the right derivative does not vanish in the limit where a symmetry causes the upper and lower parts of the 2-bin loop to coincide.

Our theorems imply that in the $M \to \infty$ limit of infinitely many bins, successive corners become gradually less pronounced (with ever smaller derivative discontinuities), because the left asymptote of the Pareto frontier simply approaches the horizontal line $I_* = I(Y, Z)$.

3.2.1. Approximating $w(X)$

For our toy example, we knew the conditional probability distribution $P(Y|X)$ and could therefore compute $W = w(X) = P(Y = 1|X)$ exactly. For practical examples where this is not the case, we can instead train a neural network to implement a function $\hat{w}(X)$ that approximates $P(Y = 1|X)$. For our toy example, we train a fully connected feedforward neural network to predict Y from X using cross-entropy loss; it has two hidden layers, each with 256 neurons with ReLU activation, and a final linear layer with softmax activation, whose first neuron defines $\hat{w}(X)$. As illustrated in Figure 5, the network prediction for the conditional probability $\hat{w}(X) \equiv P(Y = 1)$ is fairly accurate, but slightly over-confident, tending to err on the side of predicting more extreme probabilities (further from 0.5). The average KL-divergence between the predicted and actual conditional probability distribution $P(Y|X)$ is about 0.004, which causes negligible loss of information about Y.

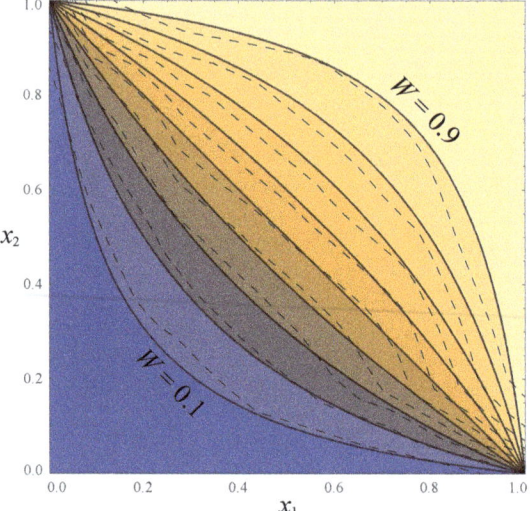

Figure 5. Contour plot of the function $W(x_1, x_2)$ computed both exactly using Equation (27) (solid curves) and approximately using a neural network (dashed curves).

3.2.2. Approximating $f_1(W)$

For practical examples where the conditional joint probability distribution $P(W, Y)$ cannot be computed analytically, we need to estimate it from the observed distribution of W-values output by the neural network. For our examples, we do this by fitting each probability distribution by a beta-distribution times the exponential of a polynomial of degree d:

$$f(w, \mathbf{a}) \equiv \exp\left[\sum_{k=0}^{d} a_k x^k\right] x^{a_{d+1}} (1-x)^{a_{d+2}}, \tag{31}$$

where the coefficient a_0 is fixed by the normalization requirement $\int_0^1 f(w, \mathbf{a}) dw = 1$. We use this simple parametrization because it can fit any smooth distribution arbitrarily well for sufficiently large d, and provides accurate fits for the probability distributions in our examples using quite modest d; for example, $d = 3$ gives $d_{KL}[f_1(w), f(w, \mathbf{a})] \approx 0.002$ for

$$\begin{aligned}\mathbf{a} &\equiv \operatorname*{argmin}_{\mathbf{a}'} d_{\mathrm{KL}}[f_1(w), f(w, \mathbf{a}')] \\ &= (-1.010, 2.319, -5.579, 4.887, 0.308, -0.307),\end{aligned} \tag{32}$$

which causes rather negligible loss of information about Y. For our examples below where we do not know the exact distribution $f_1(w)$ and merely have samples W_i drawn from it, one for each element of the data set, we instead perform the fitting by the standard technique of minimizing the cross entropy loss, i.e.,

$$\mathbf{a} \equiv \operatorname*{argmin}_{\mathbf{a}'} -\sum_{k=1}^{n} \log f(W_k, \mathbf{a}'). \tag{33}$$

Table 2 lists the fitting coefficients used, and Figure 6 illustrates the fitting accuracy.

Table 2. Fits to the conditional probability distributions $P(W|Y)$ for our experiments, in terms of the parameters a_i defined by Equation (31).

Experiment	Y	a_0	a_1	a_2	a_3	a_4	a_5	a_6
Analytic	1	0.0668	−4.7685	16.8993	−25.0849	13.758	0.5797	−0.2700
	2	0.4841	−5.0106	5.7863	−1.5697	−1.7180	−0.3313	−0.0030
Fashion-MNIST	Pullover	0.2878	−12.9596	44.9217	−68.0105	37.3126	0.3547	−0.2838
	Shirt	1.0821	−23.8350	81.6655	−112.2720	53.9602	−0.4068	0.4552
CIFAR-10	Cat	0.9230	0.2165	0.0859	6.0013	−1.0037	0.8499	
			0.6795	0.0511	0.6838	−1.0138	0.9061	
	Dog	0.8970	0.2132	0.0806	6.0013	−1.0039	0.8500	
			0.7872	0.0144	0.7974	−0.9440	0.7237	
MNIST	One	3.1188	−65.224	231.4	−320.054	150.779	1.1226	−0.6856
	Seven	−1.0325	−47.5411	189.895	−269.28	127.363	−0.8219	0.1284

3.3. MNIST, Fashion-MNIST and CIFAR-10

The MNIST database consists of 28 × 28 pixel greyscale images of handwritten digits: 60,000 training images and 10,000 testing images [19]. We use the digits 1 and 7, since they are the two that are most frequently confused, relabeled as $Y = 1$ (ones) and $Y = 2$ (sevens). To increase difficulty, we inject 30% of pixel noise, i.e., randomly flip each pixel with 30% probability (see examples in Figure 2). For easy comparison with the other cases, we use the same number of samples for each class.

The Fashion-MNIST database has the exact same format (60,000 + 10,000 28 × 28 pixel greyscale images), depicting not digits but 10 classes of clothing [20]. Here we again use the two most easily confused classes: Pullovers ($Y = 1$) and shirts ($Y = 2$); see Figure 2 for examples.

We train a neural network classifier on our datasets using the architecture from https://github.com/pytorch/examples/blob/master/mnist/main.py, changing the number of outpiut neurons from 10 to 2. We use two convolutional layers (kernel size 5, stride 1, ReLU activation) with 20 and 50 features, respectively, each of which is followed by max-pooling with kernel size 2. This is followed by a fully connected layer with 500 ReLU neurons and finally a softmax layer that produces the predicted probabilities for the two classes. After training, we apply the trained model to the test set to obtain $W_i = P(Y|X_i)$ for each dataset.

CIFAR-10 [21] is one of the most widely used datasets for machine learning research, and contains 60,000 32 × 32 color images in 10 different classes. We use only the cat ($Y = 1$) and dog ($Y = 2$) classes, which are the two that are empirically hardest to discriminate; see Figure 2 for examples. We use a ResNet18 architecture adapted from https://github.com/kuangliu/pytorch-cifar, for which we use its ResNet18 model [22]; the only difference in architecture is that we use 2 rather than 10 output neurons. We train with a learning rate of 0.01 for the first 150 epochs, 0.001 for the next 100, and 0.0001 for the final 100 epochs; we keep all other settings the same as in the original repository.

Figure 6 shows observed cumulative distribution functions $F_i(w)$ (solid curves) for the $W_i = P(Y = 1|X_i)$ generated by the neural network classifiers, together with our above-mentioned analytic fits (dashed curves). Figure 7 shows the corresponding conditional probability curves $P(Y = 1|W)$ after remapping W to have a uniform distribution as described above. Figure 6 shows that the original W-distributions are strongly peaked around $W \approx 0$ and $W \approx 1$, so this remapping stretches the W-axis so as to shift probability toward more central values.

In the case of CIFAR-10, the observed distribution $f(w)$ was so extremely peaked near the endpoints that we replaced Equation (31) by the more accurate fit

$$f(w) \equiv F'(w), \tag{34}$$

$$F(w) \equiv \begin{cases} \mathbf{a}_0^A F_*[w, \mathbf{a}^A] & \text{if } w < 1/2, \\ 1 - (1 - \mathbf{a}_0^A) F_*[1 - w, \mathbf{a}^B]] & \text{otherwise,} \end{cases} \tag{35}$$

$$F_*(x) \equiv G\left[\frac{(2x)^{a_1}}{2}\right], \tag{36}$$

$$G(x) \equiv \left[\left(\frac{x}{a_2}\right)^{a_3 a_4} + (a_5 + a_6 x)^{a_4}\right]^{1/a_4}, \tag{37}$$

$$a_6 \equiv 2\left[(1 - (2a_2)^{-a_3 a_4})^{1/a_4} - a_5\right], \tag{38}$$

where the parameters vectors $\mathbf{a}^A$ and $\mathbf{a}^B$ are given in Table 2 for both cats and dogs. For the cat case, this fit gives not $f(w)$ but $f(1-w)$. Note that $F_*(0) = 0$, $F_*(1/2) = 1$.

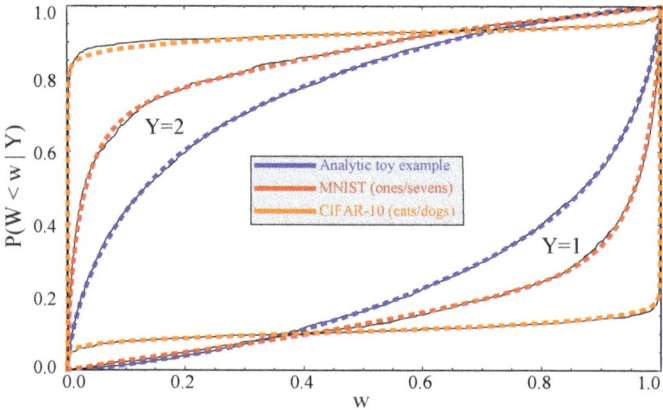

Figure 6. Cumulative distributions $F_i(w) \equiv P(W < w | Y = i)$ are shown for the analytic (blue/dark grey), Fashion-MNIST (red/grey) and CIFAR-10 (orange/light grey) examples. Solid curves show the observed cumulative histograms of W from the neural network, and dashed curves show the fits defined by Equation (31) and Table 2.

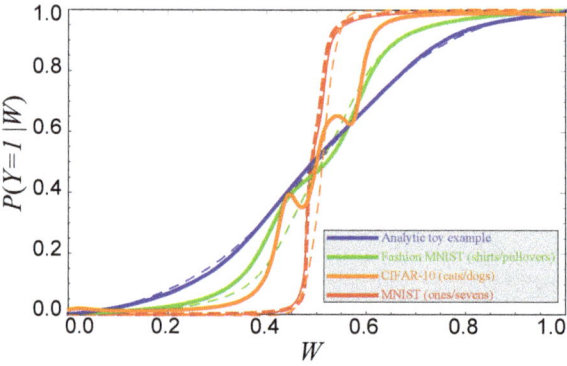

Figure 7. The solid curves show the actual conditional probability $P(Y = 1 | W)$ for CIFAR-10 (where the labels Y = 1 and 2 correspond to "cat" and "dog") and MNIST with 20% label noise (where the labels Y = 1 and 2 correspond to "1" and "7"), respectively. The color-matched dashed curves show the conditional probabilities predicted by the neural network; the reason that they are not diagonal lines $P(Y = 1 | W) = W$ is that W has been reparametrized to have a uniform distribution. If the neural network classifiers were optimal, then solid and dashed curves would coincide.

The final result of our calculations is shown in Figure 8: The Pareto frontiers for our four datasets, computed using our method.

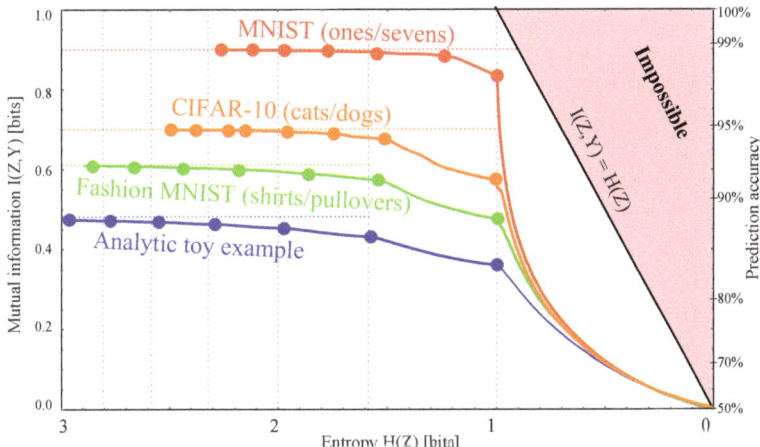

Figure 8. The Pareto frontier for compressed versions $Z = g(X)$ of our four datasets X, showing the maximum attainable class information $I(Z, Y)$ for a given entropy $H(Z)$. The "corners" (dots) correspond to the maximum $I(Z, Y)$ attainable when binning the likelihood W into a given number of bins (2, 3, ..., 8 from right to left). The horizontal dotted lines show the maximum available information $I(X, Y)$ for each case, reflecting that there is simply less to learn in some examples than in others.

3.4. Interpretation of Our Results

To build intuition for our results, let us consider our CIFAR-10 example of images X depicting cats ($Y = 1$) and dogs ($Y = 2$) as in Figure 2 and ask what aspects $Z = g(X)$ of an image X capture the most information about the species Y. Above, we estimated that $I(X, Y) \approx 0.69692$ bits, so what Z captures the largest fraction of this information for a fixed entropy? Given a good neural network classifier, a natural guess might be the single bit Z containing its best guess, say "it's probably a cat". This corresponds to defining $Z = 1$ if $P(Y = 1|X) > 0.5$, $Z = 2$ otherwise, and gives the joint distribution $P_{ij} \equiv P(Y = i, Z = j)$

$$\mathbf{P} = \begin{pmatrix} 0.454555 & 0.045445 \\ 0.042725 & 0.457275 \end{pmatrix}$$

corresponding to $(Z, Y) \approx 0.56971$ bits. However, our results show that we can improve things in two separate ways.

First of all, if we only want to store one bit Z, then we can do better, corresponding to the first "corner" in Figure 8: moving the likelihood cutoff from 0.5 to 0.51, i.e., redefining $Z = 1$ if $P(Y|X) > 0.51$, increases the mutual information to $I(Z, Y) \approx 0.56974$ bits.

More importantly, we are still falling far short of the 0.69692 bits of information we had without data compression, capturing only 88% of the available species information. Our Theorem 1 showed that we can retain all this information if we instead define Z as the cat probability itself: $Z \equiv W \equiv P(Y|X)$. For example, a given image might be compressed not into "It's probably a cat" but into "I'm 94.2477796% sure it's a cat". However, it is clearly impractical to report the infinitely many decimals required to retain all the species information, which would make $H(Z)$ infinite. Our results can be loosely speaking interpreted as the optimal way to round Z, so that the information $H(Z)$ required to store it becomes finite. We found that simply rounding to a fixed number of decimals is suboptimal; for example, if we pick 2 decimals and say "I'm 94.25% sure it's a cat", then we have effectively binned the probability W into 10,000 bins of equal size, even though we can often do much better with bins of unequal size, as illustrated in the bottom panel of Figure 1. Moreover, when the probability W is approximated by a neural network, we found that what should be optimally binned is not W but the conditional probability $P(Y=1|W)$ illustrated in Figure 7 ("vertical binning").

It is convenient to interpret our Pareto-optimal data compression $X \mapsto Z$ as clustering, i.e., as a method of grouping our images or other data X_i into clusters based on what information they contain about Y. For example, Figure 2 illustrates CIFAR-10 images clustered by their degree of "cattiness" into 5 groups $Z = 1, ..., 5$ that might be nicknamed "1.9% cat", "11.8% cat", "31.4% cat", "68.7% cat" and "96.7% cat". This gives the joint distribution $P_{ij} \equiv P(Y=i, Z=j)$ where

$$\mathbf{P} = \begin{pmatrix} 0.350685 & 0.053337 & 0.054679 & 0.034542 & 0.006756 \\ 0.007794 & 0.006618 & 0.032516 & 0.069236 & 0.383836 \end{pmatrix}$$

and gives $I(Z,Y) \approx 0.6882$, thus increasing the fraction of species information retained from 82% to 99%.

This is a striking result: We can group the images into merely five groups and discard all information about all images except which group they are in, yet retain 99% of the information we cared about. Such grouping may be helpful in many contexts. For example, given a large sample of labeled medical images of potential tumors, they can be used to define say five optimal clusters, after which future images can be classified into five degrees of cancer risk that collectively retain virtually all the malignancy information in the original images.

Given that the Pareto Frontier is continuous and corresponds to an infinite family of possible clusterings, which one is most useful in practice? Just as in more general multi-objective optimization problems, the most interesting points on the frontier are arguably its "corners", indicated by dots in Figure 8, where we do notably well on both criteria. This point was also made in the important paper [23] in the context of the DIB-frontier discussed below. We see that the parts of the frontier between corners tend to be convex and thus rather unappealing, since any weighted average of $-H(Z)$ and $I(Z,Y)$ will be maximized at a corner. Our results show that these corners can conveniently be computed without numerically tedious multiobjective optimization, by simply maximizing the mutual information $I(Z,Y)$ for $m = 2, 3, 4, ...$ bins. The first corner, at $H(Z) = 1$ bit, corresponds to the learnability phase transition for DIB, i.e., the largest β for which DIB is able to learn a non-trivial representation. In contrast to the IB learnability phase transition [24,25] where $I(Z,Y)$ increases continuously from 0, here the $I(Y;Z)$ has a jump from 0 to a positive value, due to the non-concave nature of the Pareto frontier.

Moreover, all the examples in Figure 8 are seen to get quite close to the $m \to \infty$ asymptote $I(Z,Y) \to I(X,Y)$ for $m \gtrsim 5$, so the most interesting points on the Pareto frontier are simply the first handful of corners. For these examples, we also see that the greater the mutual information is, the fewer bins are needed to capture most of it.

An alternative way if interpreting the Pareto plane in Figure 8 is as a traveoff between two evils:

Information bloat: $H(Z|Y) \equiv H(Z) - I(Z,Y) \geq 0$,
Information loss: $\Delta I \equiv I(X,Y) - I(Z,Y) \geq 0$.

What we are calling the "information bloat" has also been called "causal waste" [26]. It is simply the conditional entropy of Z given Y, and represents the excess bits we need to store in order to retain the desired information about Y. Geometrically, it is the horizontal distance to the impossible region to the right in Figure 8, and we see that for MNIST, it takes local minima at the corners for both 1 and 2 bins. The information loss is simply the information discarded by our lossy compression of X. Geometrically, it is the vertical distance to the impossible region at the top of Figure 1 (and, in Figure 8, it is the vertical distance to the corresponding dotted horizontal line). As we move from corner to corner adding more bins, we typically reduce the information loss at the cost of increased information bloat. For the examples in Figure 8, we see that going beyond a handful of bins essentially just adds bloat without significantly reducing the information loss.

3.5. Real-World Issues

We just discussed how lossy compression is a tradeoff between information bloat and information loss. Let us now elaborate on the latter, for the real-world situation where $W \equiv P(Y = 1|X)$ is approximated by a neural network.

If the neural network learns to become perfect, then the function w that it implements will be such that $W \equiv w(X)$ satisfies $P(Y = 1|W) = W$, which corresponds to the dashed curves in Figure 7 being identical to the solid curves. While we see that this is close to being the case for the analytic and MNIST examples, the neural networks are further from optimal for Fashion-MNIST and CIFAR-10. The figure illustrates that the general trend is for these neural networks to overfit and therefore be overconfident, predicting probabilities that are too extreme.

This fact that $P(Y = 1|W) \ne W$ probably indicates that our Fashion-MNIST and CIFAR-10 classifiers $W = w(X)$ destroy information about X, but it does not prove this, because if we had a perfect lossless classifier $W \equiv w(X)$ satisfying $P(Y = 1|W) = W$, then we could define an overconfident lossless classifier by an invertible (and hence information-preserving) reparameterization such as $W' \equiv W^2$ that violates the condition $P(Y = 1|W') = W'$.

So how much information does X contain about Y? One way to lower-bound $I(X;Y)$ uses the classification accuracy: if we have a classification problem where $P(Y = 1) = 1/2$ and compress X into a single classification bit Z (corresponding to a binning of W into two bins), then we can write the joint probability distribution for Y and the guessed class Z as

$$P = \begin{pmatrix} \frac{1}{2} - \epsilon_1 & \epsilon_1 \\ \epsilon_2 & \frac{1}{2} - \epsilon_2 \end{pmatrix}.$$

For a fixed total error rate $\epsilon \equiv \epsilon_1 + \epsilon_2$, Fano's Inequality implies that the mutual information takes a minimum

$$I(Z, Y) = 1 + \epsilon \log \epsilon + (1 - \epsilon) \log(1 - \epsilon) \tag{39}$$

when $\epsilon_1 = \epsilon_2 = \epsilon/2$, so if we can train a classifier that gives an error rate ϵ, then the right-hand-side of Equation (39) places a lower bound on the mutual information $I(X, Y)$. The prediction accuracy $1 - \epsilon$ is shown for reference on the right side of Figure 8. Note that getting close to one bit of mutual information requires extremely high accuracy; for example, 99% prediction accuracy corresponds to only 0.92 bits of mutual information.

We can obtain a stronger estimated lower bound on $I(X, Y)$ from the cross-entropy loss function $\mathcal{L}$ used to train our classifiers:

$$\langle \mathcal{L} \rangle = - \langle \log P(Y = Y_i | X = X_i) \rangle = H(Y|X) + d_{\text{KL}}, \tag{40}$$

where $d_{\text{KL}} \ge 0$ denotes the average KL-divergence between true and predicted conditional probability distributions, and $\langle \cdot \rangle$ denotes ensemble averaging over data points, which implies that

$$\begin{aligned} I(X, Y) &= H(Y) - H(Y|X) = H(Y) - \langle \mathcal{L} \rangle - d_{\text{KL}} \\ &\ge H(Y) - \langle \mathcal{L} \rangle. \end{aligned} \tag{41}$$

If $P(Y = 1|W) \ne W$ as we discussed above, then d_{KL} and hence the loss can be further reduced be recalibrating W as we have done, which increases the information bound from Equation (41) up to the the value computed directly from the observed joint distribution $P(W, Y)$.

Unfortunately, without knowing the true probability $p(Y|X)$, there is no rigorous and practically useful upper bound on the mutual information other than the trivial inequality $I(X, Y) < H(Y) = 1$ bit, as the following simple counterexample shows: Suppose our images X are encrypted with some encryption algorithm that is extremely time-consuming to crack, rendering the images for all practical purposes indistinguishable from random noise. Then any reasonable neural network will produce a useless classifier giving $I(W, Y) \approx 0$ even though the true mutual information $I(X, Y)$ could be as large as one bit. In other words, we generally cannot know the true information loss caused by compressing $X \mapsto W$, so the best we can do in practice is to pick a corner reasonably close to the upper asymptote in Figure 8.

3.6. Performance Compared with Blahut–Arimoto Method

The most commonly used technique to date for finding the Pareto frontier is the Blahut–Arimoto (BA) method [27,28] applied to the DIB objective of Equation (3) as described in [12]. Figure 9 and Table 3 shows the BA method implemented as in [23], applied to our above-mentioned analytic toy example, after binning using 2000 equispaced W-bins and $Z \in 1, ..., 8$, scanning the β-parameter from Equation (3) from 10^{-10} to 1 in 20,000 logarithmically equispaced steps. Our method is seen to improve on the BA method in two ways. First, our method finds the entire continuous frontier, whereas the BA method finds only six discrete disconnected points. This is because the BA-method tries to maximize the the DIB-objective from Equation (3) and thus cannot discover points where the Pareto frontier is convex as discussed above. Second, our method finds the exact frontier, whereas the BA-method finds only approximations, which are seen to generally lie below the true frontier.

Table 3. The approximate Pareto frontier points for our analytic example computed with the Blahut–Arimoto (BA) method compared with the points for those same six H-values computed with our exact method.

$H(Z)$	$I(Z, Y)$	
	BA-Method	Our Method
0.0000	0.0000	0.0000
0.9652	0.3260	0.3421
0.9998	0.3506	0.3622
1.5437	0.4126	0.4276
1.5581	0.4126	0.4298
1.5725	0.4141	0.4314

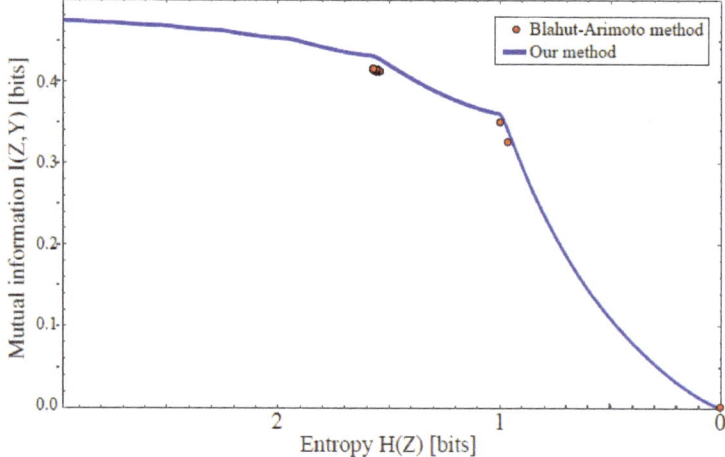

Figure 9. The Pareto frontier our analytic example is computed exactly with our method (solid curve) and approximately with the Blahut–Arimoto method (dots).

4. Conclusions and Discussion

We have presented a method for mapping out the Pareto frontier for classification tasks (as in Figure 8), reflecting the tradeoff between retained entropy and class information. In other words, we have generalized the quest for maximizing raw classification accuracy to that of mapping the full Pareto frontier corresponding to the accuracy–complexity tradeoff. The optimal soft classifiers that we have studied (corresponding to points on the Pareto frontier) are useful for the same reason that the DIB method is useful, e.g., overfitting less and therefore generalizing better.

We first showed how a random variable X (an image, say) drawn from a class $Y \in \{1, ..., n\}$ can be distilled into a vector $W = f(X) \in \mathbb{R}^{n-1}$ losslessly, so that $I(W, Y) = I(X, Y)$. For the $n = 2$ case of binary classification, we then showed how the Pareto frontier is swept out by a one-parameter family of binnings of W into a discrete variable $Z = g_\beta(W) \in \{1, ..., m_\beta\}$ that corresponds to binning W into $m_\beta = 2, 3, ...,$ bins, such that $I(Z, Y)$ is maximized for each fixed entropy $H(Z)$. Our method efficiently finds the exact Pareto frontier, significantly outperforming the Blahut–Arimoto (BA) method [27,28]. Our MATLAB code for computing the Pareto frontier is freely available here: https://github.com/tailintalent/distillation.

4.1. Relation to Information Bottleneck

As mentioned in the introduction, the discrete information bottleneck (DIB) method [12] maximizes a linear combination $I(Z, Y) - \beta H(Z)$ of the two axes in Figure 8. We have presented a method solving a generalization of the DIB problem. The generalization lies in switching the objective from Equation (3) to Equation (1), which has the advantage of discovering the full Pareto frontier in Figure 8 instead of merely the corners and concave parts (as mentioned, the DIB objective cannot discover convex parts of the frontier). The solution lies in our proof that the frontier is spanned by binnings of the likelihood into 2, 3, 4, etc., bins, which enables it to be computed more efficiently than with the iterative/variational method of [12].

The popular original Information Bottleneck (IB) method [10] generalizes DIB by allowing the compression function $g(X)$ to be non-deterministic, thus adding noise that is independent of X. Starting with a Pareto-optimal $Z \equiv g(X)$ and adding such noise will simply shift us straight to the left in Figure 8, away from the frontier (which is by definition monotonically decreasing) and into the Pareto-suboptimal region in the $I(Y; Z)$ vs. $H(Z)$ plane. As shown in [12], IB-compressions tend to altogether avoid the rightmost part of Figure 8, with an entropy $H(Z)$ that never drops below some fixed value independent of β.

4.2. Relation to Phase Transitions in DIB Learning

Recent work has revealed interesting phase transitions that occur during information bottleneck learning [12,24,25,29], as well as phase transitions in other objectives, e.g., β-VAE [30], infoDropout [31]. Specifically, when the β-parameter that controls the tradeoff between information retention and model simplicity is continuously adjusted, the resulting point in the IB-plane can sometimes "get stuck" or make discontinuous jumps. For the DIB case, our results provide an intuitive understanding of these phase transitions in terms of the geometry of the Pareto frontier.

Let us consider Figure 1 as an example. The DIB maximiziation of $I(Z, Y) - \beta H(Z)$ geometrically corresponds to finding a tangent line of the Pareto frontier of slope $-\beta$.

If the Pareto frontier $I_*(H)$ were everywhere continuous and concave, so that $I_*''(H) < 0$, then its slope would range from some steepest value $-\beta_*$ at the right endpoint $H = 0$ and continuously flatten out as we move leftward, asymptotically approaching zero slope as $H \to \infty$. The learnability phase transition studied in [24,25] would then occur when $\beta = \beta_*$: for any $\beta \geq \beta_*$, the DIB method learns nothing, e.g., discovers as optimal the point $(H, I) = (0, 0)$ where Z retains no information whatsoever about Y. As $\beta \leq \beta_*$ is continuously reduced, the DIB-discovered point would then continuously move up and to the left along the Pareto frontier.

This was for the case of an everywhere concave frontier, but Figures 1 and 8 show that actual Pareto frontiers need not be concave—indeed, none of the frontiers that we have computed are concave. Instead, they are seen to consist of long convex segments joint together by short concave pieces near the "corners". This means that as β is continuously increased, the DIB solution exhibits first-order phase transitions, making discontinuous jumps from corner to corner at certain critical β-values; these phase transitions correspond to increasing the number of clusters into which the data X is grouped.

4.3. Outlook

Our results suggest a number of opportunities for further work, ranging from information theory to machine learning, neuroscience and physics.

As to information theory, it will be interesting to try to generalize our method from binary classification into classification into more than two classes. Moreover, one can ask if there is a way of pushing the general information distillation problem all the way to bits. It is easy to show that a discrete random variable $Z \in \{1, ..., m\}$ can always be encoded as $m - 1$ independent random bits (Bernoulli variables) $B_1, ..., B_{m-1} \in \{0, 1\}$, defined by

$$P(B_k = 1) = P(Z = k+1)/P(Z \leq k+1), \tag{42}$$

while this generically requires some information bloat. The mapping z from bit strings $\mathbf{B}$ to integers $Z \equiv z(\mathbf{B})$ is defined so that $z(\mathbf{B})$ is the position of the last bit that equals one when $\mathbf{B}$ is preceded by a one. For example, for $m = 4$, the mapping from length-3 bit strings $\mathbf{B} \in \{0,1\}^3$ to integers $Z \in \{1, ..., 4\}$ is $z(001) = z(011) = z(101) = z(111) = 4$, $z(010) = z(110) = 3$, $z(100) = 2$, $z(000) = 1$. So in the spirit of the introduction, is there some useful way of generalizing PCA, autoencoders, CCA and/or the method we have presented so that the quantities Z_i and Z'_i in Table 1 are not real numbers but bits?

As to neural networks, it is interesting to explore novel classifier architectures that reduce the overfitting and resulting overconfidence revealed by Figure 7, as this might significantly increase the amount of information we can distill into our compressed data. It is important not to complacently declare victory just because classification accuracy is high; as mentioned, even 99% binary classification accuracy can waste 8% of the information.

As to neuroscience, our discovery of optimal "corner" binnings begs the question of whether evolution may have implemented such categorization in brains. For example, if some binary variable Y that can be inferred from visual imagery is evolutionarily important for a given species (say, whether potential food items are edible), might our method help predict how many distinct colors m their brains have evolved to classify hues into? In this example, X might be a triplet of real numbers corresponding to light intensity recorded by three types of retinal photoreceptors, and the integer Z might end up corresponding so some definitions of yellow, orange, etc. A similar question can be asked for other cases where brains define finite numbers of categories, for example categories defined by distinct words.

As to physics, it has been known even since the introduction of Maxwell's Demon that a physical system can use information about its environment to extract work from it. If we view an evolved life form as an intelligent agent seeking to perform such work extraction, then it faces a tradeoff between retaining too little relevant infomation (consequently extrating less work) and retaining too much (wasting energy on information processing and storage). Susanne Still recently proved the remarkable physics result [32] that the lossy data compression optimizing such work extraction efficiency is precisely that prescribed by the above-mentioned information bottleneck method [10]. As she puts it, an intelligent data representation strategy emerges from the optimization of a fundamental physical limit to information processing. This derivation made minimal and reasonable seeming assumptions about the physical system, but did not include an energy cost for information encoding. We conjecture that this can be done such that an extra Shannon coding term proportional to $H(Z)$ gets added to the loss function, which means that when this term dominates, the generalized Still criterion would instead prefer the Deterministic Information Bottleneck or one of our Pareto-optimal data compressions.

While noise-adding IB-style data compression may turn out to be commonplace in many biological settings, it is striking that the types of data compression we typically associate with human perception intelligence appears more deterministic, in the spirit of DIB and our work. For example, when we compress visual input into "this is a probably a cat", we do not typically add noise by deliberately flipping our memory to "this is probably a dog". Similarly, the popular jpeg image compression algorithm dramatically reduces image sizes while retaining essentially all information that we humans find relevant, and does so deterministically, without adding noise.

It is striking that simple information-theoretical principles such as IB, DIB and Pareto-optimality appear relevant across the spectrum of known intelligence, ranging from extremely simple physical systems as in Still's work all the way up to high-level human perception and cognition. This motivates further work on the exciting quest for a deeper understanding of Pareto-optimal data compression and its relation to neuroscience and physics.

Author Contributions: Conceptualization, resources, supervision, project administration, funding acquisition, M.T.; methodology, software, validation, formal analysis, investigation, writing—original draft preparation, writing—review and editing, visualization, M.T. and T.W. All authors have read and agreed to the published version of the manuscript.

Funding. This work was supported by The Casey and Family Foundation, the Ethics and Governance of AI Fund, the Foundational Questions Institute, the Rothberg Family Fund for Cognitive Science and by Theiss Research through TWCF grant #0322. The opinions expressed in this publication are those of the authors and do not necessarily reflect the views of the funders.

Acknowledgments: The authors wish to thank Olivier de Weck for sharing the AWS multiobjective optimization software.

Conflicts of Interest: The authors declare no conflict of interest. The funders had no role in the design of the study; in the collection, analyses, or interpretation of data; in the writing of the manuscript, or in the decision to publish the results.

Appendix A. Binning Can Be Practically Lossless

If the conditional probability distribution $p_1(w) \equiv P(Y = 1|W = w)$ is a slowly varying function and the range of W is divided into tiny bins, then $p_1(w)$ will be almost constant within each bin and so binning W (discarding information about the exact position of W within a bin) should destroy almost no information about Y. This intuition is formalized by the following theorem, which says that a random variable W can be binned into a finite number of bins at the cost of losing arbitrarily little information about Y.

Theorem A1. *Binning can be practically lossless: Given a random variable $Y \in \{1, 2\}$ and a uniformly distributed random variable $W \in [0, 1]$ such that the conditional probability distribution $p_1(w) \equiv P(Y = 1|W = w)$ is monotonic, there exists for any real number $\epsilon > 0$ a vector $\mathbf{b} \in \mathbb{R}^{N-1}$ of bin boundaries such that the information reduction*

$$\Delta I \equiv I[W, Y] - I[B(W, \mathbf{b}), Y] < \epsilon,$$

where B is the binning function defined by Equation (17).

Proof. The binned bivariate probability distribution is

$$P_{ij} \equiv P(Z = j, Y = i) = \int_{b_{j-1}}^{b_j} p_i(w) dw \qquad (A1)$$

with marginal distribution

$$P_j^Z \equiv P(Z = j) = b_j - b_{j-1}. \qquad (A2)$$

Let $\bar{p}_i(w)$ denote the piecewise constant function that in the jth bin $b_{j-1} < w \le b_j$ takes the average value of $p_i(w)$ in that bin, i.e.,

$$\bar{p}_i(w) \equiv \frac{1}{b_j - b_{j-1}} \int_{b_{j-1}}^{b_j} p_i(w) dw = \frac{P_{ij}}{P_j^Z}. \tag{A3}$$

These definitions imply that

$$-\sum_{j=1}^{N} P_{ij} \log \frac{P_{ij}}{P_j^Z} = \int_0^1 h\left[\bar{p}_i(w)\right] dw, \tag{A4}$$

where $h(x) \equiv -x \log x$. Since $h(x)$ vanishes at $x = 0$ and $x = 1$ and takes its intermediate maximum value at $x = 1/e$, the function

$$h_*(x) \equiv \begin{cases} h(x) & \text{if } x < e^{-1}, \\ 2h(e^{-1}) - h(x) & \text{if } x \ge e^{-1} \end{cases} \tag{A5}$$

is continuous and increases monotonically for $x \in [0,1]$, with $h'_* = |h'(x)|$. This means that if we define the non-negative monotonic function

$$h_+(w) \equiv h_*[p_1(w)] - h_*[p_2(w)],$$

it changes at least as fast as either of its terms, so that for any $w_1, w_2 \in [0, 1]$, we have

$$\begin{aligned} |h[p_i(w_2)] - h[p_i(w_1)]| &\le |h_*[p_i(w_2)] - h_*[p_i(w_1)]| \\ &\le |h_+(w_2) - h_+(w_1)|. \end{aligned} \tag{A6}$$

We will exploit this bound to limit how much $h[p_i(w)]$ can vary within a bin. Since $h_+(0) \ge 0$ and $h_+(1) \le 2h_*(1) = 4/e \ln 2 \approx 2.12 < 3$, we pick $N - 1$ bins boundaries b_k implicitly defined by

$$h_+(b_j) = h_+(0) + [h_+(1) - h_+(0)] \frac{j}{N} \tag{A7}$$

for some integer $N \gg 1$. Using Equation (A6), this implies that

$$|h[\bar{p}_i(w)] - h[p_i(w)]| \le \frac{h_+(1) - h_+(0)}{N} < \frac{3}{N}. \tag{A8}$$

The mutual information between two variables is given by $I(Y, U) = H(Y) - H(Y|U)$, where the second term (the conditional entropy is given by the following expressions in the cases that we need:

$$H(Y|Z) = -\sum_{i=1}^{N} \sum_{j=1}^{2} P_{ij} \log \frac{P_{ij}}{P_i}, \tag{A9}$$

$$H(Y|W) = -\sum_{i=1}^{2} \int_0^1 p_i(w) \log p_i(w) dw. \tag{A10}$$

The information loss caused by our binning is therefore

$$\begin{aligned}
\Delta I &= I(W,Y) - I(Z,Y) = H(Y|Z) - H(Y|W) \\
&= -\sum_{i=1}^{2}\left(\sum_{j=1}^{N} P_{ij}\log\frac{P_{ij}}{P_{j}^{Z}} + \int_{0}^{1} h\left[p_{i}(w)\right] dw\right) \\
&= \sum_{i=1}^{2}\int_{0}^{1}\left(h\left[\bar{p}_{i}(w)\right] - h\left[p_{i}(w)\right]\right) dw \\
&\leq \sum_{i=1}^{2}\int_{0}^{1}\left|h\left[\bar{p}_{i}(w)\right] - h\left[p_{i}(w)\right]\right| dw \\
&< \sum_{i=1}^{2}\int_{0}^{1}\frac{3}{N} = \frac{6}{N}, \quad \quad \quad \quad \quad \text{(A11)}
\end{aligned}$$

where we used Equation (A4) to obtain the third row and Equation (A8) to obtain the last row. This means that however small an information loss tolerance ϵ we want, we can guarantee $\Delta I < \epsilon$ by choosing $N > 6/\epsilon$ bins placed according to Equation (A7), which completes the proof. □

Note that the proof still holds if the function $p_i(w)$ is not monotonic, as long as the number of times M that it changes direction is finite: In that case, we can simply repeat the above-mentioned binning procedure separately in the $M+1$ intervals where $p_i(w)$ is monotonic, using $N > 6/\epsilon$ bins in each interval, i.e., a total of $N > 6M/\epsilon$ bins.

Appendix B. More Varying Conditional Probability Boosts Mutual Information

Mutual information is loosely speaking a measure of how far a probability distribution P_{ij} is from being separable, i.e., a product of its two marginal distributions. (specifically, the mutual information is the Kullback–Leibler divergence between the bivariate probability distribution and the product of its marginals.) If all conditional probabilities for one variable Y given the other variable Z are identical, then the distribution is separable and the mutual information $I(Z,Y)$ vanishes, so one may intuitively expect that making conditional probabilities more different from each other will increase $I(Z,Y)$. The following theorem formalizes this intuition in a way that enables Theorem 2.

Theorem A2. *Consider two discrete random variables $Z \in \{1,...,n\}$ and $Y \in \{1,2\}$ and define $P_i \equiv P(Z=i)$, $p_i \equiv P(Y=1|Z=i)$, so that the joint probability distribution $P_{ij} \equiv P(Z=i, Y=j)$ is given by $P_{i1} = P_i p_i$, $P_{i2} = P_i(1-p_i)$. If two conditional probabilities p_k and p_l differ, then we increase the mutual information $I(Y,Z)$ if we bring them further apart by adjusting P_{kj} and P_{lj} in such a way that both marginal distributions remain unchanged.*

Proof. The only such change that keeps the marginal distributions for both Z and Y unchanged takes the form

$$\begin{pmatrix} P_1 p_1 & \cdots & P_k p_k - \epsilon & \cdots & P_l p_l + \epsilon & \cdots \\ P_1(1-p_1) & \cdots & P_k(1-p_k) + \epsilon & \cdots & P_l(1-p_l) - \epsilon & \cdots \end{pmatrix}$$

where the parameter ϵ must be kept small enough for all probabilities to remain non-negative. Without loss of generality, we can assume that $p_k < p_l$, so that we make the conditional probabilities

$$P(Y=1|Z=k) = \frac{P_{k1}}{P_k} = p_k - \epsilon/P_k, \quad \quad \text{(A12)}$$

$$P(Y=1|Z=l) = \frac{P_{l1}}{P_l} = p_l + \epsilon/P_l \quad \quad \text{(A13)}$$

more different when increasing ϵ from zero. Computing and differentiating the mutual information with respect to ϵ, most terms cancel and we find that

$$\left.\frac{\partial I(Z,Y)}{\partial \epsilon}\right|_{\epsilon=0} = \log\left[\frac{1/p_k - 1}{1/p_l - 1}\right] > 0 \qquad (A14)$$

which means that adjusting the probabilities with a sufficiently tiny $\epsilon > 0$ will increase the mutual information, completing the proof. □

References

1. Pearson, K. LIII. On lines and planes of closest fit to systems of points in space. *Lond. Edinb. Dublin Philos. Mag. J. Sci.* **1901**, *2*, 559–572. [CrossRef]
2. Vincent, P.; Larochelle, H.; Bengio, Y.; Manzagol, P.A. Extracting and composing robust features with denoising autoencoders. In Proceedings of the 25th International Conference on Machine Learning, Helsinki, Finland, 5–9 July 2008; pp. 1096–1103.
3. Goodfellow, I.; Pouget-Abadie, J.; Mirza, M.; Xu, B.; Warde-Farley, D.; Ozair, S.; Courville, A.; Bengio, Y. Generative adversarial nets. In Proceedings of the Neural Information Processing Systems 2014, Montreal, QC, Canada, 8–13 December 2014; pp. 2672–2680.
4. Hotelling, H. Relation between two sets of variates. *Biometrica* **1936**, *28*, 321–377. [CrossRef]
5. Eckart, C.; Young, G. The approximation of one matrix by another of lower rank. *Psychometrika* **1936**, *1*, 211–218. [CrossRef]
6. van den Oord, A.; Li, Y.; Vinyals, O. Representation learning with contrastive predictive coding. *arXiv* **2018**, arXiv:1807.03748.
7. Clark, D.G.; Livezey, J.A.; Bouchard, K.E. Unsupervised Discovery of Temporal Structure in Noisy Data with Dynamical Components Analysis. *arXiv* **2019**, arXiv:1905.09944.
8. Tegmark, M. Optimal Latent Representations: Distilling Mutual Information into Principal Pairs. *arXiv* **2019**, arXiv:1902.03364.
9. Kurkoski, B.M.; Yagi, H. Quantization of binary-input discrete memoryless channels. *IEEE Trans. Inf. Theory* **2014**, *60*, 4544–4552. [CrossRef]
10. Tishby, N.; Pereira, F.C.; Bialek, W. The information bottleneck method. *arXiv* **2000**, arXiv:physics/0004057.
11. Tan, A.; Meshulam, L.; Bialek, W.; Schwab, D. The renormalization group and information bottleneck: A unified framework. In Proceedings of the APS Meeting Abstracts, Boston, MA, USA, 4–8 March 2019.
12. Strouse, D.; Schwab, D.J. The deterministic information bottleneck. *Neural Comput.* **2017**, *29*, 1611–1630. [CrossRef]
13. Alemi, A.A.; Fischer, I.; Dillon, J.V.; Murphy, K. Deep variational information bottleneck. *arXiv* **2016**, arXiv:1612.00410.
14. Chalk, M.; Marre, O.; Tkacik, G. Relevant sparse codes with variational information bottleneck. In Proceedings of the Neural Information Processing Systems 2016, Barcelona, Spain, 5–10 December 2016; pp. 1957–1965.
15. Fischer, I. The Conditional Entropy Bottleneck. 2018. Available online: https://openreview.net/forum?id=rkVOXhAqY7 (accessed on 11 December 2019).
16. Amjad, R.A.; Geiger, B.C. Learning representations for neural network-based classification using the information bottleneck principle. *IEEE Trans. Pattern Anal. Mach. Intell.* **2019**. [CrossRef] [PubMed]
17. Kim, I.Y.; de Weck, O.L. Adaptive weighted-sum method for bi-objective optimization: Pareto front generation. *Struct. Multidiscip. Optim.* **2005**, *29*, 149–158. [CrossRef]
18. Krizhevsky, A.; Nair, V.; Hinton, G. The CIFAR-10 Dataset. 2014. Available online: https://www.cs.toronto.edu/~kriz/cifar.html (accessed on 11 December 2019).
19. LeCun, Y.; Cortes, C.; Burges, C. MNIST Handwritten Digit Database. 2010. Available online: http://yann.lecun.com/exdb/mnist (accessed on 11 December 2019).
20. Xiao, H.; Rasul, K.; Vollgraf, R. Fashion-mnist: A novel image dataset for benchmarking machine learning algorithms. *arXiv* **2017**, arXiv:1708.07747.
21. Krizhevsky, A.; Hinton, G. *Learning Multiple Layers of Features from Tiny Images*; Technical Report TR-2009; University of Toronto: Toronto, ON, Canada, 2009.

22. He, K.; Zhang, X.; Ren, S.; Sun, J. Deep residual learning for image recognition. In Proceedings of the IEEE Conference on Computer Vision and Pattern Recognition, Las Vegas, NV, USA, 27–30 June 2016; pp. 770–778.
23. Strouse, D.; Schwab, D.J. The information bottleneck and geometric clustering. *Neural Comput.* **2019**, *31*, 596–612. [CrossRef] [PubMed]
24. Wu, T.; Fischer, I.; Chuang, I.; Tegmark, M. Learnability for the information bottleneck. *arXiv* **2019**, arXiv:1907.07331.
25. Wu, T.; Fischer, I.; Chuang, I.; Tegmark, M. Learnability for the information bottleneck. *Entropy* **2019**, *21*, 924. [CrossRef]
26. Thompson, J.; Garner, A.J.; Mahoney, J.R.; Crutchfield, J.P.; Vedral, V.; Gu, M. Causal asymmetry in a quantum world. *Phys. Rev. X* **2018**, *8*, 031013. [CrossRef]
27. Blahut, R. Computation of channel capacity and rate-distortion functions. *IEEE Trans. Inf. Theory* **1972**, *18*, 460–473. [CrossRef]
28. Arimoto, S. An algorithm for computing the capacity of arbitrary discrete memoryless channels. *IEEE Trans. Inf. Theory* **1972**, *18*, 14–20. [CrossRef]
29. Chechik, G.; Globerson, A.; Tishby, N.; Weiss, Y. Information bottleneck for Gaussian variables. *J. Mach. Learn. Res.* **2005**, *6*, 165–188.
30. Rezende, D.J.; Viola, F. Taming VAEs. *arXiv* **2018**, arXiv:1810.00597.
31. Achille, A.; Soatto, S. Emergence of invariance and disentanglement in deep representations. *J. Mach. Learn. Res.* **2018**, *19*, 1947–1980.
32. Still, S. Thermodynamic cost and benefit of data representations. *arXiv* **2017**, arXiv:1705.00612.

© 2019 by the authors. Licensee MDPI, Basel, Switzerland. This article is an open access article distributed under the terms and conditions of the Creative Commons Attribution (CC BY) license (http://creativecommons.org/licenses/by/4.0/).

Article

The Convex Information Bottleneck Lagrangian

Borja Rodríguez Gálvez *,†, Ragnar Thobaben *,† and Mikael Skoglund *,†

Department of Intelligent Systems, Division of Information Science and Engineering (ISE), KTH Royal Institute of Technology, 11428 Stockholm, Sweden
* Correspondence: borjarg@kth.se (B.R.G.); ragnart@kth.se (R.T.); skoglund@kth.se (M.S.)
† Current address: Malvinas väg 10, 100 44 Stockholm, Sweden

Received: 9 December 2019; Accepted: 8 January 2020; Published: 14 January 2020

Abstract: The information bottleneck (IB) problem tackles the issue of obtaining relevant compressed representations T of some random variable X for the task of predicting Y. It is defined as a constrained optimization problem that maximizes the information the representation has about the task, $I(T;Y)$, while ensuring that a certain level of compression r is achieved (i.e., $I(X;T) \leq r$). For practical reasons, the problem is usually solved by maximizing the IB Lagrangian (i.e., $\mathcal{L}_{\text{IB}}(T;\beta) = I(T;Y) - \beta I(X;T)$) for many values of $\beta \in [0,1]$. Then, the curve of maximal $I(T;Y)$ for a given $I(X;T)$ is drawn and a representation with the desired predictability and compression is selected. It is known when Y is a deterministic function of X, the IB curve cannot be explored and another Lagrangian has been proposed to tackle this problem: the squared IB Lagrangian: $\mathcal{L}_{\text{sq-IB}}(T;\beta_{\text{sq}}) = I(T;Y) - \beta_{\text{sq}} I(X;T)^2$. In this paper, we (i) present a general family of Lagrangians which allow for the exploration of the IB curve in all scenarios; (ii) provide the exact one-to-one mapping between the Lagrange multiplier and the desired compression rate r for known IB curve shapes; and (iii) show we can approximately obtain a specific compression level with the convex IB Lagrangian for both known and unknown IB curve shapes. This eliminates the burden of solving the optimization problem for many values of the Lagrange multiplier. That is, we prove that we can solve the original constrained problem with a single optimization.

Keywords: information bottleneck; representation learning; mutual information; optimization

1. Introduction

Let $X \in \mathcal{X}$ and $Y \in \mathcal{Y}$ be two statistically dependent random variables with joint distribution $p_{(X,Y)}$. The information bottleneck (IB) [1] investigates the problem of extracting the relevant information from X for the task of predicting Y.

For this purpose, the IB defines a bottleneck variable $T \in \mathcal{T}$ obeying the Markov chain $Y \leftrightarrow X \leftrightarrow T$ so that T acts as a representation of X. Tishby et al. [1] define the relevant information as the information the representation keeps from Y after the compression of X (i.e., $I(T;Y)$), provided a certain level of compression (i.e., $I(X;T) \leq r$). Therefore, we select the representation which yields the value of the IB curve that best fits our requirements.

Definition 1 (IB Functional). *Let X and Y be statistically dependent variables. Let Δ be the set of random variables T obeying the Markov condition $Y \leftrightarrow X \leftrightarrow T$. Then the IB functional is*

$$F_{\text{IB,max}}(r) = \max_{T \in \Delta} \{I(T;Y)\} \text{ s.t. } I(X;T) \leq r, \ \forall r \in [0,\infty). \tag{1}$$

Definition 2 (IB Curve). *The IB curve is the set of points defined by the solutions of $F_{\text{IB,max}}(r)$ for varying values of $r \in [0,\infty)$.*

Definition 3 (Information Plane). *The plane is defined by the axes $I(T;Y)$ and $I(X;T)$.*

This method has been successfully applied to solve different problems from a variety of domains. For example:

- Supervised learning. In supervised learning, we are presented with a set of n pairs of input features and task outputs instances. We seek an approximation of the conditional probability distribution between the task outputs Y and the input features X. In classification tasks (i.e., when Y is a discrete random variable), the introduction of the variable T learned through the information bottleneck principle maintained the performance of standard algorithms based on the cross-entropy loss while providing with more adversarial attacks robustness and invariance to nuisances [2–4]. Moreover, by the nature of its definition the information bottleneck appears to be closely related with a trade-off between accuracy on the observable set and generalization to new, unseen instances (see Section 2).
- Clustering. In clustering, we are presented with a set of n pairs of instances of a random variable X and their attributes of interest Y. We seek groups of instances (or clusters T) such that the attributes of interest within the instances of each cluster are similar and the attributes of interest of the instances of different clusters are dissimilar. Therefore, the information bottleneck can be employed since it allows us to aim for attribute representative clusters (maximizing the similarity between instances within the clusters) and enforce a certain compression of the random variable X (ensuring a certain difference between instances of the different clusters). This has been successfully implemented, for instance, for gene expression analysis and word, document, stock pricing, or movie rating clustering [5–7].
- Image segmentation. In image segmentation, we want to partition an image into segments such that each pixel in a region shares some attributes. If we divide the image into very small regions X (e.g., each region is a pixel or a set of pixels defined by a grid), we can consider the problem of segmentation as that of clustering the regions X based on the region attributes Y. Hence, we can use the information bottleneck so that we seek region clusters T that are maximally informative about the attributes Y (e.g., the intensity histogram bins) and maintain a level of compression of the original regions X [8].
- Quantization. In quantization, we consider a random variable $X \in \mathcal{X}$ such that $\mathcal{X}$ is a large or continuous set. Our objective is to map X into a variable $T \in \mathcal{T}$ such that $\mathcal{T}$ is a smaller, countable set. If we fix the quantization set size to $|\mathcal{T}| = \lfloor r \rfloor$ and aim at maximizing the information of the quantized variable with another random variable Y and restrict the mapping to be deterministic, then the problem is equivalent to the information bottleneck [9,10].
- Source coding. In source coding, we consider a data source $\mathcal{S}$ which generates a signal $Y \in \mathcal{Y}$, which is later perturbed by a channel $\mathcal{C} : \mathcal{Y} \to \mathcal{X}$ that outputs X. We seek a coding scheme that generates a code $T \in \mathcal{T}$ from the output of the channel X which is as informative as possible about the original source signal Y and can be transmitted at a small rate $I(X;T) \leq r$. Therefore, this problem is equivalent to the the formulation of the information bottleneck [11].

Furthermore, it has been employed as a tool for development or explanation in other disciplines like reinforcement learning [12–14], attribution methods [15], natural language processing [16], linguistics [17] or neuroscience [18]. Moreover, it has connections with other problems such as source coding with side information (or the Wyner-Ahlswede-Körner (WAK) problem), the rate-distortion problem or the cost-capacity problem (see Sections 3, 6 and 7 from [19]).

In practice, solving a constrained optimization problem such as the IB functional is challenging. Thus, in order to avoid the non-linear constraints from the IB functional, the IB Lagrangian is defined.

Definition 4 (IB Lagrangian). *Let X and Y be statistically dependent variables. Let Δ be the set of random variables T obeying the Markov condition $Y \leftrightarrow X \leftrightarrow T$. Then we define the IB Lagrangian as*

$$\mathcal{L}_{IB}(T;\beta) = I(T;Y) - \beta I(X;T). \qquad (2)$$

Here $\beta \in [0,1]$ is the Lagrange multiplier which controls the trade-off between the information of Y retained and the compression of X. Note we consider $\beta \in [0,1]$ because (i) for $\beta \leq 0$ many uncompressed solutions such as $T = X$ maximize $\mathcal{L}_{IB}(T;\beta)$, and (ii) for $\beta \geq 1$ the IB Lagrangian is non-positive due to the data processing inequality (DPI) (Theorem 2.8.1 from Cover and Thomas [20]) and trivial solutions like $T = $ const are maximizers with $\mathcal{L}_{IB}(T;\beta) = 0$ [21].

We know the solutions of the IB Lagrangian optimization (if existent) are solutions of the IB functional by the Lagrange's sufficiency theorem (Theorem 5 in Appendix A of Courcoubetis [22]). Moreover, since the IB functional is concave (Lemma 5 of Gilad-Bachrach et al. [19]) we know they exist (Theorem 6 in Appendix A of Courcoubetis [22]).

Therefore, the problem is usually solved by maximizing the IB Lagrangian with adaptations of the Blahut-Arimoto algorithm [1], deterministic annealing approaches [23] or a bottom-up greedy agglomerative clustering [6] or its improved sequential counterpart [24]. However, when provided with high-dimensional random variables X such as images, these algorithms do not scale well and deep learning-based techniques, where the IB Lagrangian is used as the objective function, prevailed [2,25,26].

Note the IB Lagrangian optimization yields a representation T with a given performance $(I(X;T), I(T;Y))$ for a given β. However, there is no one-to-one mapping between β and $I(X;T)$. Hence, we cannot directly optimize for the desired compression level r but we need to perform several optimizations for different values of β and select the representation with the desired performance; e.g., [2]. The Lagrange multiplier selection is important since (i) sometimes even choices of $\beta < 1$ lead to trivial representations such that $p_{T|X} = p_T$, and (ii) there exist some discontinuities on the performance level w.r.t. the values of β [27].

Moreover, recently Kolchinsky et al. [21] showed how in deterministic scenarios (such as many classification problems where an input x_i belongs to a single particular class y_i) the IB Lagrangian could not explore the IB curve. Particularly, they showed that multiple β yielded the same performance level and that a single value of β could result in different performance levels. To solve this issue, they introduced the squared IB Lagrangian, $\mathcal{L}_{\text{sq-IB}}(T;\beta_{\text{sq}}) = I(T;Y) - \beta_{sq} I(X;T)^2$, which is able to explore the IB curve in any scenario by optimizing for different values of β_{sq}. However, even though they realized a one-to-one mapping between β_{sq} and the compression level existed, they did not find such mapping. Hence, multiple optimizations of the Lagrangian were still required to find the best trade-off solution.

The main contributions of this article are:

1. We introduce a general family of Lagrangians (the convex IB Lagrangians) which are able to explore the IB curve in any scenario for which the squared IB Lagrangian [21] is a particular case of. More importantly, the analysis made for deriving this family of Lagrangians can serve as inspiration for obtaining new Lagrangian families that solve other objective functions with intrinsic trade-offs such as the IB Lagrangian.
2. We show that in deterministic scenarios (and other scenarios where the IB curve shape is known) one can use the convex IB Lagrangian to obtain a desired level of performance with a single optimization. That is, there is a one-to-one mapping between the Lagrange multiplier used for the optimization and the level of compression and informativeness obtained, and we provide the exact mapping. This eliminates the need for multiple optimizations to select a suitable representation.
3. We introduce a particular case of the convex IB Lagrangians: the shifted exponential IB Lagrangian, which allows us to approximately obtain a specific compression level in any scenario. This way, we can approximately solve the initial constrained optimization problem from Equation (1) with a single optimization.

Furthermore, we provide some insight for explaining why there are discontinuities in the performance levels w.r.t. the values of the Lagrange multipliers. In a classification setting, we connect those discontinuities with the intrinsic clusterization of the representations when optimizing the IB bottleneck objective.

The structure of the article is the following: In Section 2 we motivate the usage of the IB in supervised learning settings. Then, in Section 3 we outline the important results used about the IB curve in deterministic scenarios. Later, in Section 4 we introduce the convex IB Lagrangian and explain some of its properties like the bijective mapping between Lagrange multipliers and the compression level and the range of such multipliers. After that, we support our (proved) claims with some empirical evidence on the MNIST [28] and TREC-6 [29] datasets in Section 5. Finally, in Section 6 we discuss our claims and empirical results. A PyTorch [30] implementation of the article can be found at https://github.com/burklight/convex-IB-Lagrangian-PyTorch.

In the Appendices A–F we provide with the proofs of the theoretical results. Then, in Appendix G we show some alternative families of Lagrangians with similar properties. Later, in Appendix H we provide with the precise experimental setup details to reproduce the results from the paper, and further experimentation with different datasets and neural network architectures. To conclude, in Appendix I we show some guidelines on how to set the convex information bottleneck Lagrangians for practical problems.

2. The IB in Supervised Learning

In this section, we will first give an overview of supervised learning in order to later motivate the usage of the information bottleneck in this setting.

2.1. Supervised Learning Overview

In supervised learning we are given a dataset $\mathcal{D}_n = \{(x_i, y_i)\}_{i=1}^n$ of n pairs of input features and task outputs. In this case, X and Y are the random variables of the input features and the task outputs. We assume x_i and y_i are sampled i.i.d. from the true distribution $p_{(X,Y)} = p_{Y|X} p_X$. The usual aim of supervised learning is to use the dataset $\mathcal{D}_n$ to learn a particular conditional distribution $q_{\hat{Y}|X}$ of the task outputs given the input features, parametrized by θ, which is a good approximation of $p_{Y|X}$. We use $\hat{Y}$ and $\hat{y}$ to indicate the predicted task output random variable and its outcome. We call a supervised learning task regression when Y is continuous-valued and classification when it is discrete.

Usually, supervised learning methods employ intermediate representations of the inputs before making predictions about the outputs; e.g., hidden layers in neural networks (Chapter 5 from Bishop [31]) or transformations in a feature space through the kernel trick in kernel machines like SVMs or RVMs (Sections 7.1 and 7.2 from Bishop [31]). Let T be a possibly stochastic function of the input features X with a parametrized conditional distribution $q_{T|X}$, then, T obeys the Markov condition $Y \leftrightarrow X \leftrightarrow T$. The mapping from the representation to the predicted task outputs is defined by the parametrized conditional distribution $q_{\hat{Y}|T}$. Therefore, in representation-based machine learning methods, the full Markov Chain is $Y \leftrightarrow X \leftrightarrow T \leftrightarrow \hat{Y}$. Hence, the overall estimation of the conditional probability $p_{Y|X}$ is given by the marginalization of the representations; i.e., $q_{\hat{Y}|X} = \mathbb{E}_{t \sim q_{T|X}} \left[q_{\hat{Y}|T=t} \right]$ (The notation $q_{\hat{Y}|T=t}$ represents the probability distribution $q_{\hat{Y}|T}(\cdot|t;\theta)$. For the rest of the text, we will use the same notation to represent conditional probability distributions where the conditioning argument is given).

In order to achieve the goal of having a good estimation of the conditional probability distribution $p_{Y|X}$, we usually define an instantaneous cost function $j : \mathcal{X} \times \mathcal{Y} \to \mathbb{R}$. The value of this function $j(x,y;\theta)$ serves as a heuristic to measure the loss of our algorithm, parametrized by θ, obtains when trying to predict the realization of the task output y with the input realization x.

Clearly, we can be interested in minimizing the expectation of the instantaneous cost function over all the possible input features and task outputs, which we call the cost function. However, since we only have a finite dataset $\mathcal{D}_n$ we have instead to minimize the empirical cost function.

Definition 5 (Cost Function and Empirical Cost Function). *Let X and Y be the input features and task output random variables and $x \in \mathcal{X}$ and $y \in \mathcal{Y}$ their realizations. Let also j be the instantaneous cost function, θ the parametrization of our learning algorithm, and $\mathcal{D}_n = \{(x_i, y_i)\}_{i=1}^{n}$ the given dataset. Then, we define:*

1. The cost function:
$$J(p_{(X,Y)}; \theta) = \mathbb{E}_{(x,y) \sim p_{(X,Y)}}[j(x, y; \theta)] \qquad (3)$$

2. The emprical cost function:
$$\hat{J}(\mathcal{D}_n; \theta) = \frac{1}{n} \sum_{i=1}^{n} j(x_i, y_i; \theta) \qquad (4)$$

The discrepancy between the normal and empirical cost functions is called the generalization gap or generalization error (see Section 1 of Xu and Raginsky [32], for instance) and intuitively, the smaller this gap is, the better our model generalizes; i.e., the better it will perform to new, unseen samples in terms of our cost function.

Definition 6 (Generalization Gap). *Let $J(p_{(X,Y)}; \theta)$ and $\hat{J}(\mathcal{D}_n; \theta)$ be the cost and the empirical cost functions as defined in Definition 5. Then, the generalization gap is defined as*

$$\text{gen}(\mathcal{D}_n; \theta) = J(p_{(X,Y)}; \theta) - \hat{J}(\mathcal{D}_n; \theta), \qquad (5)$$

and it represents the error incurred when the selected distribution is the one parametrized by θ when the rule $\hat{J}(\mathcal{D}_n; \theta)$ is used instead of $J(p_{(X,Y)}; \theta)$ as the function to minimize.

Ideally, we would want to minimize the cost function. Hence, we usually try to minimize the empirical cost function and the generalization gap simultaneously. The modifications to our learning algorithm which intend to reduce the generalization gap but not hurt the performance on the empirical cost function are known as regularization.

2.2. Why Do We Use the IB?

Definition 7 (Representation cross-entropy cost function). *Let X and Y be two statistically dependent variables with joint distribution $p_{(X,Y)} = p_{Y|X} p_X$. Let also T be a random variable obeying the Markov condition $Y \leftrightarrow X \leftrightarrow T$ and $q_{T|X}$ and $q_{\hat{Y}|T}$ be the encoding and decoding distributions of our model, parametrized by θ. Finally, let $\mathbb{C}(p_Z || q_Z) = -\mathbb{E}_{z \sim p_Z}[\log(q_Z(z))]$ be the cross entropy between two probability distributions p_Z and q_Z. Then, the cross-entropy cost function is*

$$J_{\text{CE}}(p_{(X,Y)}; \theta) = \mathbb{E}_{(x,t) \sim q_{T|X} p_X}\left[\mathbb{C}(q_{Y|T=t} || q_{\hat{Y}|T=t})\right] = \mathbb{E}_{(x,y) \sim p_{(X,Y)}}[j_{\text{CE}}(x, y; \theta)], \qquad (6)$$

where $j_{\text{CE}}(x, y; \theta) = -\mathbb{E}_{t \sim q_{T|X=x}}[q_{\hat{Y}|T=t}(y|t; \theta)]$ is the instantaneous representation cross-entropy cost function and $q_{Y|T} = \mathbb{E}_{x \sim p_X}[p_{Y|X=x} q_{T|X=x} / q_T]$ and $q_T = \mathbb{E}_{x \sim p_X}[q_{T|X=x}]$.

The cross-entropy is a widely used cost function in classification tasks (e.g., Teahan [8], Krizhevsky et al. [33], Shore and Gray [34]) which has many interesting properties [35]. Moreover, it is known that minimizing the $J_{\text{CE}}(p_{(X,Y)}; \theta)$ maximizes the mutual information $I(T; Y)$. That is:

Proposition 1 (Minimizing the Cross Entropy Maximizes the Mutual Information). *Let $J_{\text{CE}}(p_{(X,Y)}; \theta)$ be the representation cross-entropy cost function as defined in Definition 7. Let also $I(T; Y)$ be the mutual information between random variables T and Y in the setting from Definition 7. Then, minimizing $J_{\text{CE}}(p_{(X,Y)}; \theta)$ implies maximizing $I(T; Y)$.*

The proof of this proposition can be found in Appendix A.

Definition 8 (Nuisance). *A nuisance is any random variable that affects the observed data X but is not informative to the task we are trying to solve. That is, Ξ is a nuisance for Y if $Y \perp \Xi$ or $I(\Xi, Y) = 0$.*

Similarly, we know that minimizing $I(X;T)$ minimizes the generalization gap for restricted classes when using the cross-entropy cost function (Theorem 1 of Vera et al. [36]), and when using $I(T;Y)$ directly as an objective to maximize (Theorem 4 of Shamir et al. [37]). Furthermore, Achille and Soatto [38] in Proposition 3.1 upper bound the information of the input representations, T, with nuisances that affect the observed data, Ξ, with $I(X;T)$. Therefore, minimizing $I(X;T)$ helps generalization by not keeping useless information of Ξ in our representations.

Thus, jointly maximizing $I(T;Y)$ and minimizing $I(X;T)$ is a good choice both in terms of performance in the available dataset and in new, unseen data, which motivates studies on the IB.

3. The Information Bottleneck in Deterministic Scenarios

Kolchinsky et al. [21] showed that when Y is a deterministic function of X (i.e., $Y = f(X)$), the IB curve is piecewise linear. More precisely, it is shaped as stated in Proposition 2.

Proposition 2 (The IB Curve is Piecewise Linear in Deterministic Scenarios). *Let X be a random variable and $Y = f(X)$ be a deterministic function of X. Let also T be the bottleneck variable that solves the IB functional. Then the IB curve in the information plane is defined by the following equation:*

$$\begin{cases} I(T;Y) = I(X;T) & \text{if } I(X;T) \in [0, I(X;Y)) \\ I(T;Y) = I(X;Y) & \text{if } I(X;T) \geq I(X;Y) \end{cases} \quad (7)$$

Furthermore, they showed that the IB curve could not be explored by optimizing the IB Lagrangian for multiple β because the curve was not strictly concave. That is, there was not a one-to-one relationship between β and the performance level.

Theorem 1 (In Deterministic Scenarios, the IB Curve cannot be Explored Using the IB Lagrangian). *Let X be a random variable and $Y = f(X)$ be a deterministic function of X. Let also Δ be the set of random variables T obeying the Markov condition $Y \leftrightarrow X \leftrightarrow T$. Then:*

1. *Any solution $T \in \Delta$ such that $I(X;T) \in [0, I(X;Y))$ and $I(T;Y) = I(X;T)$ solves $\arg\max_{T\in\Delta}\{\mathcal{L}_{\text{IB}}(T;\beta)\}$ for $\beta = 1$. That is, many different compression and performance levels can be achieved for $\beta = 1$.*
2. *Any solution $T \in \Delta$ such that $I(X;T) > I(X;Y)$ and $I(T;Y) = I(X;Y)$ solves $\arg\sup_{T\in\Delta}\{\mathcal{L}_{\text{IB}}(T;\beta)\}$ for $\beta = 0$. That is, many compression levels can be achieved with the same performance for $\beta = 0$.*

   Note we use the supremum in this case since for $\beta = 0$ we have that $I(X;T)$ could be infinite and then the search set from Equation (1); i.e., $\{T : Y \leftrightarrow X \leftrightarrow T\} \cap \{T : I(X;T) < \infty\}$ is not compact anymore.
3. *Any solution $T \in \Delta$ such that $I(X;T) = I(T;Y) = I(X;Y)$ solves $\arg\max_{T\in\Delta}\{\mathcal{L}_{\text{IB}}(T;\beta)\}$ for all $\beta \in (0,1)$. That is, many different β achieve the same compression and performance level.*

An alternative proof for this theorem can be found in Appendix B.

4. The Convex IB Lagrangian

4.1. Exploring the IB Curve

Clearly, a situation like the one depicted in Theorem 1 is not desirable, since we cannot aim for different levels of compression or performance. For this reason, we generalize the effort from Kolchinsky et al. [21] and look for families of Lagrangians which are able to explore the IB curve. Inspired by the squared IB Lagrangian, $\mathcal{L}_{\text{sq-IB}}(T;\beta_{\text{sq}}) = I(T;Y) - \beta_{\text{sq}}I(X;T)^2$, we look at the conditions a function of $I(X;T)$ requires in order to be able to explore the IB curve. In this way, we realize that any monotonically increasing and strictly convex function will be able to do so, and we call the family of Lagrangians with these characteristics the convex IB Lagrangians, due to the nature of the introduced function.

Theorem 2 (Convex IB Lagrangians). *Let Δ be the set of r.v. T obeying the Markov condition $Y \leftrightarrow X \leftrightarrow T$. Then, if u is a monotonically increasing and strictly convex function, the IB curve can always be recovered by the solutions of $\arg\max_{T \in \Delta}\{\mathcal{L}_{IB,u}(T; \beta_u)\}$, with*

$$\mathcal{L}_{IB,u}(T; \beta_u) = I(T;Y) - \beta_u u(I(X;T)). \tag{8}$$

That is, for each point $(I(X;T), I(T;Y))$ s.t. $dI(T;Y)/dI(X;T) > 0$ there is a unique β_u for which maximizing $\mathcal{L}_{IB,u}(T; \beta_u)$ achieves this solution. Furthermore, β_u is strictly decreasing w.r.t. $I(X;T)$. We call $\mathcal{L}_{IB,u}(T; \beta_u)$ the convex IB Lagrangian.

The proof of this theorem can be found in Appendix C. Furthermore, by exploiting the IB curve duality (Lemma 10 of Gilad-Bachrach et al. [19]) we were able to derive other families of Lagrangians which allow for the exploration of the IB curve (Appendix G).

Remark 1. *Clearly, we can see how if u is the identity function (i.e., $u(I(X;T)) = I(X;T)$) then we end up with the normal IB Lagrangian. However, since the identity function is not strictly convex, it cannot ensure the exploration of the IB curve.*

During the proof of this theorem we observed a relationship between the Lagrange multipliers and the solutions obtained of the normal IB Lagrangian $\mathcal{L}_{IB}(T; \beta)$ and the convex IB Lagrangian $\mathcal{L}_{IB,u}(T; \beta_u)$. This relationship is formalized in the following corollary.

Corollary 1 (IB Lagrangian and IB convex Lagrangian connection). *Let $\mathcal{L}_{IB}(T; \beta)$ be the IB Lagrangian and $\mathcal{L}_{IB,u}(T; \beta_u)$ the convex IB Lagrangian. Then, maximizing $\mathcal{L}_{IB}(T; \beta)$ and $\mathcal{L}_{IB,u}(T; \beta_u)$ can obtain the same point in the IB curve if $\beta_u = \beta/u'(I(X;T))$, where u' is the derivative of u.*

This corollary allows us to better understand why the addition of u allows for the exploration of the IB curve in deterministic scenarios. If we note that for $\beta = 1$ we can obtain any point in the increasing region of the curve, then we clearly see how evaluating u' for different values of $I(X;T)$ define different values of β_u that obtain such points. Moreover, it lets us see how if for $\beta = 0$ maximizing the IB Lagrangian could obtain any point $(I(X;Y); I(X;T))$ with $I(X;T) > I(X;Y)$, then the same happens for the IB convex Lagrangian.

4.2. Aiming for a Specific Compression Level

Let B_u denote the domain of Lagrange multipliers β_u for which we can find solutions in the IB curve with the convex IB Lagrangian. Then, the convex IB Lagrangians do not only allow us to explore the IB curve with different β_u. They also allow us to identify the specific β_u that obtains a given point $(I(X;T), I(T;Y))$, provided we know the IB curve in the information plane. Conversely, the convex IB Lagrangian allows finding the specific point $(I(X;T), I(T;Y))$ that is obtained by a given β_u.

Proposition 3 (Bijective Mapping between IB Curve Point and Convex IB Lagrange multiplier). *Let the IB curve in the information plane be known; i.e., $I(T;Y) = f_{IB}(I(X;T))$ is known. Then there is a bijective mapping from Lagrange multipliers $\beta_u \in B_u \setminus \{0\}$ from the convex IB Lagrangian to points in the IB curve $(I(X;T), f_{IB}(I(X;T))$. Furthermore, these mappings are:*

$$\beta_u = \frac{df_{IB}(I(X;T))}{dI(X;T)} \frac{1}{u'(I(X;T))} \quad \text{and} \quad I(X;T) = (u')^{-1}\left(\frac{df_{IB}(I(X;T))}{dI(X;T)} \frac{1}{\beta_u}\right), \tag{9}$$

where u' is the derivative of u and $(u')^{-1}$ is the inverse of u'.

This is especially interesting since in deterministic scenarios we know the shape of the IB curve (Theorem 2) and since the convex IB Lagrangians allow for the exploration of the IB curve (Theorem 2). A proof for Proposition 3 can be found in Appendix D.

Remark 2. *Note that the definition from Tishby et al. [1] $\beta = df_{\text{IB}}(I(X;T))/dI(X;T)$ only allows for a bijection between β and $I(X;T)$ if f_{IB} is a strictly convex, and known function, and we have seen this is not the case in deterministic scenarios (Theorem 1).*

A direct result derived from this proposition is that we know the domain of Lagrange multipliers, B_u, which allows for the exploration of the IB curve if the shape of the IB curve is known. Furthermore, if the shape is not known we can at least bound that range.

Corollary 2 (Domain of Convex IB Lagrange Multiplier with Known IB Curve Shape). *Let the IB curve in the information plane be $I(T;Y) = f_{\text{IB}}(I(X;T))$ and let $I_{\max} = I(X;Y)$. Let also $I(X;T) = r_{\max}$ be the minimum mutual information s.t. $f_{\text{IB}}(r_{\max}) = I_{\max}$; i.e., $r_{\max} = \arg\inf_r \{f_{\text{IB}}(r)\}$ s.t. $f_{\text{IB}}(r) = I_{\max}$. Then, the range of Lagrange multipliers that allow the exploration of the IB curve with the convex IB Lagrangian is $B_u = [\beta_{u,\min}, \beta_{u,\max}]$, with*

$$\beta_{u,\min} = \lim_{r \to r_{\max}^-} \left\{ \frac{f'_{\text{IB}}(r)}{u'(r)} \right\} \quad \text{and} \quad \beta_{u,\max} = \lim_{r \to 0^+} \left\{ \frac{f'_{\text{IB}}(r)}{u'(r)} \right\}, \tag{10}$$

where $f'_{\text{IB}}(r)$ and $u'(r)$ are the derivatives of $f_{\text{IB}}(I(X;T))$ and $u(I(X;T))$ w.r.t. $I(X;T)$ evaluated at r respectively. Also, note that there are some scenarios where $r_{\max} \to \infty$ (see, e.g., [39]), in these scenarios $\beta_{u,\min} = \lim_{r \to \infty} \{f'_{\text{IB}}(r)/u'(r)\} \geq 0$.

Corollary 3 (Domain of Convex IB Lagrange Multiplier Bound). *The range of the Lagrange multipliers that allow the exploration of the IB curve is contained by $[0, \beta_{u,\text{top}}]$ which is also contained by $[0, \beta_{u,\text{top}}^+]$, where*

$$\beta_{u,\text{top}} = \frac{(\inf_{\Omega_x \subset \mathcal{X}} \{\beta_0(\Omega_x)\})^{-1}}{\lim_{r \to 0^+} \{u'(r)\}}, \quad \text{and} \quad \beta_{u,\text{top}}^+ = \frac{1}{\lim_{r \to 0^+} \{u'(r)\}}, \tag{11}$$

where $u'(r)$ is the derivative of $u(I(X;T))$ w.r.t. $I(X;T)$ evaluated at r, $\mathcal{X}$ is the set of possible realizations of X and β_0 and Ω_x are defined as in [27] (Note in [27] they consider the dual problem (see Appendix G), so when they refer to β^{-1} it translates to β in this article). That is, $B_u \subseteq [0, \beta_{u,\text{top}}] \subseteq [0, \beta_{u,\text{top}}^+]$.

Corollaries 2 and 3 allow us to reduce the range search for β when we want to explore the IB curve. Practically, $\inf_{\Omega_x \subset \mathcal{X}} \{\beta_0(\Omega_x)\}$ might be difficult to calculate so Wu et al. [27] derived an algorithm to approximate it. However, we still recommend setting the numerator to 1 for simplicity. The proofs for both corollaries are found in Appendices E and F.

5. Experimental Support

In order to showcase our claims we use the MNIST [28] and the TREC-6 [29] datasets. We modify the nonlinear-IB method [26], which is a neural network that minimizes the cross-entropy while also minimizing a differentiable kernel-based estimate of $I(X;T)$ [40]. Then, we used this technique to maximize a lower bound on the convex IB Lagrangians by applying the functions u to the $I(X;T)$ estimate.

The network structure is the following: first, a stochastic encoder $T = f_{\text{enc}}(X;\theta) + W$ with $p_W = \mathcal{N}(0, I_d)$ such that $T \in \mathbb{R}^d$, where d is the dimension of the bottleneck variable (Note that the encoder needs to be stochastic to (i) ensure a finite and well-defined mutual information [21,41] and (ii) make gradient-based optimization methods over the IB Lagrangian useful [41]). Second, a deterministic decoder $q_{\hat{Y}|T} = f_{\text{dec}}(T; \theta)$. For the MNIST dataset both the encoder and the decoder are fully-connected networks, for a fair comparison with [26]. For the TREC-6 dataset, the encoder is a set of convolutions of word embeddings followed by a fully-connected network and the decoder is also a fully-connected network. For further details about the experiment setup, additional results for different values of α and η and supplementary experimental results for different datasets and network architectures, please refer to Appendix H.

In Figure 1 we show our results for two particularizations of the convex IB Lagrangians:

1. the power IB Lagrangians: $\mathcal{L}_{\text{IB,pow}}(T; \beta_{\text{pow}}, \alpha) = I(T;Y) - \beta_{\text{pow}} I(X;T)^{(1+\alpha)}$, $\alpha > 0$ (Note when $\alpha = 1$ we have the squared IB functional from Kolchinsky et al. [21]).
2. the exponential IB Lagrangians: $\mathcal{L}_{\text{IB,exp}}(T; \beta_{\text{exp}}, \eta) = I(T;Y) - \beta_{\text{exp}} \exp(\eta I(X;T))$, $\eta > 0$.

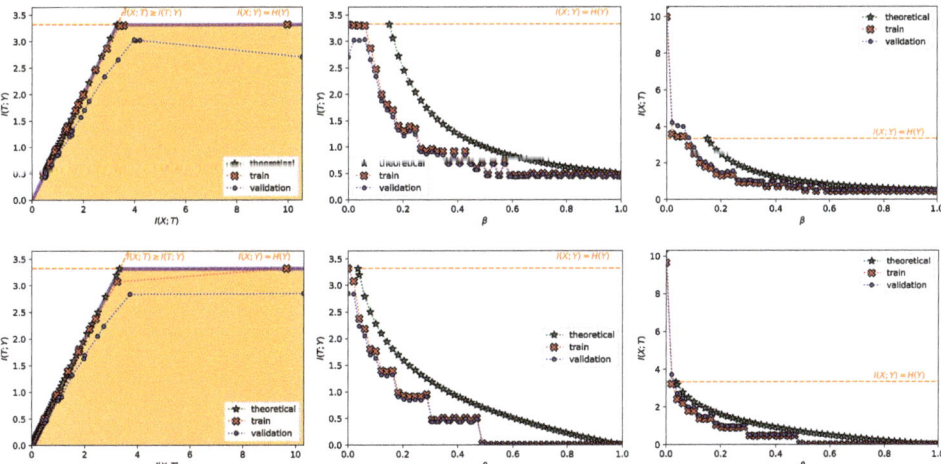

Figure 1. The top row shows the results for the power information bottleneck (IB) Lagrangian with $\alpha = 1$, and the bottom row for the exponential IB Lagrangian with $\eta = 1$, both in the MNIST dataset. In each row, from left to right it is shown (i) the information plane, where the region of possible solutions of the IB problem is shadowed in light orange and the information-theoretic limits are the dashed orange line; (ii) $I(T;Y)$ as a function of β_u; and (iii) the compression $I(X;T)$ as a function of β_u. In all plots, the red crosses joined by a dotted line represent the values computed with the training set, the blue dots the values computed with the validation set and the green stars the theoretical values computed as dictated by Proposition 3. Moreover, in all plots, it is indicated $I(X;Y) = H(Y) = \log_2(10)$ in a dashed, orange line. All values are shown in bits.

We can clearly see how both Lagrangians are able to explore the IB curve (first column from Figure 1) and how the theoretical performance trend of the Lagrangians matches the experimental results (second and third columns from Figure 1). There are small mismatches between the theoretical and experimental performance. This is because using the nonlinear-IB, as stated by Kolchinsky et al. [21], does not guarantee that we find optimal representations due to factors like (i) inaccurate estimation of $I(X;T)$, (ii) restrictions on the structure of T, (iii) use of an estimation of the decoder instead of the real one and (iv) the typical non-convex optimization issues that arise with gradient-based methods. The main difference comes from the discontinuities in performance for increasing β, which cause is still unknown (cf. Wu et al. [27]). It has been observed, however, that the bottleneck variable performs an intrinsic clusterization in classification tasks (see, for instance, [21,26,42] or Figure 2b). We observed how this clusterization matches with the quantized performance levels observed (e.g., compare Figure 2a with the top center graph in Figure 1); with maximum performance when the number of clusters is equal to the cardinality of Y and reducing performance with a reduction of the number of clusters, which is in line with the concurrent work from Wu and Fischer [43]. We do not have a mathematical proof for the exact relationship between these two phenomena; however, we agree with Wu et al. [27] that it is an interesting matter and hope this observation serves as motivation to derive new theory.

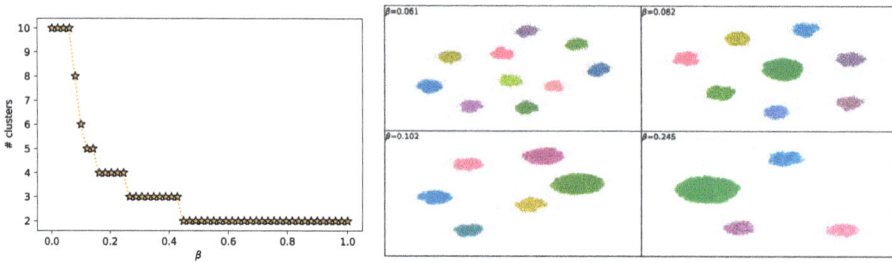

(a) Number of clusters for different β_{pow}. (b) Example of clusters for different β_{pow}.

Figure 2. Depiction of the clusterization behavior of the bottleneck variable for the power IB Lagrangian in the MNIST dataset with $\alpha = 1$. The clusters were obtained using the DBSCAN algorithm [44,45].

In practice, there are different criteria for choosing the function u. For instance, the exponential IB Lagrangian could be more desirable than the power IB Lagrangian when we want to draw the IB curve since it has a finite range of β_u. This is $B_u = [(\eta \exp(\eta I_{\max}))^{-1}, \eta^{-1}]$ for the exponential IB Lagrangian vs. $B_u = [((1+\alpha)I_{\max}^{\alpha})^{-1}, \infty)$ for the power IB Lagrangian. Furthermore, there is a trade-off between (i) how much the selected u function resembles a linear function in our region of interest; e.g., with α or η close to zero, since it will suffer from similar problems as the original IB Lagrangian; and (ii) how fast it grows in our region of interest; e.g., higher values of α or η, since it will suffer from value convergence; i.e., optimizing for separate values of β_u will achieve similar levels of performance (Figure 3). Please, refer to Appendix I for a more thorough explanation of these two phenomena.

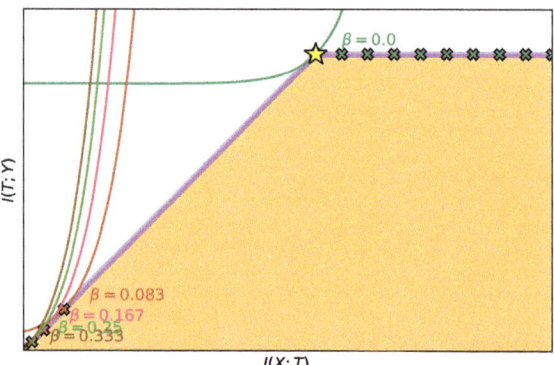

Figure 3. Example of value convergence with the exponential IB Lagrangian with $\eta = 3$. We show the intersection of the isolines of $\mathcal{L}_{\text{IB,exp}}(T; \beta_{\text{exp}})$ for different $\beta_{\text{exp}} \in B_{\text{exp}} \approx [1.56 \times 10^{-5}, 3^{-1}]$ using Corollary 2.

Particularly, the value convergence phenomenon can be exploited in order to approximately obtain a particular level of compression r^*, both for known and unkown IB curves (see Appendix I or the example in Figure 4). For known IB curves, we also know the achieved predictability $I(T;Y)$ since it is the same as the level of compression $I(X;T)$. For this exploitation, we can employ the shifted version of the exponential IB Lagrangian (which is also a particular case of the convex IB Lagrangian):

- the shifted exponential IB Lagrangians:

$$\mathcal{L}_{\text{IB,sh-exp}}(T; \beta_{\text{sh-exp}}, \eta, r^*) = I(T;Y) - \beta_{\text{sh-exp}} \exp(\eta(I(X;T) - r^*)), \eta > 0, r^* \in [0, \infty).$$

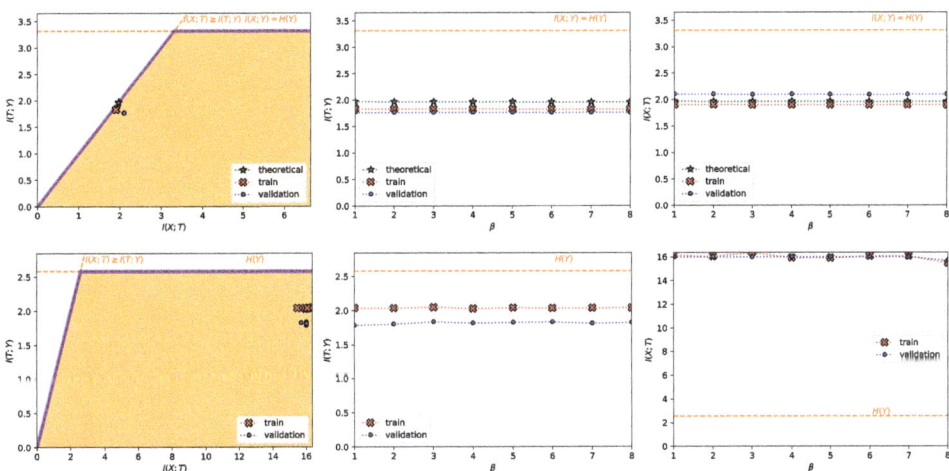

Figure 4. Example of value convergence exploitation with the shifted exponential Lagrangian with $\eta = 200$. In the top row, for the MNIST dataset aiming for a compression level $r^* = 2$ and in the bottom row, for the TREC-6 dataset aiming for a compression level of $r^* = 16$. In each row, from left to right it is shown (i) the information plane, where the region of possible solutions of the IB problem is shadowed in light orange and the information-theoretic limits are the dashed orange line; (ii) $I(T;Y)$ as a function of β_u; and (iii) the compression $I(X;T)$ as a function of β_u. In all plots, the red crosses joined by a dotted line represent the values computed with the training set, the blue dots the values computed with the validation set and the green stars the theoretical values computed as dictated by Proposition 3. Moreover, in all plots, it is indicated $H(Y)$ in a dashed, orange line. All values are shown in bits.

For this Lagrangian, the optimization procedure converges to representations with approximately the desired compression level r^* if the hyperparameter η is set to a large value.

In Figure 4 we show the results of aiming for a compression level of $r^* = 2$ bits in the MNIST dataset and of $r^* = 16$ bits in the TREC-6 dataset, both with $\eta = 200$. We can see how for different values of $\beta_{\text{sh-exp}}$ we can obtain the same desired compression level, which makes this method stable to variations in the Lagrange multiplier selection.

To sum up, in order to achieve a desired level of performance with the convex IB Lagrangian as an objective one should:

1. In a deterministic or close to a deterministic setting (see ϵ-deterministic definition in Kolchinsky et al. [21]): Use the adequate β_u for that performance using Proposition 3. Then if the performance is lower than desired, i.e., we are placed in the wrong performance plateau, gradually reduce the value of β_u until reaching the previous performance plateau. Alternatively, exploit the value convergence phenomenon with, for instance, the shifted exponential IB Lagrangian.
2. In a stochastic setting: exploit the value convergence phenomenon with, for instance, the shifted exponential IB Lagrangian. Alternatively, draw the IB curve with multiple values of β_u on the range defined by Corollary 3 and select the representations that best fit their interests.

6. Conclusions

The information bottleneck is a widely used and studied technique. However, it is known that the IB Lagrangian cannot be used to achieve varying levels of performance in deterministic scenarios. Moreover, in order to achieve a particular level of performance, multiple optimizations with different Lagrange multipliers must be done to draw the IB curve and select the best traded-off representation.

In this article we introduced a general family of Lagrangians which allow to (i) achieve varying levels of performance in any scenario, and (ii) pinpoint a specific Lagrange multiplier β_u to optimize for a specific performance level in known IB curve scenarios; e.g., deterministic. Furthermore, we showed the β_u domain when the IB curve is known and a β_u domain bound for exploring the IB curve when it is unknown. This way we can reduce and/or avoid multiple optimizations and, hence, reduce the computational effort for finding well traded-off representations. Moreover, (iii) when the IB curve is not known, we saw how we can exploit the value convergence issue of the convex IB Lagrangian to approximately obtain a specific compression level for both known and unknown IB curve shapes. Finally, (iv) we provided some insight into the discontinuities on the performance levels w.r.t. the Lagrange multipliers by connecting those with the intrinsic clusterization of the bottleneck variable.

Author Contributions: Conceptualization, B.R.G. and R.T.; formal analysis, B.R.G.; funding acquisition, M.S.; methodology, B.R.G. and R.T.; resources, M.S.; software, B.R.G.; supervision, R.T. and M.S.; visualization, B.R.G.; writing—original draft, B.R.G.; writing—review and editing, B.R.G., R.T. and M.S. All authors have read and agreed to the published version of the manuscript.

Funding: This work was supported in part by the Swedish Research Council.

Acknowledgments: We want to thank the anonymous reviewers for their insightful comments.

Conflicts of Interest: The authors declare no conflict of interest.

Appendix A. Proof of Proposition 1

Proof. We can easily prove this statement by finding $I(T;Y)$ is lower bounded by the $\gamma J_{CE}(p_{(X,Y)};\theta) + C$ where $\gamma < 0$ and C does not depend on T. This way maximizing such lower bound would be equivalent to minimizing $J_{CE}(p_{(X,Y)};\theta)$ and, moreover, it would imply maximizing $I(T;Y)$.

We can find such an expression as follows:

$$I(T;Y) = \mathbb{E}_{(y,t)\sim q_{Y|T}q_T}\left[\log\left(\frac{q_{Y|T=t}(y|t;\theta)}{p_Y(y)}\right)\right] = H(Y) + \mathbb{E}_{(y,t)\sim q_{Y|T}q_T}\left[\log(q_{Y|T=t}(y|t;\theta))\right] \tag{A1}$$

$$= H(Y) + \mathbb{E}_{t\sim q_T}\left[D_{KL}\left(q_{Y|T=t}||q_{\hat{Y}|T=t}\right)\right] + \mathbb{E}_{(y,t)\sim q_{Y|T}q_T}\left[\log(q_{\hat{Y}|T}(y|t;\theta))\right] \tag{A2}$$

$$\geq H(Y) + \mathbb{E}_{(x,y,t)\sim q_{Y|T}q_{T|X}p_X}\left[\log(q_{\hat{Y}|T=t}(y|t,\theta))\right] = H(Y) - \mathbb{E}_{(x,t)\sim q_{T|X}p_X}\left[\mathbb{C}(q_{Y|T=t}||q_{\hat{Y}|T=t})\right] \tag{A3}$$

$$= H(Y) - J_{CE}(p_{(X,Y)};\theta). \tag{A4}$$

Here, in Equation (A1) we just used the definition of the mutual information between two random variables, and then we decoupled it using the definition of the entropy of a variable (Note we used $H(\cdot)$ which is usually employed for discrete variables. However, in this setting $H(\cdot)$ could also refer to the differential entropy $h(\cdot)$ of a continuous random variable since we employed the general definition using the expectation). Then, in Equation (A2) we only multiplied and divided by $q_{\hat{Y}|T}$ inside the logarithm and employed the definition of the Kullback–Leibler divergence. Finally, in Equation (A3) we first used the fact the Kullback–Leibler divergence is always positive (Theorem 2.6.3 from Cover and Thomas [20]) and then the properties of the Markov chain $T \leftrightarrow X \leftrightarrow Y$.

Therefore, since $H(Y)$ does not depend on T and we have a negative multiplicative term on $J_{CE}(p_{(X,Y)};\theta)$ the proposition is proved. □

Appendix B. Alternative Proof of Theorem 1

Proof. We will proof all the enumerated statements sequentially, since the third one requires from the two first ones to be proved.

1. Proposition 2 states that the IB curve in the information plane follows the equation $I(T;Y) = I(X;T)$ if $I(X;T) \in [0, I(X;Y))$. Then, since $\beta = dI(T;Y)/dI(X;T)$ [1], we know $\beta = 1$ in all these points. Therefore, for $\beta = 1$ all points $(I(X;T), I(X;T))$ such that $I(X;T) \in [0, I(X;Y))$ are solutions of optimizing the IB Lagrangian.
2. Similarly, Proposition 2 states that the IB curve follows the equation $I(T;Y) = I(X;Y)$ if $I(X;T) \geq I(X;Y)$. Then, since $\beta = dI(T;Y)/dI(X;T)$ [1], we know $\beta = 0$ in all points such that $I(X;T) > I(X;Y)$. We cannot ensure it at $I(X;T) = I(X;Y)$ since $\beta = 1$ for $I(X;T) = \lim_{\epsilon \to 0^+}\{I(X;Y) - \epsilon\}$.
3. Finally, in order to prove the last statement we will first prove that if $\beta \in (0,1)$ achieves a solution, it is $(I(X;Y), I(X;Y))$. Then, we will prove that if the solution $(I(X;Y), I(X;Y))$ exists, this can be yield by any $\beta \in (0,1)$. Hence, the solution $(I(X;Y), I(X;Y))$ is achieved $\forall \beta \in (0,1)$ and it is the only solution achievable.

   (a) Since the IB curve is concave we know β is non-increasing in $I(X;T) \in \mathbb{R}^+$. We also know $\beta = 1$ at the points in the IB curve where $I(X;T) \leq \lim_{\epsilon \to 0^+}\{I(X;Y) - \epsilon\}$ and $\beta = 1$ at the points in the IB curve where $I(X;T) \geq \lim_{\epsilon \to 0^+}\{I(X;Y) + \epsilon\}$. Hence, if we achieve a solution with $\beta \in (0,1)$, this solution is $I(X;T) = I(T;Y) = I(X;Y)$.

   (b) We can upper bound the IB Lagrangian by

   $$\mathcal{L}_{\text{IB}}(T;\beta) = I(T;Y) - \beta I(X;T) \leq (1-\beta)I(T;Y) \leq (1-\beta)I(X;Y), \tag{A5}$$

   where the first and second inequalities use the DPI (Theorem 2.8.1 from Cover and Thomas [20]).

   Then, we can consider the point of the IB curve $(I(X;Y), I(X;Y))$. Since the function is concave a tangent line to $(I(X;Y), I(X;Y))$ exists such that all other points in the curve lie below this line. Let β be the slope of this curve (which we know it is from Tishby et al. [1]). Then,

   $$I(X;Y) - \beta I(X;Y) = (1-\beta)I(X;Y) \geq F_{\text{IB,max}}(r) - \beta r, \; \forall r \in [0, \infty). \tag{A6}$$

   As we see, by the upper bound on the IB Lagrangian from Equation (A5), if the point $(I(X;Y), I(X;Y))$ exists, any β can be the slope of the tangent line to $(I(X;Y), I(X;Y))$ that ensures concavity.

□

Appendix C. Proof of Theorem 2

Proof. We start the proof by remembering the optimization problem at hand (Definition 1):

$$F_{\text{IB,max}}(r) = \max_{T \in \Delta}\{I(T;Y)\} \text{ s.t. } I(X;T) \leq r \tag{A7}$$

We can modify the optimization problem by

$$\max_{T \in \Delta}\{I(T;Y)\} \text{ s.t. } u(I(X;T)) \leq u(r) \tag{A8}$$

iff u is a monotonically non-decreasing function since otherwise $u(I(X;T)) \leq u(r)$ would not hold necessarily. Now, let us assume $\exists T^* \in \Delta$ and β_u^* s.t. T^* maximizes $\mathcal{L}_{\text{IB},u}(T;\beta_u^*)$ over all $T \in \Delta$, and $I(X;T^*) \leq r$. Then, we can operate as follows:

$$\max_{\substack{T \in \Delta \\ u(I(X;T)) \le u(r)}} \{I(T;Y)\} = \max_{\substack{T \in \Delta \\ u(I(X;T)) \le u(r)}} \{I(T;Y) - \beta_u^*(u(I(X;T)) - u(r) + \xi)\} \quad \text{(A9)}$$

$$\le \max_{T \in \Delta} \{I(T;Y) - \beta_u^*(u(I(X;T)) - u(r) + \xi)\} \quad \text{(A10)}$$

$$= I(T^*;Y) - \beta_u^*(u(I(X;T^*) - u(r) + \xi) = I(T^*;Y). \quad \text{(A11)}$$

Here, the equality from Equation (A9) comes from the fact that since $I(X;T) \le r$, then $\exists \xi \ge 0$ s.t. $u(I(X;T)) - u(r) + \xi = 0$. Then, the inequality from Equation (A10) holds since we have expanded the optimization search space. Finally, in Equation (A11) we use that T^* maximizes $\mathcal{L}_{\text{IB},u}(T;\beta_u^*)$ and that $I(X;T^*) \le r$.

Now, we can exploit that $u(r)$ and ξ do not depend on T and drop them in the maximization in Equation (A10). We can then realize we are maximizing over $\mathcal{L}_{\text{IB},u}(T;\beta_u^*)$; i.e.,

$$\max_{\substack{T \in \Delta \\ u(I(X;T)) \le u(r)}} \{I(T;Y)\} \le \max_{T \in \Delta} \{I(T;Y) - \beta_u^*(u(I(X;T)) - u(r) + \xi)\} \quad \text{(A12)}$$

$$= \max_{T \in \Delta} \{I(T;Y) - \beta_u^*(I(X;T))\} = \max_{T \in \Delta} \{\mathcal{L}_{\text{IB},u}(T;\beta_u^*)\}. \quad \text{(A13)}$$

Therefore, since $I(T^*;Y)$ satisfies both the maximization with $T^* \in \Delta$ and the constraint $I(X;T^*) \le r$, maximizing $\mathcal{L}_{\text{IB},u}(T;\beta_u^*)$ obtains $F_{\text{IB,max}}(r)$.

Now, we know if such β_u^* exists, then the solution of the Lagrangian will be a solution for $F_{\text{IB,max}}(r)$. Then, if we consider Theorem 6 from the Appendix of Courcoubetis [22] and consider the maximization problem instead of the minimization problem, we know if both $I(T;Y)$ and $-u(I(X;T))$ are concave functions, then a set of Lagrange multipliers S_u^* exists with these conditions. We can make this consideration because f is concave if $-f$ is convex and $\max\{f\} = \min\{-f\}$. We know $I(T;Y)$ is a concave function of T for $T \in \Delta$ (Lemma 5 of Gilad-Bachrach et al. [19]) and $I(X;T)$ is convex w.r.t. T given p_X is fixed (Theorem 2.7.4 of Cover and Thomas [20]). Thus, if we want $-u(I(X;T))$ to be concave we need u to be a convex function.

Finally, we will look at the conditions of u so that for every point $(I(X;T), I(T;Y))$ in the IB curve, there exists a unique β_u^* s.t. $\mathcal{L}_{\text{IB,u}}(T;\beta_u^*)$ is maximized. That is, the conditions of u s.t. $|S_u^*| = 1$. For this purpose we will look at the solutions of the Lagrangian optimization:

$$\frac{d\mathcal{L}_{\text{IB},u}(T;\beta_u)}{dT} = \frac{d(I(T;Y) - \beta_u u(I(X;T)))}{dT} = \frac{dI(T;Y)}{dT} - \beta_u \frac{du(I(X;T))}{dI(X;T)} \frac{dI(X;T)}{dT} = 0 \quad \text{(A14)}$$

Now, if we integrate both sides of Equation (A14) over all $T \in \Delta$ we obtain

$$\beta_u = \frac{dI(T;Y)}{dI(X;T)} \left(\frac{du(I(X;T))}{dI(X;T)}\right)^{-1} = \frac{\beta}{u'(I(X;T))}, \quad \text{(A15)}$$

where β is the Lagrange multiplier from the IB Lagrangian [1] and $u'(I(X;T))$ is $\frac{du(I(X;T))}{dI(X;T)}$. Also, if we want to avoid indeterminations of β_u we need $u'(I(X;T))$ not to be 0. Since we already imposed u to be monotonically non-decreasing, we can solve this issue by strengthening this condition. That is, we will require u to be monotonically increasing.

We would like β_u to be continuous, this way there would be a unique β_u for each value of $I(X;T)$. We know β is a non-increasing function of $I(X;T)$ (Lemma 6 of Gilad-Bachrach et al. [19]). Hence, if we want β_u to be a strictly decreasing function of $I(X;T)$, we will require u' to be a strictly increasing function of $I(X;T)$. Therefore, we will require u to be a strictly convex function.

Thus, if u is a strictly convex and monotonically increasing function, for each point $(I(X;T), I(T;Y))$ in the IB curve s.t. $dI(T;Y)/dI(X;T) > 0$ there is a unique β_u for which maximizing $\mathcal{L}_{\mathrm{IB},u}(T;\beta_u)$ achieves this solution. □

Appendix D. Proof of Proposition 3

Proof. In Theorem 2 we showed how each point of the IB curve $(I(X;T), I(T;Y))$ can be found with a unique β_u maximizing $\mathcal{L}_{\mathrm{IB},u}(T;\beta_u)$. Therefore, since we also proved $\mathcal{L}_{\mathrm{IB},u}(T;\beta_u)$ is strictly concave w.r.t. T we can find the values of β_u that maximize the Lagrangian for fixed $I(X;T)$.

First, we look at the solutions of the Lagrangian maximization:

$$\frac{d\mathcal{L}_{\mathrm{IB},u}(T;\beta_u)}{dT} = \frac{d(f_{\mathrm{IB}}(I(X;T)) - \beta_u u(I(X;T)))}{dT} = \frac{df_{\mathrm{IB}}(I(X;T))}{dT} - \beta_u \frac{du(I(X;T))}{dI(X;T)} \frac{dI(X;T)}{dT} = 0. \tag{A16}$$

Then as before we can integrate at both sides for all $T \in \Delta$ and solve for β_u:

$$\beta_u = \frac{df_{\mathrm{IB}}(I(X;T))}{dI(X;T)} \frac{1}{u'(I(X;T))}. \tag{A17}$$

Moreover, since u is a strictly convex function it's derivative u' is strictly increasing. Hence, u' is an invertible function (since a strictly increasing function is bijective and a function is invertible iff it is bijective by definition). Now, if we consider $\beta_u > 0$ to be known and $I(X;T)$ to be the unknown we can solve for $I(X;T)$ and get:

$$I(X;T) = (u')^{-1}\left(\frac{df_{\mathrm{IB}}(I(X;T))}{dI(X;T)} \frac{1}{\beta_u}\right). \tag{A18}$$

Note we require β_u not to be 0 so the mapping is defined. □

Appendix E. Proof of Corollary 2

Proof. We will start the proof by proving the following useful Lemma.

Lemma A1. *Let $\mathcal{L}_{\mathrm{IB},u}(T;\beta_u)$ be a convex IB Lagrangian, then $\sup_{T\in\Delta}\{\mathcal{L}_{\mathrm{IB},u}(T;0)\} = I(X;Y)$.*

Proof. Since $\mathcal{L}_{\mathrm{IB},u}(T;0) = I(T;Y)$, maximizing this Lagrangian is directly maximizing $I(T;Y)$. We know $I(T;Y)$ is a concave function of T for $T \in \Delta$ (Theorem 2.7.4 from Cover and Thomas [20]); hence it has a supremum. We also know $I(T;Y) \leq I(X;Y)$. Moreover, we know $I(X;Y)$ can be achieved if, for example, Y is a deterministic function of T (since then the Markov Chain $X \leftrightarrow T \leftrightarrow Y$ is formed). Thus, $\sup_{T\in\Delta}\{\mathcal{L}_{\mathrm{IB},u}(T;0)\} = I(X;Y)$. □

For $\beta_u = 0$ we know maximizing $\mathcal{L}_{\mathrm{IB},u}(T;0)$ we can obtain the point in the IB curve $(r_{\max}, I_{\max})$ (Lemma A1). Moreover, we know that for every point $(I(X;T), f_{\mathrm{IB}}(I(X;T)))$ such that $df_{\mathrm{IB}}(I(X;T))/dI(X;T) > 0$, $\exists!\beta_u$ s.t. $\max\{\mathcal{L}_{\mathrm{IB},u}(T;\beta_u)\}$ achieves that point (Theorem 2). Thus, $\exists!\beta_{u,\min}$ s.t. $\lim_{r \to r_{\max}^-}(r, f_{\mathrm{IB}}(r))$ is achieved. From Proposition 3 we know this $\beta_{u,\min}$ is given by

$$\beta_{u,\min} = \lim_{r \to r_{\max}^-}\left\{\frac{f'_{\mathrm{IB}}(r)}{u'(r)}\right\}. \tag{A19}$$

Since we know $f_{\mathrm{IB}}(I(X;T))$ is a concave non-decreasing function in $(0, r_{\max})$ (Lemma 5 of Gilad-Bachrach et al. [19]) we know it is continuous in this interval. In addition we know β_u is strictly decreasing w.r.t. $I(X;T)$ (Theorem 2). Furthermore, by definition of $r_{\max}$ and knowing $I(T;Y) \leq I(X;Y)$ we know $f'_{\mathrm{IB}}(r) = 0$, $\forall r > r_{\max}$. Therefore, we cannot ensure the exploration of the IB curve for β'_u s.t. $0 < \beta'_u < \beta_{u,\min}$.

Then, since u is a strictly increasing function in $(0, r_{\max})$, u' is positive in that interval. Hence, taking into account β_u is strictly decreasing we can find a maximum β_u when $I(X;T)$ approaches to 0. That is,

$$\beta_{u,\max} = \lim_{r \to 0^+} \left\{ \frac{f'_{IB}(r)}{u'(r)} \right\}, \qquad (A20)$$

□

Appendix F. Proof of Corollary 3

Proof. If we use Corollary 2, it is straightforward to see that $\beta_u \subseteq [L_-, L_+]$ if $\beta_{u,\min} \geq L_-$ and $\beta_{u,\max} \leq L_+$ for all IB curves f_{IB} and functions u. Therefore, we look at a domain bound dependent on the function choice. That is, if we can find $\beta_{\min} \leq f'_{IB}(r)$ and $\beta_{\max} \geq f'_{IB}(r)$ for all IB curves and all values of r, then

$$B_u \subseteq \left[\frac{\beta_{\min}}{\lim_{r \to r_{\max}^-} \{u'(r)\}}, \frac{\beta_{\max}}{\lim_{r \to 0^+} \{u'(r)\}} \right]. \qquad (A21)$$

The region for all possible IB curves regardless of the relationship between X and Y is depicted in Figure A1. The hard limits are imposed by the DPI (Theorem 2.8.1 from Cover and Thomas [20]) and the fact that the mutual information is non-negative (Corollary with Equation 2.90 for discrete and first Corollary of Theorem 8.6.1 for continuous random variables from Cover and Thomas [20]). Hence, a minimum and maximum values of f'_{IB} are given by the minimum and maximum values of the slope of the Pareto frontier. Which means

$$B_u \subseteq \left[0, \frac{1}{\lim_{r \to 0^+} \{u'(r)\}} \right]. \qquad (A22)$$

Note $0/(\lim_{r \to r_{\max}^-} \{u'(r)\}) = 0$ since u is monotonically increasing and, thus, u' will never be 0.

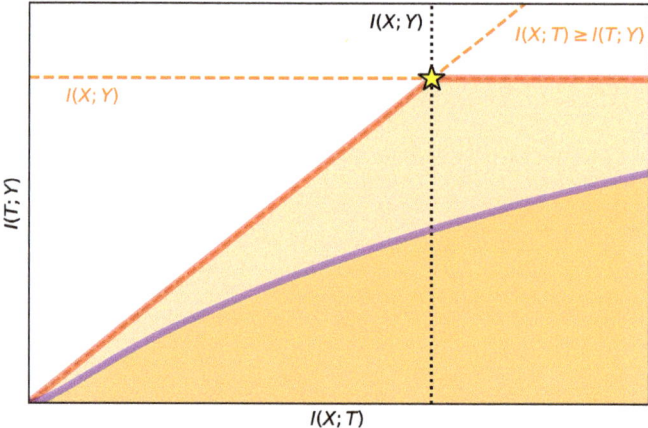

Figure A1. Graphical representation of the IB curve in the information plane. Dashed lines in orange represent tight bounds confining the region (in light orange) of possible IB curves (delimited by the red line, also known as the Pareto frontier). Black dotted lines are informative values. In blue we show an example of a possible IB curve confining a region (in darker orange) of an IB curve that does not achieve the Pareto frontier. Finally, the yellow star represents the point where the representation keeps the same information about the input and the output.

Then, we can tighten the bound using the results from Wu et al. [27], where, in Theorem 2, they showed the slope of the Pareto frontier could be bounded in the origin by $f'_{IB} \leq (\inf_{\Omega_x \subset \mathcal{X}} \{\beta_0(\Omega_x)\})^{-1}$. Finally, we know that in deterministic classification tasks $\inf_{\Omega_x \subset \mathcal{X}} \{\beta_0(\Omega_x)\} = 1$, which aligns with Kolchinsky et al. [21] and what we can observe from Figure A1. Therefore,

$$B_u \subseteq \left[0, \frac{(\inf_{\Omega_x \subset \mathcal{X}}\{\beta_0(\Omega_x)\})^{-1}}{\lim_{r \to 0^+}\{u'(r)\}}\right] \subseteq \left[0, \frac{1}{\lim_{r \to 0^+}\{u'(r)\}}\right]. \quad (A23)$$

□

Appendix G. Other Lagrangian Families

We can use the same ideas we used for the convex IB Lagrangian to formulate new families of Lagrangians that allow the exploration of the IB curve. For that, we will use the duality of the IB curve (Lemma 10 of [19]). That is:

Definition A1 (IB Dual Functional). *Let X and Y be statistically dependent variables. Let also Δ be the set of random variables T obeying the Markov condition $Y \leftrightarrow X \leftrightarrow T$. Then the IB dual functional is*

$$F_{\mathrm{IB,min}}(i) = \min_{T \in \Delta} \{I(X;T)\} \text{ s.t. } I(T;Y) \geq i, \ \forall i \in [0, I(X;Y)). \quad (A24)$$

Theorem A1 (IB Curve Duality). *Let the IB curve be defined by the solutions of $F_{\mathrm{IB,max}}(r)$ for varying $r \in [0, \infty)$. Then,*

$$\forall r \exists i \text{ s.t. } (r, F_{\mathrm{IB,max}}(r)) = (F_{\mathrm{IB,min}}(i), i) \quad (A25)$$

and

$$\forall i \exists r \text{ s.t. } (F_{\mathrm{IB,min}}(i), i) = (r, F_{\mathrm{IB,max}}(r)). \quad (A26)$$

From this definition, it follows that minimizing the dual IB Lagrangian, $\mathcal{L}_{\mathrm{IB,dual}}(T; \beta_{\mathrm{dual}}) = I(X;T) - \beta_{\mathrm{dual}}I(T;Y)$, for $\beta_{\mathrm{dual}} = \beta^{-1}$ is equivalent to maximizing the IB Lagrangian. In fact, the original Lagrangian for solving the problem was defined this way [1]. We decided to use the maximization version because the domain of useful β is bounded while it is not for β_{dual}.

Following the same reasoning as we did in the proof of Theorem 2, we can ensure the IB curve can be explored if:

1. We minimize the concave IB Lagrangian $\mathcal{L}_{\mathrm{IB},v}(T; \beta_v) = I(X;T) - \beta_v v(I(T;Y))$.
2. We maximize the dual concave IB Lagrangian $\mathcal{L}_{\mathrm{IB},v,\mathrm{dual}}(T; \beta_{v,\mathrm{dual}}) = v(I(T;Y)) - \beta_{v,\mathrm{dual}}I(X;T)$.
3. We minimize the dual convex IB Lagrangian $\mathcal{L}_{\mathrm{IB},u,\mathrm{dual}}(T; \beta_{u,\mathrm{dual}}) = u(I(X;T)) - \beta_{u,\mathrm{dual}}I(T;Y)$.

Here, u is a monotonically increasing strictly convex function, v is a monotonically increasing strictly concave function, and $\beta_v, \beta_{v,\mathrm{dual}}, \beta_{u,\mathrm{dual}}$ are the Lagrange multipliers of the families of Lagrangians defined above.

In a similar manner, one could obtain relationships between the Lagrange multipliers of the IB Lagrangian and the convex IB Lagrangian with these Lagrangian families. For instance, the convex IB Lagrangian $\mathcal{L}_{\mathrm{IB},u}(T; \beta_u)$ is related with the concave IB Lagrangian $\mathcal{L}_{\mathrm{IB},v}(T; \beta_v)$ as defined by Propositon A1.

Proposition A1 (Relationship between the convex and concave IB Lagrangians). *Consider the convex and concave IB Lagrangians $\mathcal{L}_{\mathrm{IB},u}(T; \beta_u), \mathcal{L}_{\mathrm{IB},v}(T; \beta_v)$. Let the IB curve defined as in Definition 2 be f_{IB}. Then, if we fix the functions u and v we can obtain the same point in the IB curve $(r, f_{\mathrm{IB}}(r))$ with both Lagrangians when*

$$\beta_v^{-1} = f'_{\mathrm{IB}}(r) v'\left(f_{\mathrm{IB}}\left((u')^{-1}\left(\frac{f'_{\mathrm{IB}}(r)}{\beta_u}\right)\right)\right), \quad (A27)$$

or equivalently,

$$\beta_u^{-1} = \frac{1}{f'_{\mathrm{IB}}(r)} u'\left(f_{\mathrm{IB}}^{-1}\left((v')^{-1}\left(\frac{\beta_v^{-1}}{f'_{\mathrm{IB}}(r)}\right)\right)\right). \quad (A28)$$

Proof. If we proceed like we did in the proof of Proposition 3 we can find the mapping between $I(X;T)$ and β_u and between $I(T;Y)$ and β_v. That is,

$$I(X;T) = (u')^{-1}\left(\frac{df_{\text{IB}}(I(X;T))}{dI(X;T)}\frac{1}{\beta_u}\right) \text{ and } I(T;Y) = (v')^{-1}\left(\left(\frac{df_{\text{IB}}(I(X;T))}{dI(X;T)}\right)^{-1}\frac{1}{\beta_v}\right). \quad \text{(A29)}$$

Then, if we recall that $I(T;Y) = f_{\text{IB}}(I(X;T))$, we can directly obtain that

$$f_{\text{IB}}\left((u')^{-1}\left(\frac{df_{\text{IB}}(I(X;T))}{dI(X;T)}\frac{1}{\beta_u}\right)\right) = (v')^{-1}\left(\left(\frac{df_{\text{IB}}(I(X;T))}{dI(X;T)}\right)^{-1}\frac{1}{\beta_v}\right). \quad \text{(A30)}$$

Then, if we solve Equation (A30) with a fixed point $(I(X;T) = r, I(T;Y) = f_{\text{IB}}(r))$ for β_v we obtain Equation (A27), and if we solve it for β_u we obtain Equation (A28). □

Also, one could find a range of values for these Lagrangians to allow for the IB curve exploration and define a bijective mapping between their Lagrange multipliers and the IB curve. However, (i) as mentioned in Section 2.2, $I(T;Y)$ is particularly interesting to maximize without transformations because of its meaning. Moreover, (ii) like β_{dual}, the domain of useful β_v and $\beta_{u,\text{dual}}$ is not upper bounded. These two reasons make these other Lagrangians less preferable. We only include them here for completeness. Nonetheless, we encourage the curious reader to explore these families of Lagrangians too. For example, a possible interesting research would be investigating if some particularization of the concave IB Lagrangian suffers from an issue like value convergence that can be exploited for approximately obtaining any predictability level $I(T;Y) = i^*$ for many values of β_v.

Appendix H. Experimental Setup Details and Further Experiments

In order to generate empirical support for our claims, we performed several experiments on different datasets with different neural network architectures and different ways of calculating the information bottleneck.

Appendix H.1. Information Bottleneck Calculations

The information bottleneck is calculated modifying either the nonlinear-IB [26]. This method of calculating the information bottleneck is a neural network that minimizes the cross-entropy while also minimizing an upper bound estimate of the mutual information $I_\theta \approx I(X;T)$. The nonlinear-IB relies on a kernel-based estimate of this mutual information [40]. We modify this calculation method by applying the function u to the $I(X;T)$ estimate.

For the nonlinear-IB calculations, we estimated the gradients of both $I_\theta(X;T)$ and the cross-entropy with the same mini-batch. Moreover, we did not learn the covariance of the mixture of Gaussians used for the kernel density estimation of $I_\theta(X;T)$ and we set it to $(\exp(-1))^2$.

In both methods, and for all the experiments, we assumed a Gaussian stochastic encoder $T = f_{\text{enc}}(X;\theta) + W$ with $p_W = \mathcal{N}(0, I_d)$, where d are the number of dimensions of the representations. We trained the neural networks with the Adam optimization algorithm [46] with a learning rate of 10^{-4} and a 0.6 decay rate every 10 epochs. We used a batch size of 128 samples and all the weights were initialized according to the method described by Glorot and Bengio [47] using a Gaussian distribution.

Then, we used the DBSCAN algorithm [44,45] for clustering. Particularly, we used the scikit-learn [48] implementation with $\epsilon = 0.3$ and `min_samples` = 50.

The reader can find the PyTorch [30] implementation in the following link: https://github.com/burklight/convex-IB-Lagrangian-PyTorch.

Appendix H.2. The Experiments

We performed experiments in four different datasets:

- A Classification Task on the MNIST Dataset [28] (Figures 1, 2, and A2–A4 and top row from Figure 3). This dataset contains 60,000 training samples and 10,000 testing samples of hand-written digits. The samples are 28x28 pixels and are labeled from 0 to 9; i.e., $\mathcal{X} = \mathbb{R}^{784}$ and $\mathcal{Y} = \{0, 1, ..., 9\}$. The data is pre-processed so that the input has zero mean and unit variance. This is a deterministic setting, hence the experiment is designed to showcase how the convex IB Lagrangians allow us to explore the IB curve in a setting where the normal IB Lagrangian cannot and the relationship between the performance plateaus and the clusterization phenomena. Furthermore, it intends to showcase the behavior of the power and exponential Lagrangians with different parameters of α and η. Finally, it wants to demonstrate how the value convergence can be employed to approximately obtain a specific compression value. In this experiment, the encoder f_{enc} is a three fully-connected layer encoder with 800 ReLU units on the first two layers and two linear units on the last layer ($T \in \mathbb{R}^2$), and the decoder f_{dec} is a fully-connected 800 ReLU unit layers followed by an output layer with 10 softmax units. The convex IB Lagrangian was calculated using the nonlinear-IB.

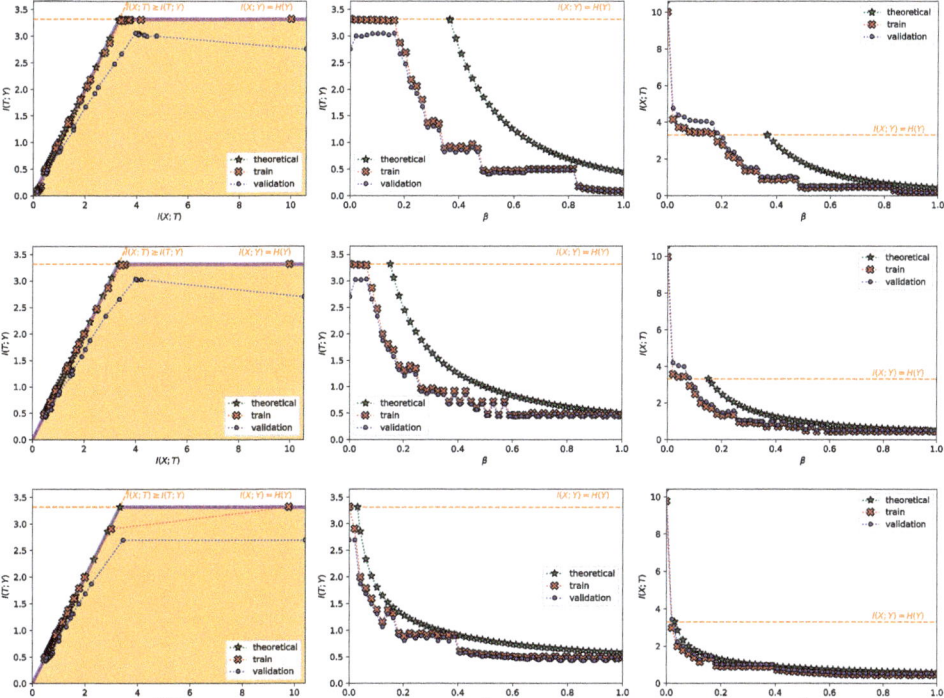

Figure A2. Results for the power IB Lagrangian in the MNIST dataset with $\alpha = \{0.5, 1, 2\}$, from top to bottom. In each row, from left to right it is shown (i) the information plane, where the region of possible solutions of the IB problem is shadowed in light orange and the information-theoretic limits are the dashed orange line; (ii) $I(T; Y)$ as a function of β_u; and (iii) the compression $I(X; T)$ as a function of β_u. In all plots, the red crosses joined by a dotted line represent the values computed with the training set, the blue dots the values computed with the validation set and the green stars the theoretical values computed as dictated by Proposition 3. Moreover, in all plots, it is indicated $I(X; Y) = H(Y) = \log_2(10)$ in a dashed, orange line. All values are shown in bits.

In Figure A2 we show how the IB curve can be explored with different values of α for the power IB Lagrangian and in Figure A3 for different values of η and the exponential IB Lagrangian.

Finally, in Figure A4 we show the clusterization for the same values of α and η as in Figure A2 and A3. In this way the connection between the performance discontinuities and the clusterization is more evident. Furthermore, we can also observe how the exponential IB Lagrangian maintains better the theoretical performance than the power IB Lagrangian (see Appendix I for an explanation of why).

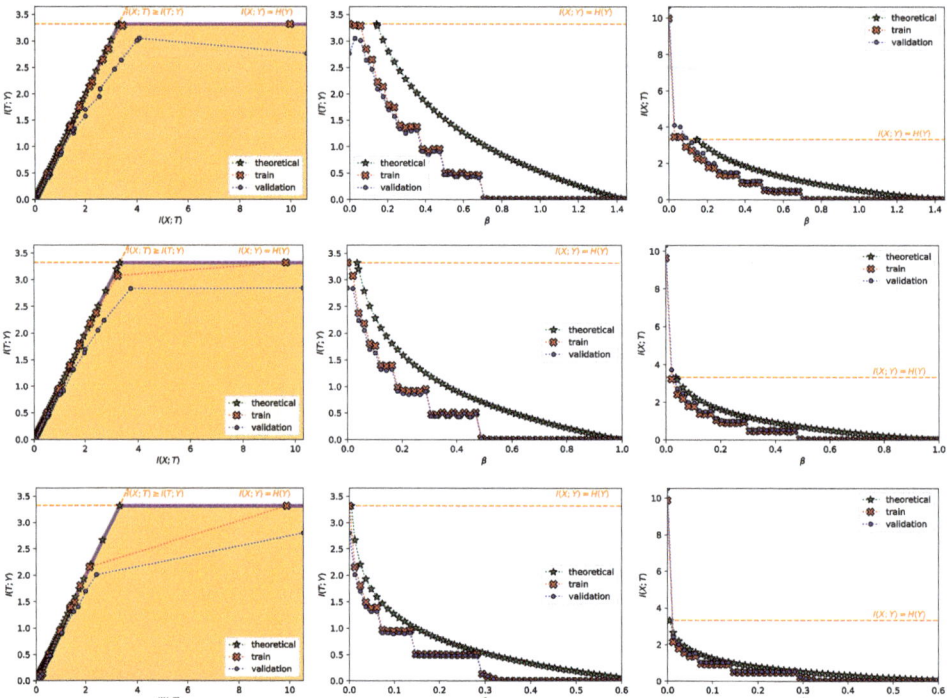

Figure A3. Results for the exponential IB Lagrangian in the MNIST dataset with $\eta = \{\log(2), 1, 1.5\}$, from top to bottom. In each row, from left to right it is shown (i) the information plane, where the region of possible solutions of the IB problem is shadowed in light orange and the information-theoretic limits are the dashed orange line; (ii) $I(T;Y)$ as a function of β_u; and (iii) the compression $I(X;T)$ as a function of β_u. In all plots, the red crosses joined by a dotted line represent the values computed with the training set, the blue dots the values computed with the validation set and the gren stars the theoretical values computed as dictated by Proposition 3. Moreover, in all plots, it is indicated $I(X;Y) = H(Y) = \log_2(10)$ in a dashed, orange line. All values are shown in bits.

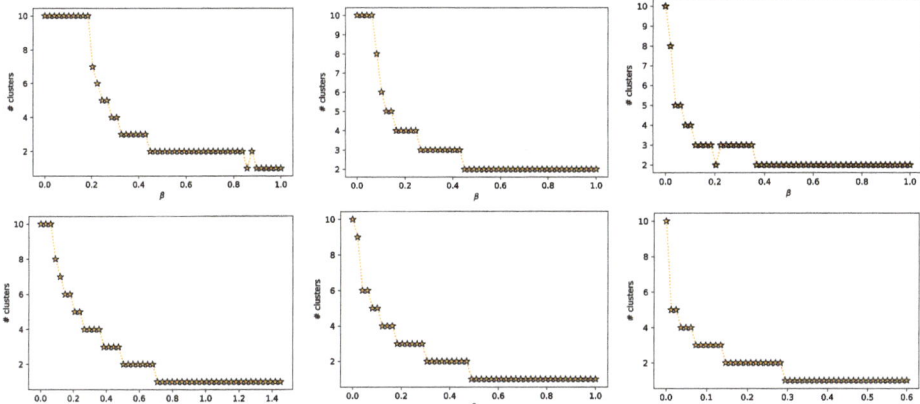

Figure A4. Depiction of the clusterization behavior of the bottleneck variable in the MNIST dataset. In the first row, from left to right, the power IB Lagrangian with different values of $\alpha = \{0.5, 1, 2\}$. In the second row, from left to right, the exponential IB Lagrangian with different values of $\eta = \{\log(2), 1, 1.5\}$.

- A Classification Task on the Fashion-MNIST Dataset [49] (Figure A5). As MNSIT, this dataset contains 60,000 training and 10,000 testing samples of 28x28 pixel images labeled from 0 to 9 and constitutes a deterministic setting. The difference is that this dataset contains fashion products instead of hand-written digits and it represents a harder classification task [49]. The data is also pre-processed so that the input has zero mean and unit variance. For this experiment, the encoder f_{enc} is composed of a two-layer convolutional neural network (CNN) with 32 filters on the first layer and 128 filters on the second with kernels of size 5 and stride 2. This CNN is followed by two fully-connected layers of 128 linear units ($T \in \mathbb{R}^{128}$). After the first convolution and the first fully-connected layer, a ReLU activation is employed. The decoder f_{dec} is a fully-connected 128 ReLU unit layer followed by an output layer with 10 softmax units. The convex IB Lagrangian was calculated using the nonlinear-IB. Therefore, this experiment intends to showcase how the convex IB Lagrangian can explore the IB curve for different neural network architectures and harder datasets.

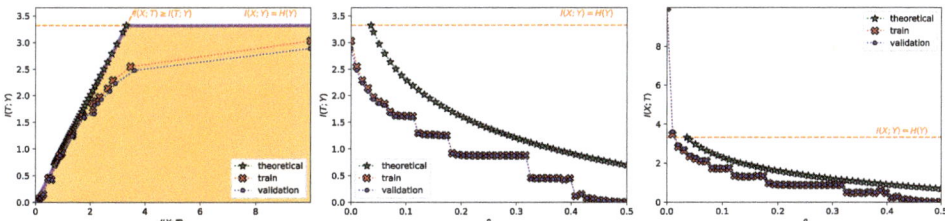

Figure A5. Results for the exponential IB Lagrangian in the Fashion MNIST dataset with $\eta = 1$. From left to right it is shown (i) the information plane, where the region of possible solutions of the IB problem is shadowed in light orange and the information-theoretic limits are the dashed orange line; (ii) $I(T;Y)$ as a function of β_u; and (iii) the compression $I(X;T)$ as a function of β_u. In all plots, the red crosses joined by a dotted line represent the values computed with the training set and the blue dots the values computed with the validation set. Moreover, in all plots, it is indicated $I(X;Y) = H(Y) = \log_2(10)$. All values are shown in bits.

- A Regression Task on the California Housing Dataset [50] (Figure A6). This dataset contains 20,640 samples of 8 real number input variables like the longitude and latitude of the house (i.e.,

$X \in \mathbb{R}^8$) and a task output real variable representing the price of the house (i.e., $Y \in \mathbb{R}$). We used the log-transformed house price as the target variable and dropped the 992 samples in which the house price was equal or greater than \$500,000 so that the output distribution was closer to a Gaussian as they did in [26]. The input variables were processed so that they had zero mean and unit variance and we randomly split the samples into a 70% training and 30% test dataset. As in [40], for regression tasks we approximate $H(Y)$ with the entropy of a Gaussian with variance $\mathrm{Var}(Y)$ and $H(Y|T)$ with the entropy of a Gaussian with variance equal to the mean-squared error (MSE). This leads to the estimate $I(T;Y) \approx 0.5\log(\mathrm{Var}(Y)/\mathrm{MSE})$. The encoder f_{enc} is a three fully-connected layer encoder with 128 ReLU units on the first two layers and 2 linear units on the last layer ($T \in \mathbb{R}^2$), and the decoder f_{dec} is a fully-connected 128 ReLU unit layers followed by an output layer with 1 linear unit. The convex IB Lagrangian was calculated using the nonlinear-IB. Hence, this experiment was designed to showcase the convex IB Lagrangian can explore the IB curve in stochastic scenarios for regression tasks.

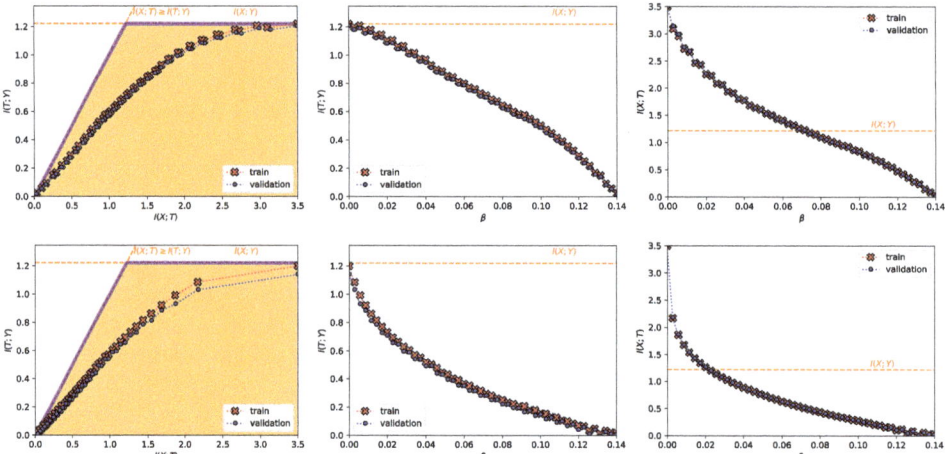

Figure A6. The top row shows the results for the normal IB Lagrangian, and the bottom row for the exponential IB Lagrangian with $\eta = 1$, both in the California housing dataset. In each row, from left to right it is shown (i) the information plane, where the region of possible solutions of the IB problem is shadowed in light orange and the information-theoretic limits are the dashed orange line; (ii) $I(T;Y)$ as a function of β_u; and (iii) the compression $I(X;T)$ as a function of β_u. In all plots, the red crosses joined by a dotted line represent the values computed with the training set and the blue dots the values computed with the validation set. Moreover, in all plots, it is indicated $I(X;Y)$ as the empirical value obtained maximizing $I(T;Y)$ without compression limitations as in [26]. All values are shown in bits.

- A Classification Task on the TREC-6 Dataset [29] (Figure A7 and bottom row from Figure 3). This dataset is the six-class version of the TREC [51] dataset. It contains 5452 training and 500 test samples of text questions. Each question is labeled within six different semantic categories based on what the answer is; namely: Abbreviation, description and abstract concepts, entities, human beings, locations, and numeric values. This dataset does not constitute a deterministic setting since there are examples that could belong to more than one class and there are examples which are wrongly labeled (e.g., "What is a fear of parasites?" could belong both to the description and abstract concept category, however it is labeled into the entity category), and hence $H(Y|X) > 0$. Following Ben Trevett's tutorial on Sentiment Analysis [52] the encoder f_{enc} is composed by a 6 billion token pre-trained 100-dimensional Glove word embedding [53], followed by a concatenation of three convolutions with kernel sizes 2–4 respectively, and finalized with a fully-connected 128 linear unit layer ($T \in \mathbb{R}^{128}$). The decoder f_{dec} is a single fully-connected

6 softmax unit layer. The convex IB Lagrangian was calculated using the nonlinear-IB. Thus, this experiment intends to show an example where the classification task does not convey a deterministic scenario, that the convex IB Lagrangian can recover the IB curve in complex stochastic tasks with complex neural network architectures and that the value convergence can be employed to obtain a specific compression value even in stochastic settings where the IB curve is unknown.

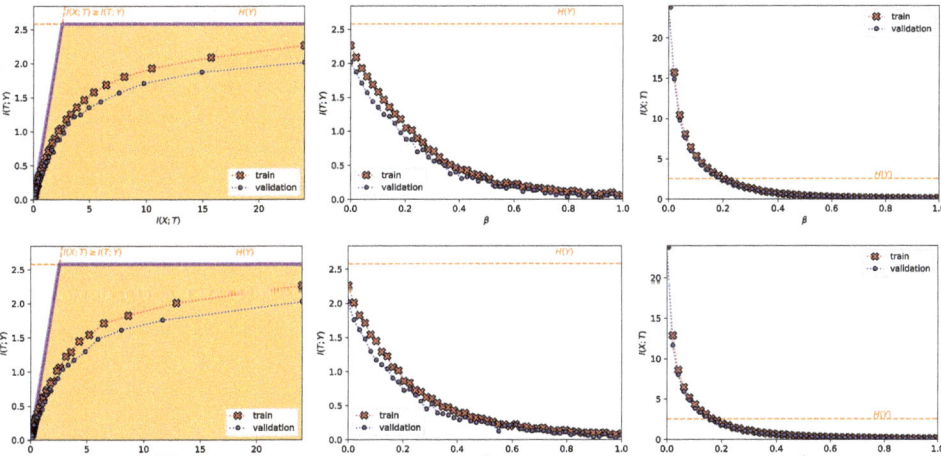

Figure A7. The top row shows the results for the normal IB Lagrangian, and the bottom row for the power IB Lagrangian with $\alpha = 0.1$, both in the TREC-6 dataset. In each row, from left to right it is shown (i) the information plane, where the region of possible solutions of the IB problem is shadowed in light orange and the information-theoretic limits are the dashed orange line; (ii) $I(T;Y)$ as a function of β_u; and (iii) the compression $I(X;T)$ as a function of β_u. In all plots, the red crosses joined by a dotted line represent the values computed with the training set and the blue dots the values computed with the validation set. Moreover, in all plots, it is indicated $H(Y) = \log_2(6)$. All values are shown in bits.

Appendix I. Guidelines for Selecting A Proper Function in the Convex IB Lagrangian

When choosing the right u function, it is important to find the right balance between avoiding value convergence and aiming for strong convexity. Practically, this balance is found by looking at how much faster u grows w.r.t. the identity function.

When the aim is not to draw the IB curve but to find a specific level of performance, we can exploit the value convergence phenomenon in order to design a stable performance targeted u function.

Appendix I.1. Avoiding Value Convergence

In order to explain this issue we are going to use the example of classification on MNIST [28], where $I(X;Y) = H(Y) = \log_2(10)$, and again the power and exponential IB Lagrangians.

If we use Proposition 3 on both Lagrangians we obtain the bijective mapping between their Lagrange multipliers and a certain level of compression in the classification setting:

1. Power IB Lagrangian: $\beta_{\text{pow}} = ((1+\alpha)I(X;T)^\alpha)^{-1}$ and $I(X;T) = ((1+\alpha)\beta_{\text{pow}})^{-\frac{1}{\alpha}}$.
2. Exponential IB Lagrangian: $\beta_{\text{exp}} = (\eta \exp(\eta I(X;T)))^{-1}$ and $I(X;T) = -\log(\eta \beta_{\text{exp}})/\eta$.

Hence, we can simply plot the curves of $I(X;T)$ vs. β_u for different hyperparameters α and η (see Figure A8). In this way, we can observe how increasing the growth of the function (e.g., increasing α or

η in this case) too much provokes that many different values of β_u converge to very similar values of $I(X;T)$. This is an issue both for drawing the curve (for obvious reasons) and for aiming for a specific performance level. Due to the nature of the estimation of the IB Lagrangian, the theoretical and practical value of β_u that yields a specific $I(X;T)$ may vary slightly (see Figure 1). Then if we select a function with too high growth, a small change in β_u can result in a big change in the performance obtained.

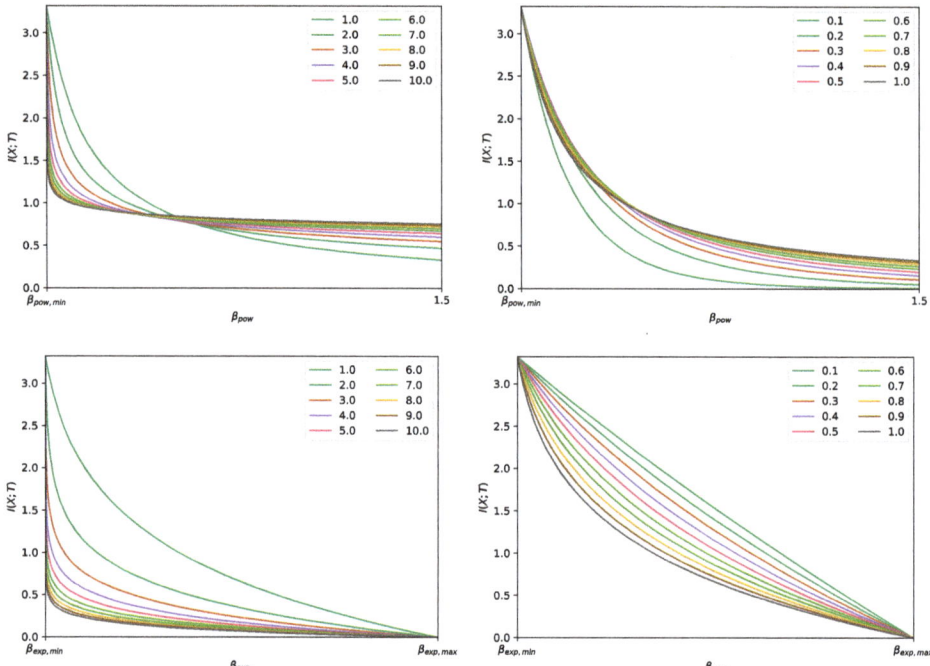

Figure A8. Theoretical bijection between $I(X;T)$ and different α from $\beta_{u,\min}$ to 1.5 in the power IB Lagrangian (**top**), and different η in the domain B_u in the exponential IB Lagrangian (**bottom**).

Appendix I.2. Aiming for Strong Convexity

Definition A2 (μ-**Strong Convexity**)**.** *If a function $f(r)$ is twice continuous differentiable and its domain is confined in the real line, then it is μ-strong convex if $f''(r) \geq \mu \geq 0 \, \forall r$.*

Experimentally, we observed when the growth of our function $u(r)$ is small in the domain of interest $r > 0$ the convex IB Lagrangian does not perform well (see first row of Figures A2 and A3). Later we realized that this was closely related to the strength of the convexity of our function.

In Theorem 2 we imposed the function u to be strictly convex to enforce having a unique β_u for each value of $I(X;T)$. Hence, since in practice we are not exactly computing the Lagrangian but an estimation of it (e.g., with the nonlinear IB [26]) we require strong convexity in order to be able to explore the IB curve.

We now look at the second derivative of the power and exponential function: $u''(r) = (1+\alpha)\alpha r^{\alpha-1}$ and $u''(r) = \eta^2 \exp(\eta r)$ respectivelly. Here we see how both functions are inherently 0-strong convex for $r > 0$ and $\alpha, \eta > 0$. However, values of $\alpha < 1$ and $\eta < 1$ could lead to low μ-strong convexity in certain domains of r. Particularly, the case of $\alpha < 1$ is dangerous because the function approaches 0-strong convexity as r increases, so the power IB Lagrangian performs poorly when low α are used to find high performances.

Appendix I.3. Exploiting Value Convergence

When the aim is not to draw or explore the IB curve, but to obtain a specific level of performance, the power of exponential IB Lagrangians aforementioned might not be the best choice due to the problems with value convergence or non-strong convexity. However, we can exploit the former in order to design a performance targeted u function.

For instance, if we look at Figure A8 we can see how a modification of the exponential IB Lagrangian could result in such a function. More precisely, a shifted exponential $u(r) = \exp(\eta(r - r^*))$, with $\eta > 0$ sufficiently large, converges to the compression level r^*. We can see this more clearly if we consider the shifted exponential IB Lagrangian $\mathcal{L}_{\text{IB,sh-exp}}(T; \beta_{\text{sh-exp}}, \eta, r^*) = I(T;Y) - \beta_{\text{sh-exp}} \exp(\eta(I(X;T) - r^*))$, since then the application of Proposition 3 results on $I(X;T) = -\log(\eta \beta_{\text{sh-exp}} / f'_{\text{IB}}(I(X;T)))/\eta + r^*$, where $f'_{\text{IB}}(I(X;T))$ is the derivative of f_{IB} evaluated at $I(X;T)$. We know $f'_{\text{IB}} = 1$ in deterministic scenarios (Theorem 2) and that $f'_{\text{IB}} < 1$ otherwise (see, e.g., [27]). Then, for large enough η, $I(X;T) \approx r^*$ regardless of the value of f'_{IB}.

For instance, if we consider a deterministic scenario like the MNIST dataset [28] with $I(X;Y) = H(Y) = \log_2(10)$, for $\eta = 200$ and $r^* = 2$ the range of the Lagrange multipliers that allow the exploration of the IB curve, according to Corollary 2, is $\beta_{\text{sh-exp}} \in [7.54 \times 10^{-178}, 2.61 \times 10^{171}]$. Furthermore, $I(X;T)$ is close to 2 for many values of $\beta_{\text{sh-exp}}$. For instance, $I(X;T) = 1.974$ for $\beta_{\text{sh-exp}} = 1$ and $I(X;T) = 1.963$ for $\beta_{\text{sh-exp}} = 8$. This ensures a stability in the performance level obtained so that small changes in the choice of $\beta_{\text{sh-exp}}$ do not result in significant changes on the performance (e.g., see top row from Figure 4).

If we now consider a stochastic scenario like the TREC-6 dataset [29] with $H(Y) = \log_2(6)$, for $\eta = 200$ and $r^* = 16$ the range of the Lagrange multipliers that allow the IB curve, according to Corollary 3, is $\beta_{\text{sh-exp}} \in [0, 2.76(\inf_{\Omega_x \subset \mathcal{X}} \{\beta_0(\Omega_x)\})^{-1} \times 10^{1287}]$, where β_0 and Ω_x are defined as in [27]. Then, unless $(\inf_{\Omega_x \subset \mathcal{X}} \{\beta_0(\Omega_x)\})^{-1}$ is of the order of 10^{-1287}, the range of possible betas is wide. Moreover, $I(X;T)$ is close to 16 for many values of $\beta_{\text{sh-exp}}$. For example, $I(X;T) = 15.939$ if $f'_{\text{IB}} = 0.001$ at that point and $I(X;T) = 15.973$ if $f'_{\text{IB}} = 0.9$ for $\beta_{\text{sh-exp}} = 1$; and $I(X;T) = 15.929$ if $f'_{\text{IB}} = 0.001$ at that point and $I(X;T) = 15.963$ if $f'_{\text{IB}} = 0.9$ for $\beta_{\text{sh-exp}} = 8$. Hence, as in the deterministic scenario, the performance level obtained is stable with changes in the choice of $\beta_{\text{sh-exp}}$ (e.g., see bottom row from Figure 4).

References

1. Tishby, N.; Pereira, F.C.; Bialek, W. The information bottleneck method. *arXiv* **2000**, arXiv:physics/0004057.
2. Alemi, A.A.; Fischer, I.; Dillon, J.V.; Murphy, K. Deep variational information bottleneck. *arXiv* **2016**, arXiv:1612.00410.
3. Peng, X.B.; Kanazawa, A.; Toyer, S.; Abbeel, P.; Levine, S. Variational Discriminator Bottleneck: Improving Imitation Learning, Inverse RL, and GANs by Constraining Information Flow. In Proceeding of the International Conference on Learning Representations (ICLR), New Orleans, LA, USA, 6–9 May 2019.
4. Achille, A.; Soatto, S. Information dropout: Learning optimal representations through noisy computation. *IEEE Trans. Pattern Anal. Mach. Intell.* **2018**, *40*, 2897–2905. [CrossRef] [PubMed]
5. Slonim, N.; Tishby, N. Document clustering using word clusters via the information bottleneck method. In Proceedings of the 23rd annual international ACM SIGIR conference on Research and development in information retrieval, Athens, Greece, 24–28 July 2000.
6. Slonim, N.; Tishby, N. Agglomerative information bottleneck. In *Advances in Neural Information Processing Systems*; MIT Press: Cambridge, MA, USA, 2000.
7. Slonim, N.; Atwal, G.S.; Tkačik, G.; Bialek, W. Information-based clustering. *Proc. Natl. Acad. Sci. USA* **2005**, *102*, 18297–18302. [CrossRef] [PubMed]
8. Teahan, W.J. Text classification and segmentation using minimum cross-entropy. In *Content-Based Multimedia Information Access*; LE CENTRE DE HAUTES ETUDES INTERNATIONALES D'INFORMATIQUE DOCUMENTAIRE: Paris, France, 2000; pp. 943–961.
9. Strouse, D.; Schwab, D.J. The deterministic information bottleneck. *Neur. Comput.* **2017**, *29*, 1611–1630. [CrossRef]

10. Nazer, B.; Ordentlich, O.; Polyanskiy, Y. Information-distilling quantizers. In Proceedings of the 2017 IEEE International Symposium on Information Theory (ISIT), Aachen, Germany, 25–30 June, 2017; pp. 96–100.
11. Hassanpour, S.; Wübben, D.; Dekorsy, A. On the equivalence of double maxima and KL-means for information bottleneck-based source coding. In Proceedings of the IEEE Wireless Communications and Networking Conference (WCNC), Barcelona, Spain, 15–18 April 2018; pp. 1–6.
12. Goyal, A.; Islam, R.; Strouse, D.; Ahmed, Z.; Botvinick, M.; Larochelle, H.; Levine, S.; Bengio, Y. Infobot: Transfer and exploration via the information bottleneck. *arXiv* **2019**, arXiv:1901.10902.
13. Yingjun, P.; Xinwen, H. Learning Representations in Reinforcement Learning:An Information Bottleneck Approach. *arXiv* **2019**, arXiv:cs.LG/1911.05695.
14. Sharma, A.; Gu, S.; Levine, S.; Kumar, V.; Hausman, K. Dynamics-Aware Unsupervised Skill Discovery. In Proceeding of the International Conference on Learning Representations (ICLR), Addis Ababa, Ethiopia, 26–30 April 2020.
15. Schulz, K.; Sixt, L.; Tombari, F.; Landgraf, T. Restricting the Flow: Information Bottlenecks for Attribution. In Proceedings of the International Conference on Learning Representations (ICLR), Addis Ababa, Ethiopia, 26–30 April 2020.
16. Li, X.L.; Eisner, J. Specializing Word Embeddings (for Parsing) by Information Bottleneck. In Proceedings of the 2019 Conference on Empirical Methods in Natural Language Processing and the 9th International Joint Conference on Natural Language Processing (EMNLP-IJCNLP), Hong Kong, China, 3–7 November 2019; pp. 2744–2754.
17. Zaslavsky, N.; Kemp, C.; Regier, T.; Tishby, N. Efficient compression in color naming and its evolution. *Proc. Natl. Acad. Sci. USA* **2018**, *115*, 7937–7942. [CrossRef]
18. Chalk, M.; Marre, O.; Tkačik, G. Toward a unified theory of efficient, predictive, and sparse coding. *Proc. Natl. Acad. Sci. USA* **2018**, *115*, 186–191. [CrossRef]
19. Gilad-Bachrach, R.; Navot, A.; Tishby, N. An information theoretic tradeoff between complexity and accuracy. In *Learning Theory and Kernel Machines*; Springer: Berlin, Germany, 2003; pp. 595–609.
20. Cover, T.M.; Thomas, J.A. *Elements of Information Theory*; John Wiley & Sons: Hoboken, NJ, USA, 2012.
21. Kolchinsky, A.; Tracey, B.D.; Van Kuyk, S. Caveats for information bottleneck in deterministic scenarios. In Proceedings of the International Conference on Learning Representations (ICLR), New Orleans, LA, USA, 6–9 May 2019.
22. Courcoubetis, C. *Pricing Communication Networks Economics, Technology and Modelling*; Wiley Online Library: Hoboken, NJ, USA, 2003.
23. Tishby, N.; Slonim, N. Data clustering by markovian relaxation and the information bottleneck method. In *Advances in Neural Information Processing Systems*; MIT Press: Cambridge, MA, USA, 2001; pp. 640–646.
24. Slonim, N.; Friedman, N.; Tishby, N. Unsupervised document classification using sequential information maximization. In Proceedings of the 25th annual international ACM SIGIR Conference on Research and Development in Information Retrieval, Tampere, Finland,11–15 August 2002.
25. Chalk, M.; Marre, O.; Tkacik, G. Relevant sparse codes with variational information bottleneck. In *Advances in Neural Information Processing Systems*; MIT Press: Cambridge, MA, USA, 2016; pp. 1957–1965.
26. Kolchinsky, A.; Tracey, B.D.; Wolpert, D.H. Nonlinear information bottleneck. *Entropy* **2019**, *21*, 1181. [CrossRef]
27. Wu, T.; Fischer, I.; Chuang, I.; Tegmark, M. Learnability for the Information Bottleneck. In Proceedings of the International Conference on Learning Representations (ICLR), New Orleans, LA, USA, 6–9 May 2019.
28. LeCun, Y.; Bottou, L.; Bengio, Y.; Haffner, P. Gradient-based learning applied to document recognition. In Proceedings of the 1998 IEEE International Frequency Control Symposium, Pasadena, CA, USA, 27–29 May 1998.
29. Li, X.; Roth, D. Learning question classifiers. In *Proceedings of the 19th international conference on Computational linguistics—Volume 1*; Association for Computational Linguistics: Stroudsburg, PA, USA, 2002; pp. 1–7.
30. Paszke, A.; Gross, S.; Chintala, S.; Chanan, G.; Yang, E.; DeVito, Z.; Lin, Z.; Desmaison, A.; Antiga, L.; Lerer, A. Automatic differentiation in pytorch. In Proceedings of the NIPS Autodiff Workshop, Long Beach, CA, USA, 9 December 2017.
31. Bishop, C.M. *Pattern Recognition and Machine Learning*; Springer Science+ Business Media: Berlin, Germany, 2006.

32. Xu, A.; Raginsky, M. Information-theoretic analysis of generalization capability of learning algorithms. In *Advances in Neural Information Processing Systems*; MIT Press: Cambridge, MA, USA, 2017; pp. 2524–2533.
33. Krizhevsky, A.; Sutskever, I.; Hinton, G.E. Imagenet classification with deep convolutional neural networks. In *Advances in Neural Information Processing Systems*; MIT Press: Cambridge, MA, USA, 2012; pp. 1097–1105.
34. Shore, J.E.; Gray, R.M. Minimum cross-entropy pattern classification and cluster analysis. *IEEE Trans. Pattern Anal. Mach. Intell.* **1982**, *1*, 11–17. [CrossRef] [PubMed]
35. Shore, J.; Johnson, R. Properties of cross-entropy minimization. *IEEE Trans. Pattern Anal. Mach. Intell.* **1981**, *27*, 472–482. [CrossRef]
36. Vera, M.; Piantanida, P.; Vega, L.R. The role of the information bottleneck in representation learning. In Proceedings of the 2018 IEEE International Symposium on Information Theory (ISIT), Vail, CO, USA, 17–22 June 2018; pp. 1580–1584.
37. Shamir, O.; Sabato, S.; Tishby, N. Learning and generalization with the information bottleneck. *Theor. Comput. Sci.* **2010**, *411*, 2696–2711. [CrossRef]
38. Achille, A.; Soatto, S. Emergence of invariance and disentanglement in deep representations. *J. Mach. Learn. Res* **2018**, *19*, 1947–1980.
39. Du Pin Calmon, F.; Polyanskiy, Y.; Wu, Y. Strong data processing inequalities for input constrained additive noise channels. *IEEE Trans. Inf. Theory* **2017**, *64*, 1879–1892. [CrossRef]
40. Kolchinsky, A.; Tracey, B. Estimating mixture entropy with pairwise distances. *Entropy* **2017**, *19*, 361. [CrossRef]
41. Amjad, R.A.; Geiger, B.C. Learning representations for neural network-based classification using the information bottleneck principle. *IEEE Trans. Pattern Anal. Mach. Intell.* **2019**, 1. [CrossRef]
42. Alemi, A.A.; Fischer, I.; Dillon, J.V. Uncertainty in the variational information bottleneck. *arXiv* **2018**, arXiv:1807.00906.
43. Wu, T.; Fischer, I. Phase Transitions for the Information Bottleneck in Representation Learning. In Proceedings of the International Conference on Learning Representations (ICLR), Addis Ababa, Ethiopia, 26–30 April 2020.
44. Ester, M.; Kriegel, H.P.; Sander, J.; Xu, X. A density-based algorithm for discovering clusters in large spatial databases with noise. In Proceedings of the Second International Conference on Knowledge Discovery and Data Mining, Menlo Park, CA, USA, 2–4 August 1996; pp. 226–231.
45. Schubert, E.; Sander, J.; Ester, M.; Kriegel, H.P.; Xu, X. DBSCAN revisited, revisited: Why and how you should (still) use DBSCAN. *ACM Trans. Database Syst. TODS* **2017**, *42*, 19. [CrossRef]
46. Kingma, D.P.; Ba, J. Adam: A method for stochastic optimization. *arXiv* **2014**, arXiv:1412.6980.
47. Glorot, X.; Bengio, Y. Understanding the difficulty of training deep feedforward neural networks. In Proceedings of the Thirteenth International Conference on Artificial Intelligence and Statistics, Sardinia, Italy, 13–15 May 2010; pp. 249–256.
48. Pedregosa, F.; Varoquaux, G.; Gramfort, A.; Michel, V.; Thirion, B.; Grisel, O.; Blondel, M.; Prettenhofer, P.; Weiss, R.; Dubourg, V.; et al. Scikit-learn: Machine learning in Python. *J. Mach. Learn. Res.* **2011**, *12*, 2825–2830.
49. Xiao, H.; Rasul, K.; Vollgraf, R. Fashion-MNIST: A Novel Image Dataset for Benchmarking Machine Learning Algorithms. *arXiv* **2017**, arXiv:1708.07747.
50. Pace, R.K.; Barry, R. Sparse spatial autoregressions. *Stat. Probab. Lett.* **1997**, *33*, 291–297. [CrossRef]
51. Voorhees, E.M.; Tice, D.M. Building a question answering test collection. In Proceedings of the 23rd Annual International ACM SIGIR Conference on Research and Development in Information Retrieval, Athens, Greece, 24–28 July 2000.
52. Trevett, Ben. Tutorial on Sentiment Analysis: 5—Multi-class Sentiment Analysis. April 2019. Available online: https://github.com/bentrevett/pytorch-sentiment-analysis (accessed on 14 January 2020).
53. Pennington, J.; Socher, R.; Manning, C. Glove: Global vectors for word representation. In Proceedings of the 2014 Conference on Empirical Methods in Natural Language Processing (EMNLP), Doha, Qatar, 25–29 October 2014; pp. 1532–1543.

© 2020 by the authors. Licensee MDPI, Basel, Switzerland. This article is an open access article distributed under the terms and conditions of the Creative Commons Attribution (CC BY) license (http://creativecommons.org/licenses/by/4.0/).

Article
Probabilistic Ensemble of Deep Information Networks

Giulio Franzese * and Monica Visintin

Electronic and Telecommunications, Politecnico di Torino, 10100 Torino, Italy; monica.visintin@polito.it
* Correspondence: giulio.franzese@polito.it

Received: 22 November 2019; Accepted: 13 January 2020; Published: 14 January 2020

Abstract: We describe a classifier made of an ensemble of decision trees, designed using information theory concepts. In contrast to algorithms C4.5 or ID3, the tree is built from the leaves instead of the root. Each tree is made of nodes trained independently of the others, to minimize a local cost function (information bottleneck). The trained tree outputs the estimated probabilities of the classes given the input datum, and the outputs of many trees are combined to decide the class. We show that the system is able to provide results comparable to those of the tree classifier in terms of accuracy, while it shows many advantages in terms of modularity, reduced complexity, and memory requirements.

Keywords: information theory; information bottleneck; classifier; decision tree; ensemble

1. Introduction

Supervised classification is at the core of many modern applications of machine learning. The history of classifiers is rich and many variants have been proposed, such as decision trees, logistic regression, Bayesian networks, and neural networks (for an overview of general methods, see [1–3]). Despite the power of modern deep learning, for many problems involving categorical structured datasets, decision trees [4–7] or Bayesian networks [8–10] usually outperform neural network based approaches.

Decision trees are particularly interesting because they can be easily interpreted. Various types of tree classifiers can be discriminated according to the metric for the iterative construction and selection of features [4]: popular tree classifiers are based on information theoretic metrics, such as ID3 and C4.5 [6,7]. However, it is known that the greedy splitting procedure at each node can be sub-optimal [11], and that decision trees are prone to overfitting when dealing with small datasets. When a classifier is not strong enough, there are, roughly speaking, two possibilities: choosing a more sophisticated classifier or ensembling multiple "weak" classifiers [12,13]. This second approach is usually called the *ensemble* method. In the performance tradeoff by using multiple classifiers simultaneously, we improve classification performance, paying with the loss of interpretability.

The so-called "information bottleneck", described by Tishby and Zaslavsky [14] and Tishby et al. [15], was proposed in [16] to build a classifier (Deep Information Network, DIN) with a tree topology that compresses the input data and generates the estimated class. DINs [16] are based on the so-called information node that, using the input samples of a feature X_{in}, generates samples of a new feature X_{out}, according to the conditional probabilities $P(X_{out} = j | X_{in} = i)$ obtained by minimizing the mutual information $\mathbb{I}(X_{in}; X_{out})$, with the constraint of a given mutual information $\mathbb{I}(X_{out}; Y)$ between X_{out} and the target/class Y (information bottleneck [14]). The outputs of two or more nodes are combined, without information loss, to generate samples of a new feature passed to a subsequent information node. The final node (root) directly outputs the class of each input datum. The tree structure of the network is thus built from the leaves, whereas C4.5 and ID3 build it from the root.

We here propose an improved implementation of the DIN scheme in [16] that only requires the propagation through the tree of small matrices containing conditional probabilities. Notice that the previous version of the DIN was stochastic, while the one we propose here is deterministic. Moreover, we use an ensemble (e.g., [12,13]) of trees with randomly permuted features and weigh their outputs to improve classification accuracy.

The proposed architecture has several advantages in terms of:

- extreme flexibility and high modularity: all the nodes are functionally equivalent and with a reduced number of inputs and outputs, which gives good opportunities for a possible hardware implementation;
- high parallelizability: each tree can be trained in parallel with the others;
- memory usage: we need to feed the network with data only at the first layer and simple incremental counters can be used to estimate the initial probability mass distribution; and
- training time and training complexity: the locality of the computed cost function allows a nodewise training that does not require any kind of information from other points of the tree apart from its feeding nodes (that are usually a very small number, e.g., 2–3).

With respect to the DINs in [16], the main difference is that samples of the random variables in the inner layers of the tree are never generated, which is an advantage in the case of large datasets. However, an assumption of statistical independence (see Section 2.3) is necessary to build the probability matrices and this might be seen as a limitation of the newly proposed method. However, experimental results (see Section 5) show that this approximation does not compromise the performance.

We underline similarities and differences of the proposed classifier with respect to the methods described in [6,7] since they are among the best performing ones. When using decision trees, as well as DINs, categorical and missing data are easily managed, but continuous random variables are not: quantization of these input features is necessary in a pre-processing phase, and it can be performed as in C4.5 [6], using other heuristics, or manually. Concerning differences, instead, the first one is that normally a hierarchical decision tree is built starting from the root and splitting at each node, whereas we here propose a way to build a tree starting from the leaves. The topology of our network implies that, once the initial ordering of the features has been set, there is no need, after each node is trained, to perform a search of the best possible next node. The second important difference is that we do not use directly mutual information as a metric for building the tree but we base our algorithm on the Information Bottleneck principle [14,15,17–21]. This allows us to extract all the relevant information (the *sufficient statistic*) while removing the redundant one, which is helpful in avoiding overfitting. As in [12,13], we use an ensemble method. We choose the simplest possible form of ensemble combination: we train independently many structurally equivalent networks, using the same single dataset but permuting the order of the features, and produce a weighted average of the outputs based on a simple rule described in Section 3.1. Notice that we use a one-shot procedure, i.e., we do not iterate more than once over the entire dataset and exploit techniques similarly to [22,23]. We leave the study of more sophisticated techniques to future works.

Sections 2 and 3 more precisely describe the structure of the DIN and how it works, Section 4 gives some insight on the theoretical properties, Section 5 comments the results obtained with standard datasets. Conclusions are finally drawn in Section 6.

2. The DIN Architecture and Its Training

The information network is made of input nodes (Section 2.1), information nodes (Section 2.2), and combiners joined together through a tree network described in Section 2.3. Moreover, an ensemble of N_{mach} trees is built, based on which the final estimated class is produced (Section 3.1). In [16], the input nodes are not present, the information node has a slightly different role, the combiners are much simpler than those described here, and just one tree was considered. As already stated, the new version of the DIN is more efficient when a large dataset with relatively few features is analyzed.

In the following, it is assumed that all the features take a finite number of discrete values; a case of continuous random variables is discussed in Section 5.2.

It is also assumed that N_{train} points are used in the training phase, N_{test} points in the testing phase, and that D features are present. The nth training point corresponds to one of N_{class} possible classes.

2.1. The Input Node

Each input node (see Figure 1) has two input vectors:

1. $\mathbf{x}_{in}$ of size N_{train}, whose elements take values in a set of cardinality N_{in}; $\mathbf{x}_{in}$ corresponds to one of the D features of the dataset (typically one column)
2. $\mathbf{y}$ of size N_{train}, whose elements take values in a set of cardinality N_{class}; $\mathbf{y}$ corresponds to the known classes of the N_{train} points

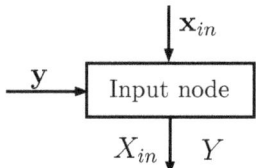

Figure 1. Schematic representation of an input node: the inputs are two vectors and the outputs are matrices that statistically describe the random variables X_{in} and Y.

The notation we use in the equations below is the following: Y, X_{in} represent random variables; $\mathbf{y}(n)$ and $\mathbf{x}_{in}(n)$ are the nth elements of vectors $\mathbf{y}$ and $\mathbf{x}_{in}$, respectively; and $\mathbf{1}(c)$ is equal to 1 if c is true, and is otherwise equal to 0. Using Laplace smoothing [2], the input node estimates the following probabilities (the probability mass function of Y in Equation (1) is common to all the input nodes: it can be evaluated only by the first one and passed to the others):

$$\hat{P}(Y = m) \simeq \frac{1 + \sum_{n=0}^{N_{train}-1} \mathbf{1}(\mathbf{y}(n) = m)}{N_{train} + N_{class}} \quad m = 0, \ldots, N_{class} - 1 \tag{1}$$

$$\hat{P}(X_{in} = i) \simeq \frac{1 + \sum_{n=0}^{N_{train}-1} \mathbf{1}(\mathbf{x}_{in}(n) = i)}{N_{train} + N_{in}}, \quad i = 0, \ldots, N_{in} - 1 \tag{2}$$

$$\hat{P}(Y = m, X_{in} = i) \simeq \frac{1 + \sum_{n=0}^{N_{train}-1} \mathbf{1}(\mathbf{y}(n) = m)\mathbf{1}(\mathbf{x}_{in}(n) = i)}{N_{train} + N_{class}N_{in}} \tag{3}$$

From basic application of probability rules, $\hat{P}(Y = m | X_{in} = i)$ and $\hat{P}(X_{in} = i | Y = m)$ are then computed. From now on, for simplicity, we denote all the estimated probabilities $\hat{P}$ simply as P.

All the above probabilities can be organized in matrices defined as follows:

$$\mathbf{P}_Y \in \mathbb{R}^{1 \times N_{class}}, \quad \mathbf{P}_Y(m) = P(Y = m) \tag{4}$$

$$\mathbf{P}_{X_{in}} \in \mathbb{R}^{1 \times N_{in}}, \quad \mathbf{P}_{X_{in}}(i) = P(X_{in} = i) \tag{5}$$

$$\mathbf{P}_{X_{in}|Y} \in \mathbb{R}^{N_{class} \times N_{in}}, \quad \mathbf{P}_{X_{in}|Y}(m, i) = P(X_{in} = i | Y = m) \tag{6}$$

$$\mathbf{P}_{Y|X_{in}} \in \mathbb{R}^{N_{in} \times N_{class}}, \quad \mathbf{P}_{Y|X_{in}}(i, m) = P(Y = m | X_{in} = i) \tag{7}$$

Note that vectors $\mathbf{x}_{in}$ and $\mathbf{y}$ are not needed by the subsequent elements in the tree; only the input nodes have access to them.

Notice also that the following equalities hold:

$$\mathbf{P}_{X_{in}} = \mathbf{P}_Y \mathbf{P}_{X_{in}|Y} \tag{8}$$

$$\mathbf{P}_Y = \mathbf{P}_{X_{in}} \mathbf{P}_{Y|X_{in}} \tag{9}$$

2.2. The Information Node

The information node is schematically shown in Figure 2: the input discrete random variable X_{in} is stochastically mapped into another discrete random variable X_{out} (see [16] for further details) through probability matrices:

- The input probability matrices $\mathbf{P}_{X_{in}}, \mathbf{P}_{X_{in}|Y}, \mathbf{P}_{Y|X_{in}}, \mathbf{P}_Y$ describe the input random variable X_{in}, with N_{in} possible values, and its relationship with class Y.
- The output matrices $\mathbf{P}_{X_{out}}, \mathbf{P}_{X_{out}|Y}, \mathbf{P}_{Y|X_{out}}, \mathbf{P}_Y$ describe the output random variable X_{out}, with N_{out} possible values, and its relationship with Y.

Compression (source encoding) is obtained by setting $N_{out} < N_{in}$.

In the training phase, the information node generates the conditional probability mass function that satisfies the following equation (see [14]):

$$P(X_{out} = j | X_{in} = i) = \frac{1}{Z(i;\beta)} P(X_{out} = j) e^{-\beta d(i,j)}, \quad i = 0, \ldots, N_{in} - 1, j = 0, \ldots, N_{out} - 1 \tag{10}$$

where

- $P(X_{out} = j)$ is the probability mass function of the output random variable X_{out}

$$P(X_{out} = j) = \sum_{i=0}^{N_{in}-1} P(X_{in} = i) P(X_{out} = j | X_{in} = i), \quad j = 0, \ldots, N_{out} - 1 \tag{11}$$

- $d(i,j)$ is the Kullback–Leibler divergence

$$d(i,j) = \sum_{m=0}^{N_{class}-1} P(Y = m | X_{in} = i) \log_2 \frac{P(Y = m | X_{in} = i)}{P(Y = m | X_{out} = j)}$$
$$= \mathbb{KL}(P(Y | X_{in} = i) || P(Y | X_{out} = j)) \tag{12}$$

and

$$P(Y = m | X_{out} = j) = \sum_{i=0}^{N_{in}-1} P(Y = m | X_{in} = i) P(X_{in} = i | X_{out} = j),$$
$$m = 0, \ldots, N_{class} - 1, j = 0, \ldots, N_{out} - 1 \tag{13}$$

- β is a real positive parameter.
- $Z(i; \beta)$ is a normalizing coefficient to get

$$\sum_{j=1}^{N_{out}-1} P(X_{out} = j | X_{in} = i) = 1. \tag{14}$$

The probabilities $P(X_{out} = j | X_{in} = i)$ can be iteratively found using the Blahut–Arimoto algorithm [14,24,25].

Equation (10) solves the information bottleneck: it minimizes the mutual information $\mathbb{I}(X_{in}; X_{out})$ under the constraint of a given mutual information $\mathbb{I}(Y; X_{out})$. In particular, Equation (10) is the solution of the minimization of the Lagrangian

$$\mathcal{L} = \mathbb{I}(X_{in}; X_{out}) - \beta \mathbb{I}(Y; X_{out}). \tag{15}$$

If the Lagrangian multiplier β is increased, then the constraint is privileged and the information node tends to maximize the mutual information between its output X_{out} and the class Y; if β is reduced, then minimization of $\mathbb{I}(X_{in}; X_{out})$ is obtained (compression). The information node must actually balance compression from X_{in} to X_{out} and propagation of the information about Y. In our implementation, the compression is also imposed by the fact that the cardinality of the output alphabet N_{out} is smaller than that of the input alphabet N_{in}.

The role of the information node is thus that of finding the conditional probability matrices

$$\mathbf{P}_{X_{out}|X_{in}} \in \mathbb{R}^{N_{in} \times N_{out}}, \quad \mathbf{P}_{X_{out}|X_{in}}(i,j) = P(X_{out} = j | X_{in} = i) \tag{16}$$

$$\mathbf{P}_{Y|X_{out}} \in \mathbb{R}^{N_{out} \times N_{class}}, \quad \mathbf{P}_{Y|X_{out}}(j,m) = P(Y = m | X_{out} = j) \tag{17}$$

$$\mathbf{P}_{X_{out}} \in \mathbb{R}^{1 \times N_{out}}, \quad \mathbf{P}_{X_{out}}(j) = P(X_{out} = j) \tag{18}$$

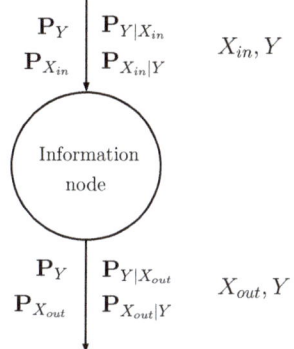

Figure 2. Schematic representation of an information node, showing the input and output matrices.

2.3. The Combiner

Consider the case depicted in Figure 3, where the two information nodes *a* and *b* feed a combiner (shown as a triangle) that generates the input of the information node *c*. The random variables $X_{out,a}$ and $X_{out,b}$, both having alphabet with cardinality N_1, are combined together as

$$X_{in,c} = X_{out,a} + N_1 X_{out,b} \tag{19}$$

that has an alphabet with cardinality $N_1 \times N_1$.

The combiner actually does not generate $X_{in,c}$; it simply evaluates the probability matrices that describe $X_{in,c}$ and Y. In particular, the information node *c* needs $\mathbf{P}_{X_{in,c}|Y}$, which can be evaluated **assuming** that $X_{out,a}$ and $X_{out,b}$ are conditionally independent given Y (notice that in implementation [16] this assumption was not necessary):

$$P(X_{in,c} = k|Y = m) = P(X_{out,a} = k_a, X_{out,b} = k_b|Y = m)$$
$$= P(X_{out,a} = k_a|Y = m)P(X_{out,b} = k_b|Y = m) \tag{20}$$

where $k = k_a + N_1 k_b$. In particular, the *m*th row of $\mathbf{P}_{X_{in,c}|Y}$ is the Kronecker product of the *m*th rows of $\mathbf{P}_{X_{out,a}|Y}$ and $\mathbf{P}_{X_{out,b}|Y}$

$$\mathbf{P}_{X_{in,c}|Y}(m,:) = \mathbf{P}_{X_{out,a}|Y}(m,:) \otimes \mathbf{P}_{X_{out,b}|Y}(m,:) \quad m = 0, \ldots, N_{class} - 1 \tag{21}$$

(here $\mathbf{A}(m,:)$ identifies the mth row of matrix $\mathbf{A}$). The probability vector $\mathbf{P}_{X_{in,c}}$ can be evaluated considering that

$$P(X_{in,c} = k) = \sum_{m=0}^{N_{class}-1} P(X_{in,c} = k, Y = m) = \sum_{m=0}^{N_{class}-1} P(X_{in,c} = k|Y = m)P(Y = m) \quad (22)$$

so that

$$\mathbf{P}_{X_{in,c}} = \mathbf{P}_Y \mathbf{P}_{X_{in,c}|Y} \quad (23)$$

At this point, matrix $\mathbf{P}_{Y|X_{in,c}}$ can be evaluated element by element since

$$P(Y = m|X_{in,c} = k) = \frac{P(X_{in,c} = k|Y = m)P(Y = m)}{P(X_{in,c} = k)},$$

$$m = 1, \ldots, N_{class} - 1, k = 0, \ldots, N_1 \times N_1 - 1 \quad (24)$$

It is straightforward to extend the equations to the case in which $X_{in,a}$ and $X_{in,b}$ have different cardinalities.

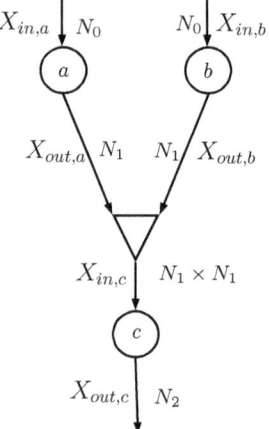

Figure 3. Sub-network: $X_{in,a}$, $X_{out,a}$, $X_{in,b}$, $X_{out,b}$, $X_{in,c}$, and $X_{out,c}$ are all random variables; N_0 is the number of values taken by $X_{in,a}$ and $X_{in,b}$; N_1 is the number of values taken by $X_{out,a}$ and $X_{out,b}$; and N_2 is the number of values taken by $X_{out,c}$.

2.4. The Tree Architecture

Figure 4 shows an example of a DIN, where we assume that the dataset has $D = 8$ features and that training is thus obtained using a matrix $\mathbf{X}_{train}$ with N_{train} rows and $D = 8$ columns, with a corresponding class vector $\mathbf{y}$. The kth column $\mathbf{x}(k)$ of matrix $\mathbf{X}_{train}$ feeds, together with vector $\mathbf{y}$, the input node $I(k)$, $k = 0, \ldots, D - 1$.

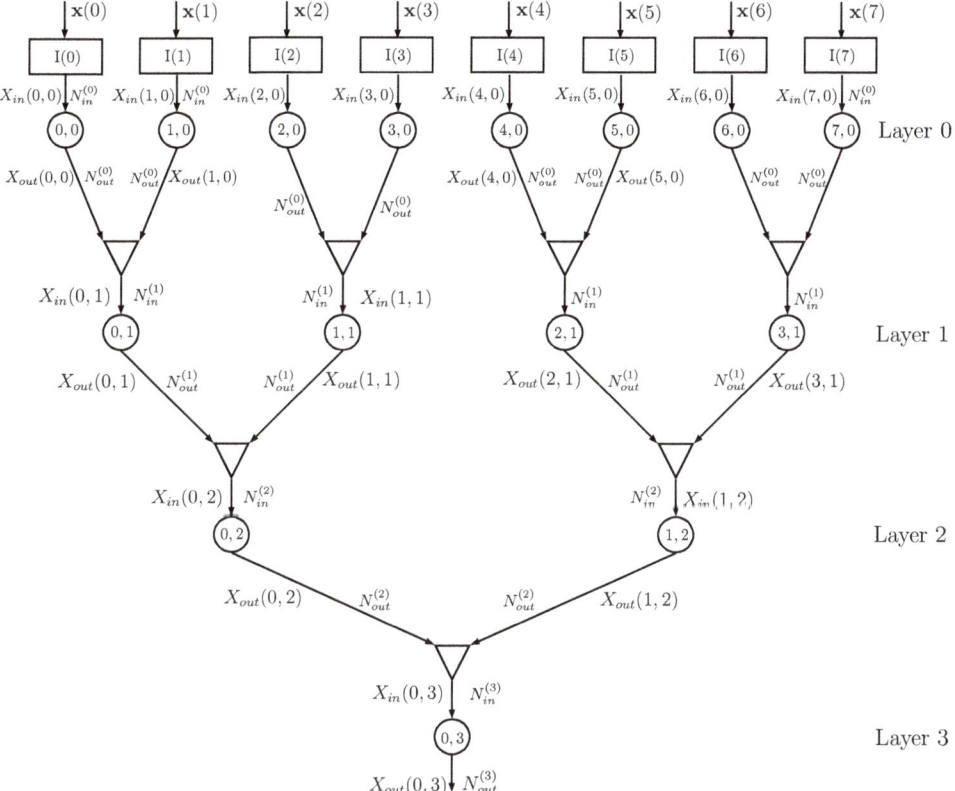

Figure 4. Example of a DIN for $D = 8$: the input nodes are represented as rectangles, the info nodes as circles, and the combiners as triangles. The numbers inside each circle identify the node (position inside the layer and layer number), $N_{in}^{(k)}$ is the number of values taken by the input of the info node at layer k, and $N_{out}^{(k)}$ is the number of values taken by the output of the info node at layer k. In this example, the info nodes at a given layer all have the same input and output cardinalities.

Information node $(k,0)$ at layer 0 processes the probability matrices generated by the input node $I(k)$, with $N_{in}^{(0)}$ possible values of $X_{in}(k,0)$, and evaluates the conditional probability matrices with $N_{out}^{(0)}$ possible values of $X_{out}(k,0)$, using the algorithm described in Section 2.2. The outputs of info nodes $(2k,0)$ and $(2k+1,0)$ are given to a combiner that outputs the probability matrices for $X_{in}(k,1)$, having alphabet of cardinality $N_{in}^{(1)} = N_{out}^{(0)} \times N_{out}^{(0)}$, using the equations described in Section 2.3. The sequence of combiners and information nodes is iterated, decreasing the number of information nodes from layer to layer, until the final root node is obtained. In the previous implementation of the DINs in [16], the root information node outputs the estimated class of the input and it is therefore necessary that the output cardinality of the root info node is equal to N_{class}. In the current implementation, this cardinality can be larger than N_{class}, since classification is based on the output probability matrix $\mathbf{P}_{Y|X_{out}}$.

For a number of features $D = 2^d$, the number of layers is d. If D is not a power of 2, then it is possible to use combiners with 3 or more inputs (the changes in the equations in Section 2.3 are straightforward, since a combiner with three inputs can be seen as two cascaded combiners with two inputs each).

The overall binary topology proposed in Figure 4 requires a number of information nodes equal to

$$N_{nodes} = D + \frac{D}{2} + \frac{D}{4} + \cdots + 2 + 1 = 2D - 1 \qquad (25)$$

and a number of combiners equal to

$$N_{comb} = \frac{D}{2} + \frac{D}{4} + \cdots + 2 + 1 = D - 1 \qquad (26)$$

All the info nodes run exactly the same algorithm and all the combiners are equal, apart from the input/output alphabet cardinalities. If the cardinalities of the alphabets are all equal, i.e., $N_{in}^{(i)}$ and $N_{out}^{(i)}$ do not depend on the layer i, then all the nodes and all the combiners are exactly equal, which might help in a possible hardware implementation; in this case, the number of parameters of the network is $(N_{out} - 1) \times N_{in} \times N_{nodes}$.

Actually, the network performance depends on how the features are coupled in subsequent layers and a random shuffling of the columns of matrix $\mathbf{X}_{train}$ provides results that might be significantly different. This property is used in Section 3.1 for building the ensemble of networks.

2.5. A Note on Computational Complexity and Memory Requirements

The modular structure of the proposed method has several advantages in terms of both memory footprint and computational cost. The considered topology in this explanation is binary, similarly to what is depicted in Figure 4. We furthermore consider for simplicity cardinalities of the D input features all equal to N_{in} and input/output cardinalities of subsequent layers information node to also be fixed constants N_{in}^* and $N_{out}^* = \frac{N_{in}^*}{2}$, respectively. As we show in the experiment (Section 5), small values for N_{in}^* and N_{out}^* such as 2, 3, or 4 are sufficient in the considered cases. Straightforward generalizations are possible when considering inhomogeneous cases.

At the first layer (the input node layer), each of the D input nodes stores the joint probabilities of the target variable Y and its input feature. Each node thus includes a simple counter that fills the probability matrix of dimension $N_{in} \times N_{class}$. Both the computational cost and the memory requirements for this first stage are the same as the Naive Bayes algorithm. Notice that, from the memory requirements point of view, it is not necessary to store all the training data but just counters with number of joint occurrences of features/classes. If after training, new data are observed, it is in fact sufficient to update the counters and properly renormalize the values to obtain the updated probability matrices. In this paper, we do not cover the topic of online learning as well as possible strategies to reduce the computational complexity in such a scenario.

At the second layer (the first information node layer), each node receives as input the joint probability matrix of feature and target variable and performs the Blahut–Arimoto algorithm. The internal memory requirement of this node is the space needed to store two probability matrices of dimensions $N_{in}^* \times N_{class}$ and $N_{in}^* \times N_{out}^*$, respectively. The cost per iteration of Blahut–Aritmoto depends on matrix multiplication of sizes $N_{in}^* \times N_{out}^*$ and $N_{in}^* \times N_{class}$, and thus obviously the complexity scales with the number of classes of the considered classification problem. To the best of our knowledge, the convergence rate for the Blahut–Arimoto algorithm applied to information bottleneck problems is unknown. In this study, however, we found empirically that, for the considered datasets, 5–6 iterations per node are sufficient, as discussed in Section 5.5.

Each combiner process the matrices generated by two information nodes: the memory requirement is zero and the computational cost is roughly N_{class} Kronecker products between rows of probability matrices. Since for ease of explanation we chose $N_{out}^* = \frac{N_{in}^*}{2}$ the output probability matrix have again dimensions $N_{in}^* \times N_{class}$.

The overall memory requirement and computational complexity (for a single DIN) is thus going to scale as D times the requirements for an input node, $2D - 1$ times the requirements for an information node, and $D - 1$ times the requirements for a combiner. To complete the discussion, we have to

3. The Running Phase

During the running phase, the columns of matrix $\mathbf{X}$ with N rows and D columns are used as inputs. Assume again that the network architecture is that depicted in Figure 4 with $D = 8$, and consider the nth input row $\mathbf{X}(n,:)$.

In particular, assume that $\mathbf{X}(n, 2k) = i$ and $\mathbf{X}(n, 2k+1) = j$. Then,

1. (a) input node $I(2k)$ passes value i to info node $(2k, 0)$;
   (b) input node $I(2k+1)$ passes value j to info node $(2k+1, 0)$;
2. (a) info node $(2k, 0)$ passes the probability vector $\mathbf{p}_a = \mathbf{P}_{X_{out}(2k,0)|X_{in}(2k,0)}(i,:)$ (ith row) to the combiner; $\mathbf{p}_a$ stores the conditional probabilities $P(X_{out}(2k, 0) = g|\mathbf{X}(n, 2k) = i)$ for $g = 0, \ldots, N_{out}^{(0)} - 1$;
   (b) info node $(2k+1, 0)$ passes the probability vector $\mathbf{p}_b = \mathbf{P}_{X_{out}(2k+1,0)|X_{in}(2k+1,0)}(j,:)$ (jth row) to the combiner; $\mathbf{p}_b$ stores the conditional probabilities $P(X_{out}(2k+1, 0) = h|\mathbf{X}(n, 2k+1) = j)$ for $h = 0, \ldots, N_{out}^{(0)} - 1$;
3. the combiner generates vector
$$\mathbf{p}_c = \mathbf{p}_a \otimes \mathbf{p}_b, \qquad (27)$$
which stores the conditional probabilities $P(X_{in}(k, 1) = s|\mathbf{X}(n, 2k) = i, \mathbf{X}(n, 2k+1) = j)$ for $s = 0, \ldots, N_{in}^{(1)} - 1$, where $N_{in}^{(1)} = N_{out}^{(0)} \times N_{out}^{(0)}$;
4. info node $(k, 1)$ generates the probability vector
$$\mathbf{p}_c \mathbf{P}_{X_{out}(k,1)|X_{in}(k,1)}, \qquad (28)$$
which stores the conditional probabilities $P(X_{out}(k, 1) = r|\mathbf{X}(n, 2k) = i, \mathbf{X}(n, 2k+1) = j)$ for $r = 0, \ldots, N_{out}^{(1)}$;
5. in the following layer, each combiner performs the Kronecker product of its two input vectors and each info node performs the product between the input vector and its conditional probability matrix $\mathbf{P}_{X_{out}|X_{in}}$;
6. the root information node at Layer 3, having the input vector $\mathbf{p}$, outputs
$$\mathbf{p}_{out}(n) = \mathbf{p} \mathbf{P}_{X_{out}(0,3)|X_{in}(0,3)} \mathbf{P}_{Y|X_{out}(0,3)}, \qquad (29)$$
which stores the estimated probabilities $P(Y = m|\mathbf{X}(n,:))$ for $m = 0, \ldots, N_{class} - 1$.

According to the MAP criterion, the estimated class of the input point $\mathbf{X}(n,:)$ is
$$\hat{Y}(n) = \arg \max \mathbf{p}_{out}(n) \qquad (30)$$
but we propose to use an improved method, as described in Section 3.1.

3.1. The DIN Ensemble

At the end of the training phase, when all the conditional matrices have been generated in each information node and combiner, the network is run with input matrix $\mathbf{X}_{train}$ (N_{train} rows and D columns) and the probability vector $\mathbf{p}_{out}$ is obtained for each input point $\mathbf{X}_{train}(n,:)$. As anticipated at the end of Section 2.4, the DIN classification accuracy depends on how the input features are combined together. By permuting the columns of $\mathbf{X}_{train}$, a different probability vector $\mathbf{p}_{out}$ is typically obtained. We thus propose to generate an ensemble of DINs by randomly permuting the columns of $\mathbf{X}_{train}$, and then combine their outputs.

Since in the training phase $\mathbf{y}(n)$ is known, it is possible to get for each DIN v the probability $\mathbf{p}_{out}^v(n)$, and ideally $\mathbf{p}_{out}^v(n, \mathbf{y}(n))$, the estimated probability corresponding to the true class $\mathbf{y}(n)$, should be equal to one. The weights

$$w^v = \frac{\sum_{n=0}^{N_{train}-1} \mathbf{p}_{out}^v(n, \mathbf{y}(n))}{\sum_{n=0}^{N_{train}-1} \sum_{j=0}^{N_{mach}-1} \mathbf{p}_{out}^j(n, \mathbf{y}(n))} \tag{31}$$

thus represent the reliability of the vth DIN.

In the running phase, feeding the N_{mach} machines each with the correctly permuted vector $\mathbf{X}(n,:)$, the final estimated probability vector is determined as

$$\hat{\mathbf{p}}_{ens}(n) = \sum_{v=0}^{N_{mach}-1} w^v \hat{\mathbf{p}}_{out}^v(n) \tag{32}$$

and the estimated class is

$$\hat{Y}(n) = \arg\max \hat{\mathbf{p}}_{ens}(n). \tag{33}$$

4. The Probabilistic Point of View

This section is intended to underline the difference in probability terms formulation between the Naive Bayes classifier [2,26] and the proposed scheme, since both use the assumption of conditional independence of the input features. Both classifiers build in a simplified way the probability matrix $\mathbf{P}_{Y|X_0,\ldots,X_D}$ with N_{class} rows and $\prod_{i=0}^{D-1} N_{in}^{(i)}$, where $N_{in}^{(i)}$ is the cardinality for the input feature X_i. In the next sections, we show the different structure of these two probability matrices.

4.1. Assumption of Conditionally Independent Features

The Naive Bayes assumption allows writing the output estimated probability of the Naive Bayes classifier as follows:

$$P(Y = m | \mathbf{x} = \mathbf{x}_0) = \frac{P(\mathbf{x} = \mathbf{x}_0 | Y = m) P(Y = m)}{P(\mathbf{x} = \mathbf{x}_0)}$$

$$= \frac{\left[\prod_{k=0}^{D-1} P(X_k = x_{k0} | Y = m)\right] P(Y = m)}{\sum_{s=0}^{N_{class}} \left[\prod_{k=0}^{D-1} P(X_k = x_{k0} | Y = s)\right] P(Y = s)} \tag{34}$$

which is very easily implemented, without the need of generating the tree network. We rewrite this output probability in a fairly complex way to show the difference between the naive Bayes probability matrix and the DIN one. Consider the nth feature $x(n)$, which can take values in the set $\{c_n^0, \ldots, c_n^{D_n-1}\}$. Define $\mathbf{p}_{x(n)|y=m} = [P(x(n) = c_n^0 | Y = m), \ldots P(x(n) = c_n^{D_n-1} | Y = m)]$; then,

$$\mathbf{P}_{X_{in}|Y}(m,:) = \otimes_{k=0}^{D-1} \mathbf{P}_{x(k)|y=m} \tag{35}$$

and thus obviously

$$\mathbf{P}_{X_{in}|Y} = \begin{bmatrix} \otimes_{k=0}^{D-1} \mathbf{P}_{x(k)|y=0} \\ \otimes_{k=0}^{D-1} \mathbf{P}_{x(k)|y=1} \\ \vdots \\ \otimes_{k=0}^{D-1} \mathbf{P}_{x(k)|y=N_{class}} \end{bmatrix} \tag{36}$$

We can write the joint probability matrix as

$$\mathbf{P}_{X_{in},Y} = diag(\mathbf{P}_Y) \mathbf{P}_{X|Y} \tag{37}$$

and the probability matrix of target class given observation as

$$\mathbf{P}_{Y|X_{in}} = (\mathbf{P}_{X_{in},Y} diag(\mathbf{P}_{X_{in}}^{\circ(-1)}))^T \qquad (38)$$

The hypothesis of conditional statistical independence of the features is not always correct and thus we can incur obvious performance degradation.

4.2. The Overall Probability Matrix

We now instead compute the output estimated probability for the DIN classifier. Consider again the sub-network in Figure 3 made of info nodes a, b, and c. Info node a is characterized by matrix $\mathbf{P}_a$, whose element $P_a(i,j)$ is $P(X_{out,a} = j|X_{in,a} = i)$; similar definitions hold for $\mathbf{P}_b$ and $\mathbf{P}_c$. Note that $\mathbf{P}_a$ and $\mathbf{P}_b$ have N_0 rows and N_1 columns, whereas $\mathbf{P}_c$ has $N_1 \times N_1$ rows and N_2 columns; the overall probability matrix between the inputs $X_{in,a}$, $X_{in,b}$ and the output $X_{out,c}$ is $\tilde{\mathbf{P}}$ with $N_0 \times N_0$ rows and N_2 columns. Then,

$$
\begin{aligned}
&P(X_{out,c} = i|X_{in,a} = j, X_{in,b} = k) \\
&= \sum_{r=0}^{N_1-1} \sum_{s=0}^{N_1-1} P(X_{out,c} = i, X_{out,a} = r, X_{out,b} = s|X_{in,a} = j, X_{in,b} = k) \\
&= \sum_{r=0}^{N_1-1} \sum_{s=0}^{N_1-1} P(X_{out,c} = i|X_{out,a} = r, X_{out,b} = s) P(X_{out,a} = r|X_{in,a} = j) P(X_{out,b} = s|X_{in,b} = k) \\
&= \sum_{r=0}^{N_1-1} \sum_{s=0}^{N_1-1} P(X_{out,c} = i|X_{out,s} = r, X_{out,b} = s) \mathbf{P}_a(j,r) \mathbf{P}_b(k,s). \qquad (39)
\end{aligned}
$$

It can be shown that

$$\tilde{\mathbf{P}} = (\mathbf{P}_a \otimes \mathbf{P}_b) \mathbf{P}_c \qquad (40)$$

where $\otimes$ identifies the Kronecker matrix multiplication; note that $\mathbf{P}_a \otimes \mathbf{P}_b$ has $N_0 \times N_0$ rows and $N_1 \times N_1$ columns. By iteratively applying the above rule, we can get the expression of the overall matrix $\tilde{\mathbf{P}}$ for the exact topology of Figure 4, with eight input nodes and four layers:

$$\tilde{\mathbf{P}} = \Big[\big\{[(\mathbf{P}_{0,0} \otimes \mathbf{P}_{1,0})\mathbf{P}_{0,1}] \otimes [(\mathbf{P}_{2,0} \otimes \mathbf{P}_{3,0})\mathbf{P}_{1,1}]\big\}\mathbf{P}_{0,2}$$
$$\otimes \big\{[(\mathbf{P}_{4,0} \otimes \mathbf{P}_{5,0})\mathbf{P}_{2,1}] \otimes [(\mathbf{P}_{6,0} \otimes \mathbf{P}_{7,0})\mathbf{P}_{3,1}]\big\}\mathbf{P}_{1,2}\Big]\mathbf{P}_{0,3}. \qquad (41)$$

The overall output probability matrix $\mathbf{P}_{Y|X}$ can finally be computed as

$$\mathbf{P}_{Y|X_{in}} = \tilde{\mathbf{P}} \mathbf{P}_{Y|X_{out}(0,3)}. \qquad (42)$$

The DIN then behaves as a one-layer system that generates the output according to matrix $\mathbf{P}_{Y|X_{in}}$, whose size might be impractically large. It is also possible to interpret the system as a sophisticated way of factorizing and approximating the exponentially large true probability matrix. In fact, the proposed layered structure needs smaller probability matrices, which makes the system computationally efficient. The equivalent probability matrix is thus different in the DIN (Equation (42)) and Naive Bayes (Equation (38)) cases.

5. Experiments

In this section, we analyze the results obtained with benchmark datasets. In particular, we consider the DIN ensemble when: (a) each DIN is based on the probability matrices (the scheme described in this paper); and (b) each information node of the DIN randomly generates the symbols, as described in the previous work [16]. We refer to these two variants in captions and labels as DIN(Prob) and DIN(Gen), respectively. The reason for this comparison is that conditional statistical independence is not required in the case DIN(Gen), and the classification accuracy could be different in the two cases. Note that Franzese and Visintin [16] considered just one DIN, not an ensemble of DINs. In the following, we introduce three datasets on which we tested the method (Sections 5.1– 5.3) and propose some examples of DINs architectures. Complete analysis of numerical results is described in Section 5.4. Sections 5.5 and 5.6 analyze the impact of changing the maximum number of iterations of Blahut–Arimoto algorithm and Lagrangian coefficient β, respectively. Finally, a synthetic multiclass experiment is described in Section 5.7. In all experiments, the value of β was optimized similarly to what is described in Section 5.6 using the training set.

5.1. UCI Congressional Voting Records Dataset

The first experiment on real data was conducted on the UCI Congressional Voting Records dataset [27], which collects the votes given by each of the U.S. House of Representatives Congressmen on 16 key laws (in 1985). Each vote can take three values corresponding to (roughly, see [27] for more details) yes, no, and missing value; each datum belongs to one of two classes (Democrats or Republican). The aim of the network is, given the list of 16 votes, decide if the voter is Republican or Democratic. In this dataset, we thus have $D = 16$ features and 435 data split into N_{train} data for training and $N_{test} = 435 - N_{train}$ data for testing. The architecture of the used network is the same as the one described in Section 2.4, except for the fact that there are 16 input features instead of 8 (the network has thus one more layer). The input cardinality in the first layer is $N_{in}^{(0)} = 3$ (yes/no/missing) and the output cardinality is set to $N_{out}^{(0)} = 2$. From the second layer on, the input cardinality for each information node is equal to $N_{in}^* = 4$ and $N_{out}^* = 2$. In the majority of the cases, the size of the probability matrices is therefore 4×2 or 2×2. In this example, we used $N_{mach} = 30$ and $N_{train} = 218$ (roughly 50% of the data). The value of β was set to 2.2.

5.2. UCI Kidney Disease Dataset

The second considered dataset was the UCI Kidney Disease dataset [28]. The dataset has a total of 24 medical features, consisting of mixed categorical, integer, and real values, with missing values. Quantization of non-categorical features of the dataset was performed according to the thresholds in Appendix A, agreed upon by a medical doctor.

The aim of the experiment is to correctly classify patients affected by chronic kidney disease. We performed 100 different trials training the algorithms using only $N_{train} = 50$ out of 400 samples for the training. Layer zero has 24 input nodes, and then the outputs of layer zero are mixed two at a time to get 12 information nodes at Layer 1, 6 at Layer 2, and 3 at Layer 3; the last three nodes are combined into a unique final node. The output cardinality of all nodes is equal to $N_{out}^* = 2$. The value of β was set equal to 5.6. In addition, in this case, we used an ensemble of $N_{mach} = 30$ DINs.

5.3. UCI Mushroom dataset

The last considered dataset was the UCI Mushroom dataset [29]. This dataset is comprised of 22 categorical features with different cardinalities, which describe some properties of mushrooms, and one target variable that defines whether the considered mushroom is edible or poisonous/unsafe. There are 8124 entries in the dataset. We padded the dataset with two null features to reach the cardinality of 24 and used exactly the same architecture as the kidney disease experiment. We selected $N_{train} = 50$, $\beta = 2.7$, and number of DINs equal to $N_{mach} = 15$.

5.4. Misclassification Probability Analysis

We hereafter report results in terms of misclassification probability between the proposed method and several classification methods implemented using MATLAB® Classification Learner. All datasets were randomly split 100 times into training and testing subsets, thus generating 100 different experiments. The proposed method shows competitive results in the considered cases, as can be observed in Table 1. It is interesting to compare in terms of performance the proposed algorithm with respect to the Naive Bayes classifier, i.e., Equation (34), and the Bagged Tree algorithm, which is the closest algorithm (conceptually) to the one we propose. In general, the two variants of the DINs perform similarly to the Bagged Trees, while outperforming Naive Bayes. For Bagged Trees and KNN-Ensemble, the same number of learners as DIN ensembles were used.

Table 1. Mean misclassification probability (over 100 random experiments) for the three datasets with the considered classifiers.

Classifier	Congressional Voting Records	Kidney Disease	Mushroom
Naive Bayes	0.10894	0.051 g	0.20641
Decision Tree	0.050691	0.062314	0.05505
Bagged Trees	0.043641	0.0268	0.038305
DIN Prob	0.050138	0.037229	0.020796
DIN Gen	0.049447	0.026286	0.022182
Linear Discriminant Classifier	0.059724	0.091029	0.069923
Logistic Regression	0.075161	0.096429	0.07074
Linear SVM	0.063226	0.049914	0.04513
KNN	0.08682	0.11369	0.037018
KNN-Ensemble	0.062811	0.036057	0.043967

5.5. The Impact of Number of Iterations of Blahut–Arimoto on The Performance

As anticipated in Section 2.5, the computational complexity of a single node scales with the number of iterations of Blahut–Arimoto algorithm. To the best of our knowledge, a provable convergence rate for the Blahut–Arimoto algorithm in the information nottleneck setting does not exist. We hereafter (Figure 5) present empirical results on the impact of limiting the number of iterations of Blahut–Arimoto algorithm (for simplicity, the same bound is applied to all nodes in the networks). When the number of iterations is too small, there is a drastic decrease in performance because the probability matrices in the information nodes have not yet converged, while 5–6 iterations are sufficient and a further increase in the number of iterations is not necessary in terms of performance improvements.

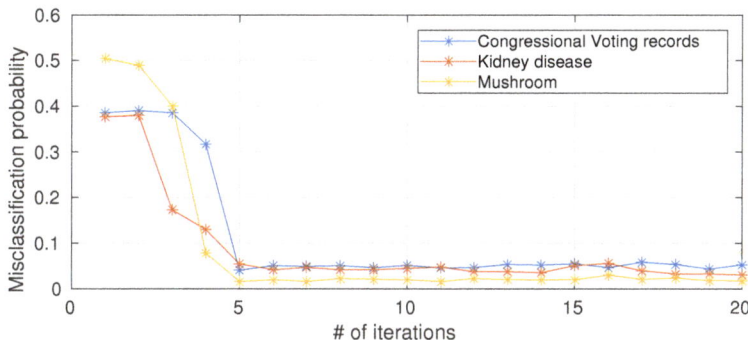

Figure 5. Misclassification probability versus number of iterations (average over 10 different trials) for the considered UCI datasets.

5.6. The Role of β: Underfitting, Optimality, and Overfitting

As usual with almost all machine learning algorithms, the choice of hyperparameters is of fundamental importance. For simplicity, in all experiments described in the previous sections, we kept the value of β constant through the network. To gain some intuition, Figure 6 shows the misclassification probability for different β for the three considered datasets (each time keeping β

constant through the network). While the three curves are quantitatively different, we can notice the same qualitative trend: when β is too small, not enough information about the target variable is propagated, and then by increasing β above a certain threshold, the misclassification probability drops. Increasing β too much however induces overfitting, as expected, and the classification error (slowly) increases again. Remember (from Equation (15)) that the Lagrangian we are minimizing is

$$\mathcal{L} = \mathbb{I}(X_{in}; X_{out}) - \beta \mathbb{I}(Y; X_{out}).$$

Information theory tells us that at every information node we should propagate only the sufficient statistic about the target variable Y. In practice, this is reflected in the role of β: when it is too small, we neglect the term $\mathbb{I}(Y; X_{out})$ and just minimize $\mathbb{I}(X_{in}; X_{out})$ (that corresponds to underfitting), while increasing β allows passing more information about the target variable through the bottleneck. It is important to remember, however, that we do not have direct access to the true mutual information values but just to an empirical estimate based on a finite dataset. Especially when the cardinalities of inputs and outputs are high, this translates into an increased probability of spotting spurious correlations that, if learned by the nodes, induce overfitting. The overall message is that β has an extremely important role in the proposed method, and its value should be chosen to modulate between underfitting and overfitting.

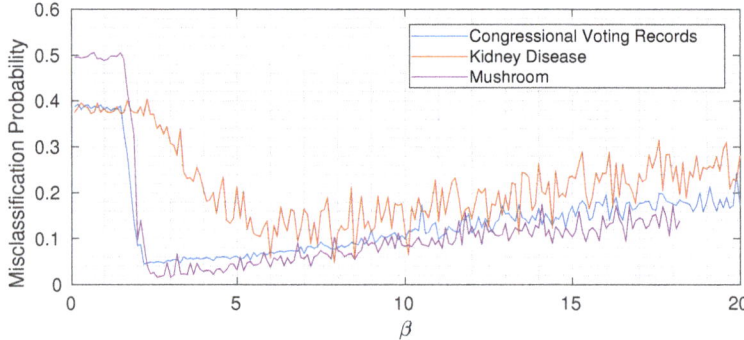

Figure 6. Misclassification probability versus β (average over 20 different trials) for the considered UCI datasets.

5.7. A Synthetic Multiclass Experiment

In this section we present results on a multiclass synthetic dataset. We generated 64-dimensional feature vectors $\mathbf{z}$ drawn from multivariate Gaussian distributions with mean and covariance depending on a target class y and a control parameter ρ:

$$p(\mathbf{z}|y=l) = |2\pi\Sigma_l|^{-\frac{1}{2}} \exp\left(-\frac{1}{2}(\mathbf{z}-\mu_l)^T(\rho\Sigma_l)^{-1}(\mathbf{z}-\mu_l)\right) \quad l=1,\cdots,N_{class} \qquad (43)$$

where for the considered experiment $N_{class} = 8$. The mean μ_l is sampled from a normal 64-dimensional random vector and Σ_l is randomly generated as $\Sigma_l = \mathbf{A}\mathbf{A}^T$ (where $\mathbf{A}$ is sampled from a matrix normal distribution) and normalized to have unit norm. The other parameter ρ is inserted to modulate the signal to noise ratio of the generated samples: a smaller value of ρ corresponds to smaller feature variances and more distinct, less overlapping, pdfs $p(\mathbf{z}|y=l)$, and an easier classification task. We then perform quantization of the result using 1 bit, i.e., the input of the ensemble of DINs is the following random vector:

$$\mathbf{x} = U(\mathbf{z}) \qquad (44)$$

where $U(\cdot)$ is the Heaviside step operator. The designed architecture has at the first layer 64 input nodes, followed by 32, 16, 4, 2, and 1. The output cardinalities are equal to 2 for the first three layers, 4 for the fourth and fifth layer, and 8 at the last layer. We selected $N_{train} = 1000$, $\beta = 7$ (constant trough the network), and number of DINs equal to $N_{mach} = 10$. Figure 7 shows the classification accuracy (on a test set of 1000 samples) for different values of ρ. As expected, when the value of ρ is small, we can reach almost perfect classification accuracy, whereas, by increasing it, the performance drops to the point where the useful signal is completely buried in noise and the classification accuracy reaches the asymptotic level of $\frac{1}{8}$ (that corresponds to random guessing when the number of classes is equal to 8).

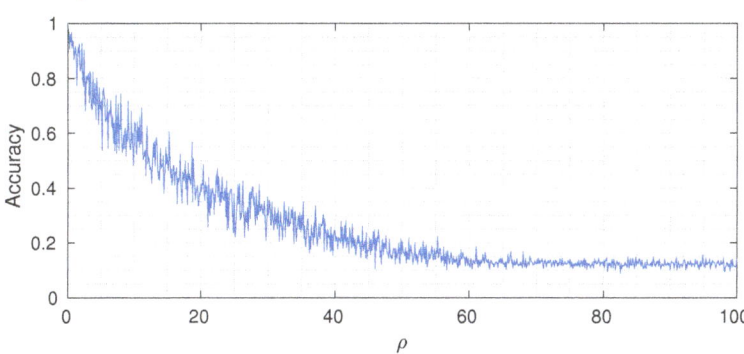

Figure 7. Varying of classification accuracy for different values of control parameter ρ.

6. Conclusions

The proposed ensemble Deep Information Network (DIN) shows good results in terms of accuracy and represents a new simple, flexible, and modular structure. The required hyperparameters are the cardinality of the alphabet at the output of each information node, the value of the Lagrangian multiplier β, and the structure of the tree itself (number of input information nodes of each combiner).

Simplistic architecture choices made for the experiments (such as equal cardinality of all node outputs, β constant through the network, etc.) performed comparably to finely tuned networks. However, we expect that, similar to what happened in neural network applications, a domain specific design of the architectures will allow for consistent improvements in terms of performance on complex datasets.

Despite the local assumption of conditionally independent features, the proposed method always outperforms Naive Bayes. As discussed in Section 4, the induced equivalent probability matrix is different in the two cases. Intuitively, we can understand the difference in performance under the point of view of probability matrix factorization. On the one side, we have the true, exponentially large, joint probability matrix of all features and target class. On the other side, we have the Naive Bayes one, which is extremely simple in terms of complexity but obviously less performing. In between, we have the proposed method, where the complexity is still reasonable but the quality of the approximation is much better. The DIN(Gen) algorithm does not require the assumption of statistical independence, but the classification accuracy is very close to that of DIN(Prob), which further suggests that the assumption can be accepted from a practical point of view.

The proposed method leaves open the possibility of devising a custom hardware implementation. Differently from classical decision trees, in fact, the execution times of all branches as well as the precise number of operations is fixed per datum and known a priori, helping in various system design choices. In fact, with classical trees, where a node's utilization depends on the datum, we are forced to design the system for the worst case, even if in the vast majority of time not all nodes are used. Instead, with DIN, there is no such a problem.

Finally, a clearly open point is related to the quantization procedure of continuous random variables. One possible self-consistent approach could be devising an information bottleneck based method (similar to the method for continuous random variables [20]).

Further studies on extremely large datasets will help understand principled ways of tuning hyperparameters and architecture choices and their relationship on performance.

Author Contributions: Conceptualization, G.F. and M.V.; methodology, G.F. and M.V.; software, G.F. and M.V.; validation, G.F. and M.V.; formal analysis, G.F. and M.V.; investigation, G.F. and M.V.; resources, G.F. and M.V.; data curation, G.F. and M.V.; writing–original draft preparation, G.F. and M.V.; writing–review and editing, G.F. and M.V.; visualization, G.F. and M.V.; All authors have read and agreed to the published version of the manuscript.

Funding: This research received no external funding.

Acknowledgments: A special thank to MD Gabriella Olmo who suggested a quantization of the continuous values of the features in the experiment in Section 5.2, which is correct from a medical point of view.

Conflicts of Interest: The authors declare no conflict of interest.

Appendix A. Quantization

Hereafter, we present the quantization scheme used for the numerical features of chronic kidney disease dataset.

- Age (Years) $\{< 10, < 18, < 45, < 70, < 120\}$
- Blood (mm/Hg) $\{< 80, < 84, < 89, < 99, < 109, \geq 110\}$
- Blood Glucose Random (mg/dl) $\{< 79, < 160, < 200, \geq 200\}$
- Blood Urea (mg/dl) $\{< 6, < 20, \geq 20\}$
- Serum Creatinine (mg/dl) $\{< 0.5, < 1.2, < 2, \geq 2\}$
- Sodium (mEq/l) $\{< 136, < 145, \geq 145\}$
- Potassium (mEq/l) $\{< 3.5, < 5, \geq 5\}$
- Haemoglobin (gm) $\{< 12, < 17, \geq 17\}$
- Packed Cell Volume $\{< 27, < 52, \geq 52\}$
- White Blood Cell Count (cells/mm^3) $\{< 3500, < 10500, \geq 10500\}$
- Red Blood Cell (millions/mm^3) $\{< 2.5, < 6, \geq 6\}$

References

1. Hastie, T.; Tibshirani, R.; Friedman, J. *The Elements Of Statistical Learning*; Springer: Berlin/Heidelberg, Germany, 2001. [CrossRef]
2. Murphy, K. *Machine Learning: A Probabilistic Perspective*; The MIT Press: Cambridge, MA, USA, 2012.
3. Bergman, M.K. *A Knowledge Representation Practionary*; Springer: Basel, Switzerland, 2018.
4. Rokach, L.; Maimon, O.Z. *Data Mining with Decision Trees: Theory and Applications*; World Scientific: Singapore, 2008; Volume 69.
5. Quinlan, J.R. Induction of decision trees. *Mach. Learn.* **1986**, *1*, 81–106. [CrossRef]
6. Quinlan, J. *C4.5: Programs for Machine Learning*; Morgan Kaufmann: Burlington, MA, USA, 1993.
7. Quinlan, J. Improved Use of Continuous Attributes in C4.5. *J. Artif. Intell. Res.* **1996**, *4*, 77–90. [CrossRef]
8. Pearl, J. *Probabilistic Reasoning in Intelligent Systems: Networks of Plausible Inference*; Elsevie: Burlington, MA, USA, 2014.
9. Barber, D. *Bayesian Reasoning and Machine Learning*; Cambridge University Press: Cambridge, UK, 2012.
10. Jensen, F.V. *Introduction to Bayesian Networks*; UCL Press: London, UK, 1996; Volume 210.
11. Norouzi, M.; Collins, M.; Johnson, M.A.; Fleet, D.J.; Kohli, P. Efficient Non-greedy Optimization of Decision Trees. In Proceedings of the 28th International Conference on Neural Information Processing Systems, Montreal, QC, Canada, 7–12 December 2015; pp. 1729–1737.
12. Breiman, L. Bagging Predictors. *Mach. Learn.* **1996**, *24*, 123–140. [CrossRef]

13. Breiman, L. Random Forests. *Mach. Learn.* **2001**, *45*, 5–32. [CrossRef]
14. Tishby, N.; Zaslavsky, N. Deep Learning and the Information Bottleneck Principle. *arXiv* **2015**, arXiv:1503.02406v1.
15. Tishby, N.; Pereira, F.; Bialek, W. The Information Bottleneck method. *arXiv* **2000**, arXiv:physics/0004057v1.
16. Franzese, G.; Visintin, M. Deep Information Networks. *arXiv* **2018**, arXiv:1803.02251v1.
17. Slonim, N.; Tishby, N. Agglomerative information bottleneck. In Proceedings of the 12th International Conference on Neural Information Processing Systems, Denver, CO, USA, 29 November–4 December 1999; pp. 617–623.
18. Still, S. Information bottleneck approach to predictive inference. *Entropy* **2014**, *16*, 968–989. [CrossRef]
19. Still, S. Thermodynamic cost and benefit of data representations. *arXiv* **2017**, arXiv:1705.00612.
20. Chechik, G.; Globerson, A.; Tishby, N.; Weiss, Y. Information bottleneck for Gaussian variables. *J. Mach. Learn. Res.* **2005**, *6*, 165–188.
21. Gedeon, T.; Parker, A.E.; Dimitrov, A.G. The mathematical structure of information bottleneck methods. *Entropy* **2012**, *14*, 456–479. [CrossRef]
22. Freund, Y.; Schapire, R. A short introduction to boosting. *Jpn. Soc. Artif. Intell.* **1999**, *14*, 1612.
23. Chen, T.; Guestrin, C. Xgboost: A scalable tree boosting system. In Proceedings of the 22nd ACM Sigkdd International Conference on Knowledge Discovery and Data Mining, San Francisco, CA, USA, 13–17 August 2016; pp. 785–794.
24. Arimoto, S. An algorithm for computing the capacity of arbitrary discrete memoryless channels. *IEEE Trans. Inf. Theory* **1972**, *18*, 14–20. [CrossRef]
25. Blahut, R. Computation of channel capacity and rate-distortion functions. *IEEE Trans. Inf. Theory* **1972**, *18*, 460–473. [CrossRef]
26. Hand, D.J.; Yu, K. Idiot's Bayes—Not so stupid after all? *Int. Stat. Rev.* **2001**, *69*, 385–398.
27. UCI Machine Learning Repository, University of California, Irvine, School of Information and Computer Sciences. Available online: http://archive.ics.uci.edu/ml (accessed on 30 September 2010).
28. Salekin, A.; Stankovic, J. Detection of chronic kidney disease and selecting important predictive attributes. In Proceedings of the IEEE International Conference on Healthcare Informatics (ICHI), Chicago, IL, USA, 4–7 October 2016; pp. 262–270.
29. Duch, W.; Adamczak, R.; Grąbczewski, K. Extraction of logical rules from neural networks. *Neural Process. Lett.* **1998**, *7*, 211–219. [CrossRef]

© 2020 by the authors. Licensee MDPI, Basel, Switzerland. This article is an open access article distributed under the terms and conditions of the Creative Commons Attribution (CC BY) license (http://creativecommons.org/licenses/by/4.0/).

Article

Convergence Behavior of DNNs with Mutual-Information-Based Regularization

Hlynur Jónsson, Giovanni Cherubini * and Evangelos Eleftheriou

IBM Research Zurich, 8803 Rüschlikon, Switzerland; hlynur4@gmail.com (H.J.); ele@zurich.ibm.com (E.E.)
* Correspondence: cbi@zurich.ibm.com; Tel.: +41-44-724-8518

Received: 17 June 2020; Accepted: 26 June 2020; Published: 30 June 2020

Abstract: Information theory concepts are leveraged with the goal of better understanding and improving Deep Neural Networks (DNNs). The information plane of neural networks describes the behavior during training of the mutual information at various depths between input/output and hidden-layer variables. Previous analysis revealed that most of the training epochs are spent on compressing the input, in some networks where finiteness of the mutual information can be established. However, the estimation of mutual information is nontrivial for high-dimensional continuous random variables. Therefore, the computation of the mutual information for DNNs and its visualization on the information plane mostly focused on low-complexity fully connected networks. In fact, even the existence of the compression phase in complex DNNs has been questioned and viewed as an open problem. In this paper, we present the convergence of mutual information on the information plane for a high-dimensional VGG-16 Convolutional Neural Network (CNN) by resorting to Mutual Information Neural Estimation (MINE), thus confirming and extending the results obtained with low-dimensional fully connected networks. Furthermore, we demonstrate the benefits of regularizing a network, especially for a large number of training epochs, by adopting mutual information estimates as additional terms in the loss function characteristic of the network. Experimental results show that the regularization stabilizes the test accuracy and significantly reduces its variance.

Keywords: deep neural networks; information bottleneck; regularization methods

1. Introduction

Deep Neural Networks (DNNs) have revolutionized several application domains of machine learning, including computer vision, natural language processing and recommender systems. Despite their success, the internal learning process of these networks is still an active field of research. One of the goals of this paper is to leverage information theoretical concepts to analyze and further improve DNNs. The analysis of DNNs through the information plane, i.e., the plane of mutual information values that each layer preserves at various learning stages on the input and the output random variables, was proposed in [1,2]. Previous approaches for the visualization of the information plane applied non-parametric estimation methods that do not work well with high dimensional data [2–4], as in this case the estimation of mutual information is nontrivial. The information plane for small fully connected networks was visualized in [2]. The results in [2] suggested that most of the training epochs of a DNN, the "compression phase", are spent on compressing the input variables after an initial "fitting phase". The existence of the compression phase was later questioned in [4–7], in case the finiteness of the mutual information between input/output and hidden-layer variables cannot be established. In this paper, we focus on Convolutional Neural Networks (CNNs) with high complexity. After briefly discussing non-parametric mutual information estimation methods, we present the convergence of mutual information on the information plane for a high-dimensional VGG-16 CNN [8] by resorting to Mutual Information Neural Estimation (MINE) [9]. The compression phase is evident from the obtained results, which confirm and extend the results previously found with low-dimensional fully connected networks, where methods with lower computational complexity were adopted to study the convergence in the information plane.

Furthermore, we consider DNNs with mutual-information-based regularization. The use of the mutual information between the input and a hidden layer of a DNN as a regularizer was suggested in [9–11]. The idea is based on the Information Bottleneck (IB) approach [12], which provides a maximally compressed version of the input random variable, while still retaining as much information as possible on a relevant random variable. Here we compare the accuracy achieved by a VGG-16 CNN, using well-known regularization techniques, such as dropout, batch normalization and data augmentation, with that of the same VGG-16 network that is enhanced by applying mutual-information-based regularization, by resorting either to MINE or the variational information bottleneck (VIB) technique [10], and demonstrate the advantages of mutual-information-based regularization, especially for a large number of training epochs.

The remainder of the paper is structured as follows. Basic definitions of mutual information and the formulation of the IB approach are recalled in Section 2. Non-parametric methods for the estimation of mutual information in DNNs are addressed in Section 3. The visualization of DNN convergence on the information plane using MINE is described in Section 4, whereas the advantages of long-term DNN regularization using mutual-information-based techniques are presented in Section 5. Finally, conclusions are given in Section 6.

2. The Information Bottleneck in DNNs

The mutual information is a measure of the mutual dependence of two random variables. In essence, it measures the relationship between them and may be regarded as the reduction in uncertainty in a random variable given the knowledge available about another one. The mutual information between two discrete random variables X and Y is defined as

$$I(X;Y) = \sum_{x \in X} \sum_{y \in Y} p(x,y) \log \frac{p(x,y)}{p(x)p(y)} \quad (1)$$

where $p(x,y)$ is the joint probability distribution and $p(x)$ and $p(y)$ are the marginal probability distributions. The mutual information may also be expressed as the Kullback–Leibler (KL) divergence between the joint distribution and the product of the marginal distributions of the random variables X and Y

$$I(X;Y) = D_{KL}\left(p(x,y) \| p(x)p(y)\right) \quad (2)$$

Computing the mutual information of two random variables is in general challenging. Exact computation can be done in the discrete setting, where the sum in Equation (1) can be calculated exhaustively, or in the continuous setting, provided the probability distributions are known in advance. Several methods to estimate the mutual information have been introduced. The most common ones are non-parametric, including discretizing the values using binning and resorting to non-parametric kernel-density estimators, as will be discussed in more detail in Section 3.

The IB method was introduced in [12] to find a maximally compressed representation of an input random variable, X, which is obtained as a function of a relevant random variable, Ψ, such that it preserves as much information as possible on Ψ. Let us denote $\hat{X}$ as the compressed version of the input random variable X parameterized by θ. What the IB method effectively does is to solve the following optimization problem

$$\min_{\theta} I_\theta(X; \hat{X})$$
$$\text{subject to:} \quad I_\theta(\hat{X}; \Psi) = I_c \leq I(X; \Psi) \quad (3)$$

where I_c is the information constraint. In a non-deterministic classification scenario, an equivalent formulation of the IB problem is obtained by introducing the Lagrangian multiplier $\beta' \geq 0$, and by minimizing the Lagrangian

$$\mathcal{L}(\theta) = I_\theta(X; \hat{X}) - \beta' I_\theta(\hat{X}; \Psi) \tag{4}$$

In a deterministic classification scenario, however, where neural networks do not exhibit any randomization of the hidden layer output variables, the two IB formulations given above are in general not equivalent, as demonstrated in [13].

The goal of any supervised learning method is to efficiently capture the relevant information about an input random variable, typically a label for classification tasks, in the output random variable [1]. Let us consider a DNN for image recognition, with input X, output $\hat{Y}$, and i-th hidden layer denoted by h_i. The classification task is related to the interpretation of an image that is generated from a relevant random variable Y.

In case the hidden layer, h_i, only processes the output of the previous layer, h_{i-1}, the layers form a Markov chain of successive internal representations of the input. By applying the Data Processing Inequality (DPI) [14] to a DNN that consists of L layers, we have

$$I(X; h_1) \geq I(X; h_2) \geq \ldots \geq I(X; h_L) \geq I(X; \hat{Y}) \tag{5}$$

As mentioned in the Introduction, DNNs may be analyzed through mutual information values on the information plane. However, estimating mutual information across layers in DNNs is a nontrivial task, as the outputs of the hidden layer neurons are continuous-valued high-dimensional random vectors.

A further difficulty arises if the input X is a continuous random variable and the neural network is deterministic. In this case, it has been shown, e.g., in [4–6], that $I(X; h_i) = \infty$ for commonly used activation functions. For the image classification problem considered here, one might argue that the input random variable is a discrete random variable, where the pixels have a discrete distribution, and h_i is also a discrete random variable given by a deterministic transformation of X via finite-precision arithmetic. The training and test sets, however, have a cardinality that is typically much lower than the alphabet size of X, thus rendering the estimation of the mutual information very difficult. To cope with the challenge of estimating the divergence in Equation (2), we will resort to MINE, as discussed in Section 4.

3. Non-Parametric Estimation of Mutual Information

As stated in Section 2, the most common methods for the estimation of mutual information are non-parametric. As our focus is on CNNs, we consider a VGG-16 network [8] to evaluate the effectiveness of non-parametric estimation methods. The block diagram of a VGG-16 network is illustrated in Figure 1. The loss function adopted to train the network is the cross-entropy loss obtained from the softmax output probabilities, that is

$$\mathcal{L}_{CE} = \sum_{m=1}^{M} y_m \log p_m \tag{6}$$

where y_m is the binary value corresponding to the m-th class of a one-hot encoded vector defined by the class labels, p_m is the softmax probability output for class m, and M is the number of classes in the classification task. In all experiments the dataset considered is CIFAR-10 [15], i.e., $M = 10$, with a batch size of 128. CIFAR-10 is a dataset consisting of 60,000 images in 10 classes. Fifty thousand of those images are used for training and 10,000 for testing. Each image of the dataset is of size 32×32 with three color channels.

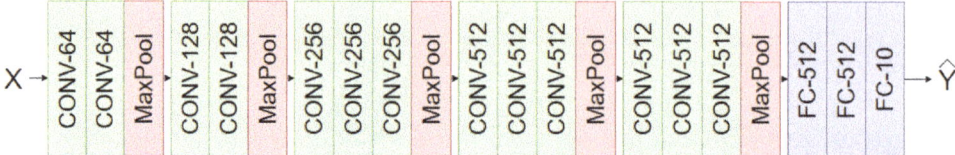

Figure 1. The Visual Geometry Group (VGG)-16 network [8] architecture from input to predicted output. CONV-64 is shorthand for a 2D convolutional layer with 64 filters. FC-512 is shorthand for a fully connected layer with 512 neurons.

3.1. Activation Binning

The mutual-information estimation method adopted in [2] resorts to binning the activations of each layer into equal-sized bins. Binning ensures that the values are discretized, which allows to calculate the joint distribution and marginal distributions by counting the activations in each bin.

Despite the promising results reported in [2], this method has a number of limitations as a general method for mutual-information estimation. Firstly, the experiments in [2] were conducted using a fully connected network with only a few neurons, which has far fewer synaptic weights and neurons than typical CNN architectures. Another shortcoming is the usage of the *tanh* function as a non-linear activation function, which bounds the activations from −1 to 1. The ReLU activation function [16] is more commonly used and allows unbounded positive activations. This limitation is pointed out in [4], where the authors showed counterexamples of the compression phase using ReLU non-linear activations. Furthermore, in [2] the input is limited to a vector of 12 binary elements and the output is also binary. In this case binning is convenient because of the low number of dimensions of the input and output random variables.

We found that the method of binning the activations does not scale well with higher dimensionality. For example, we experimented with the activations of VGG-16 trained on CIFAR-10. For the classification of a CIFAR-10 image, the input random variable has a total of 3072 dimensions as opposed to 12 in [2]. Varying the number of bins where the activations are allocated was also found not to have a significant impact on the results. The mutual-information estimates for layers with high input dimensionality turned out not to satisfy the DPI. In addition, the estimates of both $I(X; h_i)$ and $I(h_i; Y)$ for the last few CNN layers converged to values close to zero. If $I(h_i; Y)$ approximates zero the accuracy of the model should be roughly the same as for random guessing. This contrasts with the measured accuracy, which is higher than 90%, compared to 10% for random guessing.

3.2. Non-Parametric Kernel Density Estimation

We also conducted experiments on the kernel-density estimation method described in [4,17]. The assumption made is that the hidden layer activities are distributed as a mixture of Gaussian random variables with covariance matrix $\sigma^2 I$. In [4], it is further assumed for analysis purposes that Gaussian noise is added to each hidden layer T_i, which is expressed as $T_i = h_i + \epsilon$, where $\epsilon \sim \mathcal{N}(0; \sigma^2 I)$. A mutual information upper bound with respect to the input is proposed as

$$I(T_i; X) \leq -\frac{1}{P}\sum_j \log \frac{1}{P}\sum_k \exp\left(-\frac{1}{2}\frac{\|h_{ij}-h_{ik}\|_2^2}{\sigma^2}\right) \qquad (7)$$

where P denotes the number of samples and h_{ij} the hidden layer activities of layer i for sample j. Furthermore, the mutual information with respect to the output random variable is upper bounded as follows

$$I(T_i; Y) \leq -\frac{1}{P}\sum_j \log \frac{1}{P}\sum_k \exp\left(-\frac{1}{2}\frac{||h_{ij}-h_{ik}||_2^2}{\sigma^2}\right)$$
$$-\sum_m p_m \left[-\frac{1}{P_m}\sum_{j,Y_j=m} \log \frac{1}{P_m}\sum_{k,Y_k=m} \exp\left(-\frac{1}{2}\frac{||h_{ij}-h_{ik}||_2^2}{\sigma^2}\right)\right] \tag{8}$$

where P_m is the number of samples with class label m, and $p_m = P_m/P$ is the empirical probability of class m.

The same experiment as in [2] was conducted in [4] by using the non-parametric kernel density estimation.

We also tested the estimation method by a VGG-16 network, and adopted a variance of $\sigma^2 = 0.1$, as in [4]. However, as experienced with the binning method in Section 3.1, we did not find satisfactory results with a convolutional network of high complexity.

3.3. Rényi's α-Entropy

A multivariate matrix-based Rényi's α-entropy method was proposed in [3] for application to a LeNet-5 network. This approach is suitable for CNN networks, as each CNN layer has several channels, which all contain some information on the input and output random variables. However, two distinct channels of a single layer can contain the same information on the input and output random variables. Therefore, summing up the mutual information estimates between each channel and the input or output random variable only gives an upper bound on the mutual information that has little relevance to the true mutual information value. The experiments in [3] for LeNet-5 result in mutual information estimates for the various layers that satisfy the DPI. However, our experiments for VGG-16, which has up to 512 channels, did not yield estimates that comply with the DPI.

A method for the estimation of mutual information in complex DNNs was proposed in [18], which relies on matrix-based Rényi's entropy and tensor kernels to estimate the mutual information in a VGG-16 network. The method in [18] augments the multivariate extension of the matrix-based Rényi's α-order entropy presented in [19] by introducing tensor kernels. In that manner, the tensor-based nature of convolutional layers in DNNs is recognized and the numerical difficulties arising by a straightforward application of the multivariate extension of the matrix-based Rényi's entropy are avoided. However, the convergence in the information plane is affected by the overfitting that takes place when the training is conducted for a large number of epochs. Therefore, the compression phase needs to be limited by an early stopping technique to prevent overfitting.

4. DNN Convergence Analysis Using MINE

Our goal is to visualize the information plane for networks with high-dimensional variables, as previous work focuses on networks with much lower complexity [2,4,10]. The methods discussed in Section 3 for estimating the mutual information do not perform well with high-dimensional random variables. Furthermore, the existence of a compression phase during training has been disputed in [4] for networks where the finiteness of the mutual information between input/output and hidden-layer variables cannot be established. Therefore, to clarify these issues, we consider a VGG-16 network with ReLU activation function. For the estimation of the mutual information for all layers in the network, we resort to the MINE method [9].

MINE is a method first proposed in [9] for the estimation of mutual information between high-dimensional continuous random variables. The method takes advantage of the Donsker–Varadhan dual representation of the KL-divergence [20] and utilizes the lower bound

$$D_{KL}(\mathbb{P}\|\mathbb{Q}) \geq \sup_{T\in\mathcal{F}} \mathbb{E}_\mathbb{P}[T] - \log(\mathbb{E}_\mathbb{Q}[e^T]) \tag{9}$$

where $\mathcal{F}$ is any class of functions $T: \mathbb{R}^d \to \mathbb{R}$ such that the two expectations are finite, and $\mathbb{P}$ and $\mathbb{Q}$ are probability distributions. The main idea of MINE is to choose T as a function parameterized by a deep neural network with parameters $\theta \in \Theta$. By defining $\mathbb{P}$ as the joint probability distribution and $\mathbb{Q}$ as the product of the marginal distributions of the random vectors X and Y, by combining Equations (9) and (2) we get the MINE lower bound

$$\hat{I}(X;Y) = \sup_{\theta \in \Theta} \mathbb{E}_{p(x,y)}[T_\theta] - \log(\mathbb{E}_{p(x)p(y)}[e^{T_\theta}]) \qquad (10)$$

The lower bound given by Equation (10) is finite, even if the input X is a continuous random vector and the neural network under investigation is deterministic, for which the mutual information between the input and a hidden layer $I(X;h_i)$ is infinite, as discussed in Section 2. If the mutual information is not finite, the MINE may nevertheless be regarded as a well-defined estimate of the statistical divergence between the two probability distributions $p(X,h_i)$ and $p(X)p(h_i)$ that assume nonzero (possibly infinite) values over different support. This is analogous to the evaluation of the optimal cost in applications of the optimal transport theory, which is obtained by resorting to a dual representation of the original problem, see, e.g., [21].

The expectation over the product of the marginal distributions is estimated by shuffling the samples from which the empirical joint distribution is obtained. For the estimation of $I(X;h_i)$ the samples from h_i are shuffled, whereas for the estimation of $I(h_i;Y)$ the samples from Y are shuffled. The objective function from Equation (10) is adopted and is optimized by gradient ascent. For the visualization of the information plane for the i-th layer in the network, two estimations are needed, namely those of $I(X;h_i)$ and $I(h_i;Y)$. Each of these estimations is parameterized by a separate deep neural network. As stated in [9], more training samples are needed as the complexity of the MINE network increases. Therefore, very deep networks with high complexity are infeasible as MINE networks. Here we adopt a network and an overall training approach capable of accurately estimating mutual information while resorting to networks with moderate complexity.

For the experiments in this section, we train a VGG-16 network on CIFAR-10 images. Minor data augmentation is applied in the form of random cropping and randomly flipping the images horizontally. The size of each CIFAR-10 image is 32×32 pixels. For random cropping, we pad the original image to 40×40 pixels and randomly take a crop of size 32×32 pixels. In our experiments, each convolutional layer has a 3×3 pixel receptive field. In addition, batch-normalization is used for all convolutional layers. Furthermore, dropout regularization is applied after all convolutional layers that do not precede a pooling layer (dropout rate of 0.3) and after the first fully connected layer (dropout rate of 0.5). The ReLU activation function is adopted for all layers with the exception of the last one, which is a linear dense layer. The hyperparameters are chosen using a validation set obtained by extracting 10,000 samples from the training set. The MINE loss function used to train each MINE network is defined as

$$\mathcal{L}_{MINE} = \frac{1}{n}\sum_{i=1}^{n} T_\theta(i) - \log(\frac{1}{n}\sum_{j=1}^{n} \exp^{T_\theta(j)}) \qquad (11)$$

where n is the batch size and $T_\theta(i)$ and $T_\theta(j)$ are the individual network outputs referring to the expectation over the joint distribution and over the product of the marginal distributions (see Equation (10)), respectively. An illustration of the process, including input and output encoders and referring to the hidden layer h_3 as an example, is shown in Figure 2.

4.1. Visualization of the Information Plane

As discussed above with reference to Equation (10), each estimate of mutual information by MINE requires a separate neural network to learn both the expectations over the joint distribution and over the product of the marginal distributions. To characterize the convergence behavior of the VGG-16 on the information planes, we need to estimate both $I(X;h_i)$ and $I(h_i;Y)$ for each layer, i.e., a total of 32 networks are needed. Additionally, we adopt two encoders, which are employed across all layers, to encode the input and output random variables. Each MINE network encodes the respective hidden layer.

As illustrated in Figure 2, all hidden layer encoders and input/output encoders output a 64-dimensional vector. The hidden layer encoder output is concatenated with the corresponding encoder output for the input/output random variable, resulting in a 128-dimensional vector. A fully connected network takes the concatenated vector as input and outputs a single value, from which the expected value over either distribution is obtained, depending on whether the input is shuffled or not. The expected values obtained from the network yield the mutual information estimate by Equation (10). Further details on the architectures of the experimental MINE networks are given in [22].

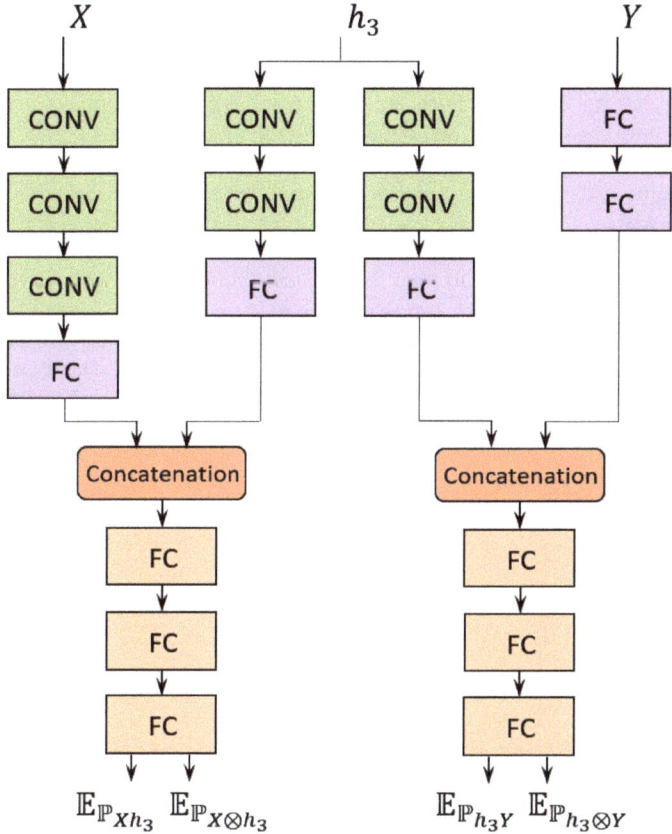

Figure 2. Example of the Mutual Information Neural Estimation (MINE) networks considered for both $I(X; h_3)$ and $I(h_3; Y)$ in layer 3 of VGG-16. The same input and output encoder are employed for all layers. The four expectations are applied as indicated in Equation (10) to estimate both $I(X; h_3)$ and $I(h_3; Y)$. FC indicates a fully connected layer, whereas CONV indicates a convolutional layer.

An information plane shows the mutual information estimates for all epochs within a certain layer. To get unbiased estimates for each epoch the training procedure is conducted as follows. Initially, the VGG-16 network is trained up to a certain epoch. Then all MINE networks are trained for a total of 1000 epochs. During MINE network training, the MINE networks use the outputs of the hidden layer neurons of the trained VGG-16 as input, without updating the weights of the VGG-16 through back-propagation. In this phase, gradient-ascent updating by back-propagation is only performed on the weights of the MINE networks. After training the MINE network for 1000 epochs, the expectations are evaluated to find the estimates of mutual information on the information plane. Therefore, each dot on the information plane of the i-th layer represents the estimates of the MINE networks for $I(X; h_i)$

and $I(h_i; Y)$, for a single epoch of VGG-16, after training the MINE network for 1000 epochs. Each mutual information estimate shown on the information plane is obtained by the same number of training iterations. To visualize the information plane, we consider the first 50 epochs of the VGG-16 training phase. The mutual information values do not provide more insight on the compression phase beyond that point. The above procedure is repeated for all 50 VGG-16 epochs shown in the information plane.

The information planes of the VGG-16 layers are shown in Figure 3. The mutual information estimates are expressed in bits. The compression phase is evident especially in the high-order layers, which is consistent with previous work presented in [2], however for the first time shown in a CNN with such a high complexity. One further difference with respect to [2] is that for the VGG-16 network trained on CIFAR-10, the compression phase appears earlier in the training process. We see that $I(X; h_i)$ starts decreasing after the first VGG-16 epoch for the high-order layers and continues to exhibit a decreasing trend until convergence. The estimation of $I(h_i; Y)$, for all layers $i = 2, ..., 16$, converges towards the upper bound equal to $\log_2 10$, which is the desired value of the mutual information in bits as CIFAR-10 contains 10 classes. An exception is constituted by the first layer, which seems to slightly underestimate the mutual information with the output. It can also be seen how the input is compressed successively in each layer. This behavior is more clear from layer 7 onward, as the estimates of $I(X; h_i)$ decrease between subsequent layers, as demanded by the DPI. While training the MINE networks, it was observed that the mutual information estimates converged to zero during training for some VGG-16 epochs. The occurrence of such events was significantly mitigated by lowering the learning rate. The lower learning rate, however, slows down the training process. Therefore, training over 1000 epochs was needed for MINE to allow the mutual information estimates to reliably converge. Figure 4 shows the decrease of the mutual information estimates $I(X; h_i)$ and $I(h_i; Y)$ as a function of the layer index, for the 1st and the 40th epoch, i.e., towards beginning and end of the considered training interval, respectively, thus indicating that the DPI is well approximated by the MINE.

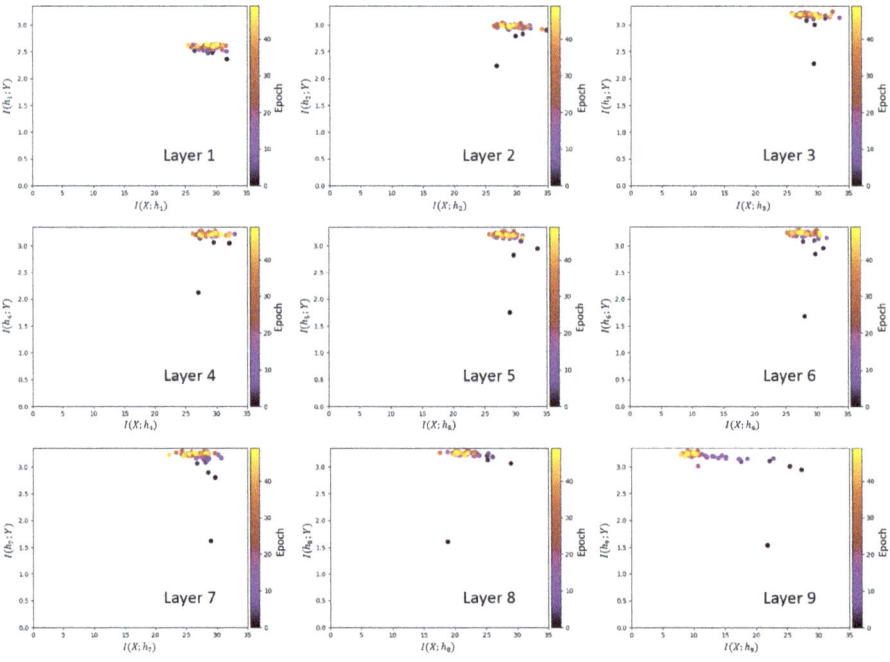

Figure 3. *Cont.*

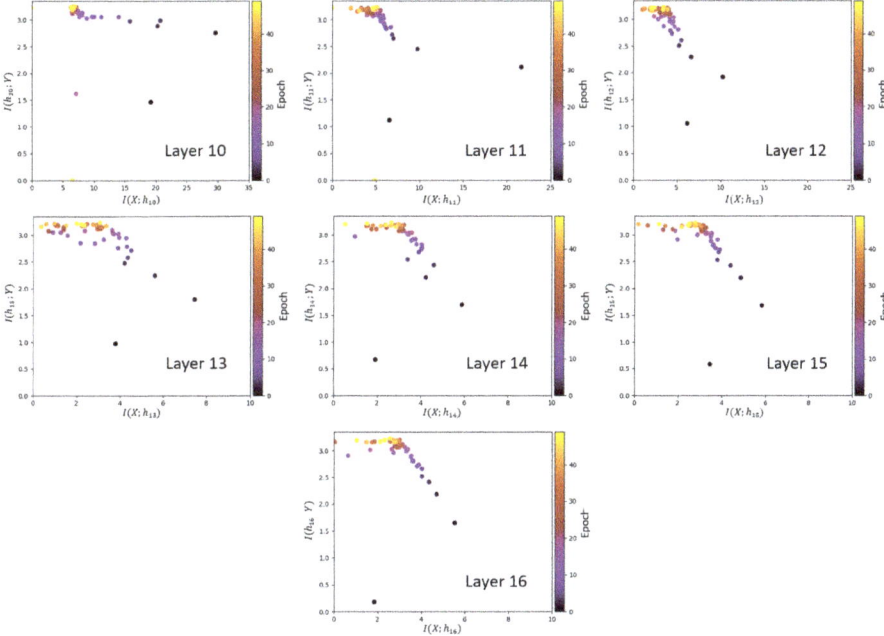

Figure 3. Information planes for the VGG-16 layers trained on Canadian Institute for Advanced Research (CIFAR)-10 image set.

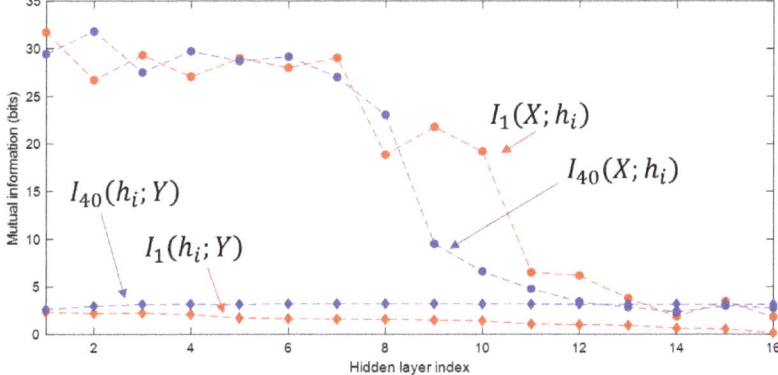

Figure 4. Mutual information estimates as function of the layer index, for the 1st and the 40th epoch. The subscript n in $I_n(h_i; Y)$ indicates the epoch.

We remark that the results presented in this section are qualitative. Proper quantitative assessment of the variance in the trajectories and a comparative study of the convergence of DNNs having different architectures will be the subject of further investigation.

5. Long-Term DNN Regularization

5.1. MINE-Based Regularization

Using MINE as a regularizer was proposed in [9] for a small fully connected network trained on MNIST. The authors replaced the variational approximation of the mutual information in [10] with a

MINE network for the mutual information estimate. We also consider a MINE network to estimate the mutual information, however with a VGG-16 network of significantly higher complexity. In our experiments, we estimated the mutual information of two layers by applying MINE networks. We trained a VGG-16 network for a total of 10,000 epochs to investigate how the MINE-based regularizer affects the test accuracy over the entire training period. An additional loss term was included in the objective function of the network that represents the estimates of $I(X; h_{14})$ and $I(X; h_{15})$ by MINE. The network parameters of the MINE networks were the same as described in Section 4.1. We performed gradient descent on the cross-entropy loss with a regularization term that equals the sum of the mutual information estimations of $I(X; h_{14})$ and $I(X; h_{15})$ multiplied by a regularization coefficient, which was chosen as $\beta = 10^{-3}$. The overall loss function is defined as

$$\mathcal{L} = \mathcal{L}_{CE} + \sum_{l=1}^{L} \mathcal{L}_{MINE}(l) \tag{12}$$

where $\mathcal{L}_{CE}$ and $\mathcal{L}_{MINE}$ are defined in Equations (6) and (11), respectively, and L is the number of layers in the VGG-16 over which the regularization takes place.

The results without and with MINE-based regularizer are shown in Figure 6a,b, respectively. The test accuracy increases and the variance decreases with respect to the experiments without MINE-based regularization. The maximum test accuracy achieved with the MINE-based regularizer is 93.9%, whereas a baseline accuracy for a VGG-16 network is measured as 93.25% in [23], which is similar to our results shown in Figure 6a.

5.2. VIB-based Regularization

As an alternative mutual-information-based estimation method between consecutive layers in CNNs, a Variational Information Bottleneck (VIB) method was proposed in [10] for fully connected networks with low complexity. The VIB technique was also used in [24] to reduce network complexity. Here we extend VIB-based regularization to CNNs with substantially higher complexity. We investigate the performance of the regularizer when training a VGG-16 for a large number of training epochs, up to 10,000, in which case overfitting is a common issue.

We adopt the same formulation of the VIB as in [24]. In a feed-forward neural network like the VGG-16 network, each hidden layer, h_i, takes as input the previous output of the hidden layer, h_{i-1}. Therefore, each layer only extracts information from the previous layer. The previous layer typically contains some information that is not relevant to the output. The aim of a VIB-based regularizer is therefore to reduce the amount of redundant information extracted from the previous layer. This is accomplished by minimizing the estimated mutual information between subsequent layers, $I(h_i; h_{i-1})$. The information bottleneck objective then becomes

$$\mathcal{L} = \sum_{i=1}^{L} \beta_i I(h_i; h_{i-1}) - I(h_i; Y) \tag{13}$$

where the coefficient $\beta_i \geq 0$ represents the strength of the VIB-based regularization in the i th layer, and L is the number of layers in the network over which the regularization takes place. An upper bound on the term given by Equation (13) can be derived as

$$\hat{\mathcal{L}} = \sum_{i=1}^{L} \beta_i \mathbb{E}_{h_{i-1} \sim p(h_{i-1})} [D_{KL} [p(h_i|h_{i-1}) \| q(h_i)]] - \mathbb{E}_{x,y \sim \mathcal{D}} [\log q(y|h_L)] \tag{14}$$

where $\mathcal{D}$ denotes the input data distribution and $q(h_i)$ and $q(y|h_L)$ are variational distributions that approximate $p(h_i)$ and $p(y|h_L)$, respectively. To optimize the network, a parametric form of the distributions $p(h_i|h_{i-1})$, $q(h_i)$ and $q(y|h_L)$ is specified. In [24], it is assumed that each conditional distribution $p(h_i|h_{i-1})$ is defined via the relation

$$h_i = (\mu_i + \epsilon_i \odot \sigma_i) \odot f_i(h_{i-1}) \tag{15}$$

where the parameters μ_i and σ_i are learnable for each layer where VIB is applied and $\epsilon_i \sim \mathcal{N}(0,I)$. The function f_i represents the network processing that takes place at the i-th layer, consisting of a linear transformation or a convolution operation for convolutional layers, plus batch normalization and a non-linear activation function. Furthermore, the distribution $q(h_i)$ is specified as a Gaussian distribution, such that

$$q(h_i) = \mathcal{N}(h_i; 0, \text{diag}[\xi_i]) \tag{16}$$

where ξ_i is a vector of variances learned from the data.

The process is illustrated in Figure 5. The element-wise multiplications in Equation (15) are applied differently in fully connected layers and convolutional layers, as the convolutional layers have several channels. For each convolutional layer the learned parameters, μ_i and σ_i, are vectors with dimensionality equal to the number of channels in the layer. Therefore, we obtain a learned Gaussian distribution for each channel. The matrix, which is adopted for the element-wise multiplications, has the same dimensions as the convolutional layer output, and is generated by sampling from the distribution of each channel n^2 times, where $n \times n$ is the feature map size. For the fully connected layers, the vectors of the learned parameters have a dimensionality equal to the number of neurons in the layer. Accordingly, each element is associated with a separate learned Gaussian distribution. Thus, the loss function is expressed as

$$\hat{\mathcal{L}} = \gamma \sum_{i=1}^{L} \beta_i \sum_{j=1}^{r_i} \log\left(1 + \frac{\mu_{i,j}^2}{\sigma_{i,j}^2}\right) - \mathbb{E}_{x,y \sim \mathcal{D}}[\log q(y|h_L)] \tag{17}$$

where r_i denotes the number of channels for convolutional layers and neurons for fully connected layers. The coefficient γ is used to scale the regularizing term. Scaling is crucial in deep networks as the accumulated loss from every layer may otherwise become too large.

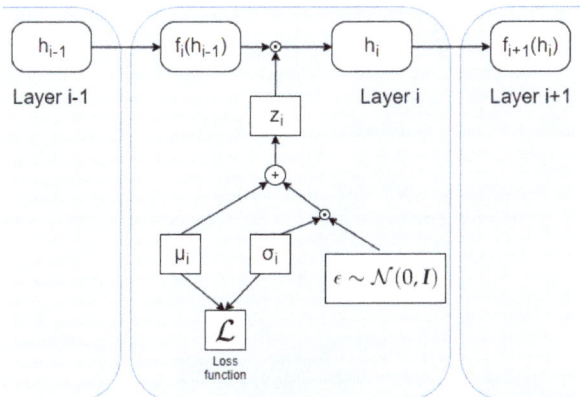

Figure 5. Illustration of how the VIB is incorporated into each layer using the formulation from [24] using Equation (15), where $z_i = \mu_i + \epsilon_i \odot \sigma_i$. The noise variable, ϵ_i, is sampled randomly from a Gaussian distribution with zero mean and unit variance.

As in the previous sections, the network adopted in our experiments was a VGG-16, trained on the CIFAR-10 dataset with the same data augmentation described in Section 4. The regularization constants were chosen as $\gamma = 10^{-5}$, $\{\beta_i\}_{i=1,\ldots,15} = \{2^{-5}, 2^{-5}, 2^{-4}, 2^{-4}, 2^{-3}, 2^{-3}, 2^{-3}, 2^{-2}, 2^{-2}, 2^{-2}, 2^{-1}, 2^{-1}, 2^{-1}, 1, 1\}$.

We trained the VGG-16 network without and with the VIB objective and compared the results. The output of the i-th layer in the experiment with the VIB was calculated as shown in Equation (15). Both models were trained for a total of 10,000 epochs on the CIFAR-10 dataset. To update the weights by back-propagation, we used the Adam [25] optimizer with exponential decay rate parameters $\beta_{A,1} = 0.9$, $\beta_{A,2} = 0.999$ and $\epsilon_A = 10^{-8}$. The learning rate was fixed to 0.001. We trained all models using a mini-batch size of 128. The results without and with the VIB objective are illustrated in Figure 6a,c, respectively.

The results in Figure 6c show that the test accuracy of the model increases with the additional VIB-based regularizer, achieving a value of 94.1%. We recall that the baseline accuracy for a VGG-16 network in [23] is 93.25%. Furthermore, the VIB-based regularizer prevents the model from overfitting. When trained for enough epochs, the model without the VIB-based regularizer eventually starts to overfit, even though it applies several regularization methods such as dropout, batch normalization and data augmentation, see Figure 6a. In contrast, the test accuracy exhibits substantially lower variance if the VIB objective is considered. To obtain the best accuracy from the model trained without VIB, early stopping is required. In contrast, the test accuracy of the model with VIB is much more stable and the stability is maintained even after 10,000 epochs of training.

We remark that the results relative to Figure 6a,b are obtained by using the exact same VGG-16 network architecture depicted in Figure 1. The VIB block illustrated in Figure 5 is added for regularization into each layer to get the results shown in Figure 6c, resulting in a modification of the loss function as well as of the overall network architecture. An interesting aspect related to the application of VIB for regularization is whether the observed improved performance is due to the modified loss function and architecture, or rather to the injection of noise alone. To investigate this aspect, we resort to a simpler LeNet-5 network on CIFAR-10. First, a comparison of test accuracy over 400 epochs without mutual-information-based regularizer, with MINE-based regularizer including the mutual information estimates of $I(X; h_i)$, $i = 1, ..., 4$, and with VIB-based regularizer is shown in Figure 7. The regularization coefficients for MINE and VIB are chosen similarly to the case of VGG-16. As observed in the case of VGG-16 on CIFAR-10, a VIB-based regularizer leads to better performance, albeit significantly lower than that achieved by VGG-16, as LeNet-5 is a much simpler network. For the same reason, overfitting is generally not an issue for a LeNet-5 network, and therefore the possible improvements due to regularization are marginal. Second, the performance of a VIB-based regularizer for LeNet-5 is compared with that achieved by a network where a Gaussian noise signal with zero mean and fixed standard deviation is added at each hidden layer. The accuracy obtained after 400 training epochs is reported in Table 1 for various values of the Gaussian noise standard deviation σ, and compared with that achieved by either VIB, MINE, or no regularization. It appears that the addition of noise alone is not adequate to achieve the same performance as the other regularizers.

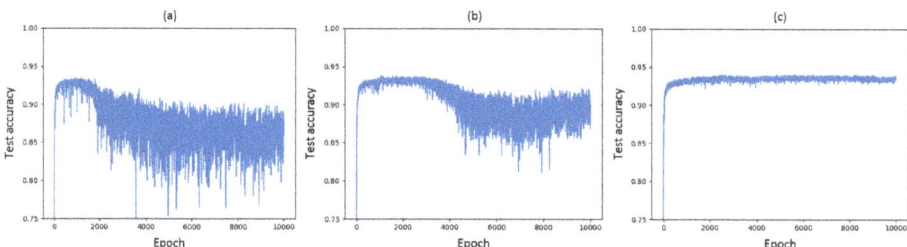

Figure 6. Test accuracies over 10,000 epochs for VGG-16 trained on CIFAR-10 (**a**) without mutual-information-based regularizer, (**b**) with MINE-based regularizer and (**c**) with Variational Information Bottleneck (VIB)-based regularizer.

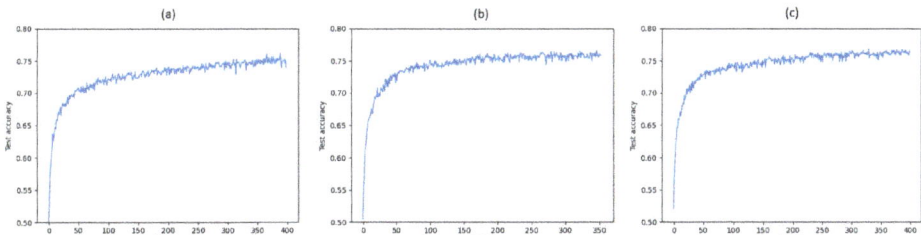

Figure 7. Test accuracies over 400 epochs for LeNet-5 trained on CIFAR-10 (**a**) without mutual-information-based regularizer, (**b**) with MINE-based regularizer and (**c**) with VIB-based regularizer.

Table 1. Test accuracies for LeNet-5 trained on CIFAR-10 with various regularization methods.

Regularization method	VIB	MINE	none	$\sigma = 0.1$	$\sigma = 0.2$	$\sigma = 0.3$	$\sigma = 0.4$
Accuracy (%)	76.9	76.7	75.9	73.8	73.9	72.3	72.3

6. Conclusions

Information theoretic concepts were adopted to analyze and improve high-complexity CNNs. We demonstrated the convergence of mutual information on the information plane and the existence of a compression phase for VGG-16, thus extending the results of [1] for fully connected networks with low-complexity. Furthermore, our experiments highlighted the advantages of regularizing DNNs by mutual-information-based additional terms in the network loss function. Specifically, mutual-information-based regularization improves and stabilizes the test accuracy, significantly reduces its variance, and prevents the model from overfitting, especially for a large number of training epochs.

Author Contributions: All authors collectively conceived the analysis of the convergence behavior of DNNs with mutual-information-based techniques. H.J. performed the experiments and derived the results under the supervision of G.C. and E.E. H.J. wrote the manuscript with input from all authors. All authors have read and agreed to the submitted version of the manuscript.

Funding: This research received no external funding.

Conflicts of Interest: The authors declare no conflict of interest.

Abbreviations

The following abbreviations are used in this manuscript:

DNN	Deep Neural Network
CNN	Convolutional Neural Network
MINE	Mutual Information Neural Estimation
ReLU	Rectified Linear Unit
VGG	Visual Geometry Group
IB	Information Bottleneck
KL	Kullback–Leibler
DPI	Data Processing Inequality
CIFAR	Canadian Institute for Advanced Research
CONV	Convolutional
FC	Fully Connected
MaxPool	Maximum Pooling
VIB	Variational Information Bottleneck

References

1. Tishby, N.; Zaslavsky, N. Deep learning and the information bottleneck principle. In Proceedings of the 2015 IEEE Information Theory Workshop (ITW), Jerusalem, Israel, 26 April–1 May 2015; pp. 1–5.
2. Shwartz-Ziv, R.; Tishby, N. Opening the black box of deep neural networks via information. *arXiv* **2017**, arXiv:1703.00810.
3. Yu, S.; Jenssen, R.; Principe, J.C. Understanding convolutional neural network training with information theory. *arXiv* **2018**, arXiv:1804.06537.
4. Saxe, A.M.; Bansal, Y.; Dapello, J.; Advani, M.; Kolchinsky, A.; Tracey, B.D.; Cox, D.D. On the information bottleneck theory of deep learning. *J. Stat. Mech. Theory Exp.* **2019**, *12*, 124020. [CrossRef]
5. Amjad, R.A.; Geiger, B.C. Learning representations for neural network-based classification using the information bottleneck principle. *IEEE Trans. Pattern Anal. Mach. Intell.* **2019**, *41*, 1–12. [CrossRef] [PubMed]
6. Geiger, B.C. On Information Plane Analyses of Neural Network Classifiers—A Review. *arXiv* **2020**, arXiv:2003.09671.
7. Goldfeld, Z.; van den Berg, E.; Greenewald, K.H.; Melnyk, I.; Nguyen, N.; Kingsbury, B.; Polyanskiy, Y. Estimating information flow in deep neural networks. *arXiv* **2014**, arXiv:1810.05728v3.
8. Simonyan, K.; Zisserman, A. Very deep convolutional networks for large-scale image recognition. *arXiv* **2014**, arXiv:1409.1556.
9. Belghazi, M.I.; Baratin, A.; Rajeswar, S.; Ozair, S.; Bengio, Y.; Courville, A.; Hjelm, R.D. MINE: Mutual information neural estimation. *arXiv* **2018**, arXiv:1801.04062.
10. Alemi, A.A.; Fischer, I.; Dillon, J.V.; Murphy, K. Deep variational information bottleneck. *arXiv* **2016**, arXiv:1612.00410.
11. Achille, A.; Soatto, S. Information dropout: Learning optimal representations through noisy computation. *IEEE Trans. Pattern Anal. Mach. Intell.* **2018**, *40*, 2897–2905. [CrossRef] [PubMed]
12. Tishby, N.; Pereira, F.C.; Bialek, W. The information bottleneck method. *arXiv* **2000**, arXiv:physics/0004057.
13. Kolchinsky, A.; Tracey, B.D.; Van Kuyk, S. Caveats for information bottleneck in deterministic scenarios. In Proceedings of the International Conference on Learning Representations (ICLR), New Orleans, LA, USA, 6–9 May 2019.
14. Cover, T.M.; Thomas, J.A. *Elements of Information Theory*; John Wiley & Sons: Hoboken, NJ, USA, 2012.
15. Krizhevsky, A. Learning Multiple Layers of Features from Tiny Images. Master's Thesis, University of Toronto, ON, Canada, 2009.
16. Nair, V.; Hinton, G.E. Rectified Linear Units Improve Restricted Boltzmann Machines. In Proceedings of the 27th International Conference on Machine Learning, Haifa, Israel, 21–24 June 2010; pp. 807–814.
17. Kolchinsky, A.; Tracey, B.D.; Wolpert, D.H. Nonlinear information bottleneck. *arXiv* **2017**, arXiv:1705.02436.
18. Wickstrøm, K.; Løkse, S.; Kampffmeyer, M.; Yu, S.; Principe, J.; Jenssen, R. Information plane analysis of deep neural networks via matrix-based Renyi's entropy and tensor kernels. *arXiv* **2019**, arXiv:1909.11396.
19. Yu, S.; Giraldo, L.; Jenssen, R.; Principe, J. Multivariate extension of matrix-based Renyi's α-order entropy functional. *IEEE Trans. Pattern Anal. Mach. Intell.* **2019**, *41*, 1–12. [CrossRef] [PubMed]
20. Donsker, M.D.; Varadhan, S.S. Asymptotic evaluation of certain Markov process expectations for large time, I. *Commun. Pure Appl. Math.* **1975**, *28*, 1–47. [CrossRef]
21. Genevay, A.; Cuturi, M.; Peyré, G.; Bach, F. Stochastic optimization for large-scale optimal transport. In Proceedings of the 2016 Conference on Neural Information Processing Systems, Barcelona, Spain, 5–10 December 2016; pp. 3440–3448.
22. Jónsson, H. Mutual-Information-Based Regularization of DNNs with Application to Abstract Reasoning. Master's Thesis, Department of Computer Science, Swiss Federal Institute of Technology, Zurich, Switzerland, 2019.
23. Li, H.; Kadav, A.; Durdanovic, I.; Samet, H.; Graf, H.P. Pruning filters for efficient convnets. *arXiv* **2017**, arXiv:1608.08710v3.
24. Dai, B.; Zhu, C.; Wipf, D. Compressing neural networks using the variational information bottleneck. *arXiv* **2018**, arXiv:1802.10399.
25. Kingma, D.P.; Ba, J. Adam: A method for stochastic optimization. *arXiv* **2014**, arXiv:1412.6980.

© 2020 by the authors. Licensee MDPI, Basel, Switzerland. This article is an open access article distributed under the terms and conditions of the Creative Commons Attribution (CC BY) license (http://creativecommons.org/licenses/by/4.0/).

Article

Variational Information Bottleneck for Semi-Supervised Classification

Slava Voloshynovskiy [1,*], Olga Taran [1], Mouad Kondah [1], Taras Holotyak [1] and Danilo Rezende [2]

1. Department of Computer Science, University of Geneva, 1227 Carouge, Switzerland; olga.taran@unige.ch (O.T.); mouad.kondah@etu.unige.ch (M.K.); taras.holotyak@unige.ch (T.H.)
2. DeepMind, London N1C 4AG, UK; danilor@google.com
* Correspondence: svolos@unige.ch; Tel.: +41(22)-379-01-58

Received: 22 July 2020; Accepted: 24 August 2020; Published: 27 August 2020

Abstract: In this paper, we consider an information bottleneck (IB) framework for semi-supervised classification with several families of priors on latent space representation. We apply a variational decomposition of mutual information terms of IB. Using this decomposition we perform an analysis of several regularizers and practically demonstrate an impact of different components of variational model on the classification accuracy. We propose a new formulation of semi-supervised IB with hand crafted and learnable priors and link it to the previous methods such as semi-supervised versions of VAE (M1 + M2), AAE, CatGAN, etc. We show that the resulting model allows better understand the role of various previously proposed regularizers in semi-supervised classification task in the light of IB framework. The proposed IB semi-supervised model with hand-crafted and learnable priors is experimentally validated on MNIST under different amount of labeled data.

Keywords: information bottleneck principle; deep networks; semi-supervised classification; latent space representation; hand crafted priors; learnable priors; regularization

---

Notations

We will denote a joint generative distribution as $p_\theta(\mathbf{x},\mathbf{z}) = p_\theta(\mathbf{z})p_\theta(\mathbf{x}|\mathbf{z})$, whereas marginal $p_\theta(\mathbf{z})$ is interpreted as a targeted distribution of latent space and marginal $p_\theta(\mathbf{x}) = \mathbb{E}_{p_\theta(\mathbf{z})}\left[p_\theta(\mathbf{x}|\mathbf{z})\right] = \int_{\mathbf{z}} p_\theta(\mathbf{x}|\mathbf{z})p_\theta(\mathbf{z})d\mathbf{z}$ as a generated data distribution with a generative model described by $p_\theta(\mathbf{x}|\mathbf{z})$, where $\mathbb{E}$ stands for the expected value. A joint data distribution $q_\phi(\mathbf{x},\mathbf{z}) = p_\mathcal{D}(\mathbf{x})q_\phi(\mathbf{z}|\mathbf{x})$, where $p_\mathcal{D}(\mathbf{x})$ denotes an empirical data distribution and $q_\phi(\mathbf{z}|\mathbf{x})$ is an inference or encoding model and marginal $q_\phi(\mathbf{z})$ denotes a "true" or "aggregated" distribution of latent space data. We will denote parameters of encoders as $\boldsymbol{\phi}_a$ and $\boldsymbol{\phi}_z$, and those of decoders as $\boldsymbol{\theta}_c$ and $\boldsymbol{\theta}_x$. The discriminators corresponding to Kullback–Leibler divergences are denoted as $\mathcal{D}_x$ where the subscript indicates the space to which this discriminator is applied to. The cross-entropy metrics are denoted as $\mathcal{D}_{x\hat{x}}$, where the subscript indicates the corresponding vectors. $\mathbf{X}$ denotes random vector, while the corresponding realization is denoted as $\mathbf{x}$.

1. Introduction

The deep supervised classifiers demonstrate an impressive performance when the amount of labeled data is large. However, their performance significantly deteriorates with the decrease of labeled samples. Recently, semi-supervised classifiers based on deep generative models such as VAE (M1 + M2) [1], AAE [2], CatGAN [3], etc., along with several other approaches based on multi-view and contrastive metrics just to mention the most recent ones [4,5], are considered to be a solution to the above problem. Besides the remarkable reported results, the information theoretic analysis of semi-supervised classifiers based on generative models and the role of different priors aiming to fulfil the gap in the lack of labeled data remain little studied. Therefore, in this paper we will try to address these issues using IB principle [6] and practically compare different priors on the same architecture of classifier.

Instead of considering the latent space of generative models such as VAE (M1 + M2) [1] and AAE [2] trained in the unsupervised way as suitable features for the classification, we will depart from the IB formulation of supervised classification, where we consider an encoder-decoder formulation of classifier and impose priors on its latent space. Thus, we study an approach to semi-supervised classification based on an IB formulation with a variational decomposition of IB compression and classification mutual information terms. To deeper understand the role and impact of different elements of variational IB on the classification accuracy, we consider two types of priors on the latent space of classifier: (i) hand-crafted and (ii) learnable priors. *Hand-crafted* latent space priors impose constraints on a distribution of latent space by fitting it to some targeted distribution according to the variational decomposition of the compression term of the IB. This type of latent space priors is well known as an information dropout [7]. One can also apply the same variational decomposition to the classification term of the IB, where the distribution of labels is supposed to follow some targeted class distribution to maximize the mutual information between inferred labels and targeted ones. This type of class label space regularization reflects an adversarial classification used in AAE [2] and CatGAN [3]. In contrast, *learnable* latent space priors aim at minimizing the need in human expertise in imposing priors on the latent space. Instead, the learnable priors are learned directly from unlabeled data using auto-encoding (AE) principle. In this way, the learnable priors are supposed to compensate the lack of labeled data in the semi-supervised learning yet minimizing the need in the hand-crafted control of the latent space distribution.

We demonstrate that several state-of-the-art models such as AAE [2], CatGAN [3], VAE (M1 + M2) [1], etc., can be considered to be instances of the variational IB with the learnable priors. At the same time, the role of different regularizers in the hand-crafted semi-supervised learning is generalized and linked to known frameworks such as information dropout [7].

We evaluate our model using standard dataset MNIST on both hand-crafted and learnable features. Besides revealing the impact of different components of variational IB factorization, we demonstrate that the proposed model outperforms prior works on this dataset.

Our main contribution is three-fold: (i) We propose a new formulation of IB for the semi-supervised classification and use a variational decomposition to convert it into a practically tractable setup with learnable parameters. (ii) We develop the variational IB for two classes of hand-crafted and learnable priors on the latent space of classifier and show its link to the state-of-the-art semi-supervised methods. (iii) We investigate the role of these priors and different regularizers in the classification, latent and reconstruction spaces for the same fixed architecture under the different amount of training data.

2. Related Work

Regularization techniques in semi-supervised learning: Semi-supervised learning tries to find a way to benefit from a large number of unlabeled samples available for training. The most common way to leverage unlabeled data is to add a special regularization term or some mechanism to better generalize to unseen data. The recent work [8] identifies three ways to construct such a regularization: (i) entropy minimization, (ii) consistency regularization and (iii) generic regularization. The entropy minimization [9,10] encourages the model to output confident predictions on unlabeled data. In addition, more recent work [3] extends this concept to adversarially generated samples or fakes for which the entropy of class label distribution was suggested to be maximized. Finally, the adversarial regularization of label space was considered in [2], where the discriminator was trained to ensure the labels produced by the classifier follow a prior distribution, which was defined to be a categorical one. The consistency regularization [11,12] encourages the model to produce the same output distribution when its inputs are perturbed. Finally, the generic regularization encourages the model to generalize well and avoid overfitting the training data. It can be achieved by imposing regularizers and corresponding priors on the model parameters or feature vectors.

In this work, we implicitly use the concepts of all three forms of considered regularization frameworks. However, instead of adding additional regularizers to the baseline classifier as suggested by the framework in [8], we will try to derive the corresponding counterparts from a semi-supervised IB framework. In this way, we will try to justify their origin and investigate their impact on overall classification accuracy for the same system architecture.

Information bottleneck: In the recent years, the IB framework [6] is considered to be a theoretical framework for analysis and explanation of supervised deep learning systems. However, as shown in [13], the original IB framework faces several practical issues: (i) for the deterministic deep networks, either the IB functional is infinite for network parameters, that leads to the ill-posed optimization problem, or it is piecewise constant, hence not admitting gradient-based optimization methods, and (ii) the invariance of the IB functional under bijections prevents it from capturing properties of the learned representation that are desirable for classification. In the same work, the authors demonstrate that these issues can be partly resolved for stochastic deep networks, networks that include a (hard or soft) decision rule, or by replacing the IB functional with related, but more well-behaved cost functions. It is important to mention that the same authors also note that rather than trying to repair the inherent problems in the IB functional, a better approach may be to design regularizers on latent representation enforcing the desired properties directly.

In our work, we extend these ideas using variational approximation approach suggested in [14] and that was also applied to unsupervised models in the previous work [15,16]. More particularly, we extend the IB framework to the semi-supervised classification and as discussed above we will consider two different ways of regularization of the latent space of classifier, i.e., either using traditional hand-crafted priors or suggested learnable priors. Although we do not consider the semi-supervised clustering and conditional generation in this work, the proposed findings can be extended to these problems in a way similar to prior works such as AAE [2], ADGM [17] and SeGMA [18].

The closest works: The proposed framework is closely related to several families of semi-supervised classifiers based on generative models. VAE (M1 + M2) [1] combines latent-feature discriminative model M1 and generative semi-supervised model M2. A new latent representation is learned using the generative model from M1 and subsequently a generative semi-supervised model M2 is trained using embeddings from the first latent representation instead of the raw data. Semi-supervised AAE classifier [2] is based on the AE architecture, where the encoder of AE outputs two latent representations: one representing class and another style. The latent class representation is regularized by an adversarial loss forcing it to follow categorical distribution. It is claimed that it plays an essential role for the overall classification performance. The latent style representation is regularized to follow Gaussian distribution. In both cases of VAE and AAE, the mean square error (MSE) metric is used for the reconstruction space loss. CatGAN [3] is an extension of GAN and is based on an objective function that trades-off mutual information between observed examples and their predicted categorical class distribution, against robustness of the classifier to an adversarial generative model.

In contrast to the above approaches and following the IB framework, we formulate the semi-supervised classification problem as a training of classifier that aims at compressing the input **x** to some latent data **a** via an encoding that is supposed to retain only class relevant information that is controlled by a decoder as shown in Figure 1. If the amount of labeled data is sufficiently large, the supervised classifier can achieve this goal. However, when the amount of labeled examples is small such an encoder-decoder pair representing an IB-driven classifier is regularized by a latent space and adversarial label space regularizers to fill the gap in training data. The adversarial label space regularization was already used in AAE and CatGAN. The latent space regularization in the scope of IB framework was reported in [7]. In this paper, we demonstrate that both label and latent space regularizations are instances of the generalized IB formulation developed in Section 3. At the same time, in contrast to the hypothesis that the considered label space and latent space regularizations are the driving factors behind the success of semi-supervised classifiers, we demonstrate that the hand-crafted priors considered in these models cannot completely fulfil the lack of labelled data and lead to relatively poor performance in comparison to a fully supervised system based on a sole cross-entropy metric. For these reasons, we analyze another mechanism of regularization of latent space based on learnable priors as shown in Figure 2 and developed in Section 4. Along this line, we provide an IB formulation of AAE and explain the driving mechanisms behind its success as an instance of IB with learnable priors. Finally, we present several extensions that explain the IB origin and role of adversarial regularization in the reconstruction space.

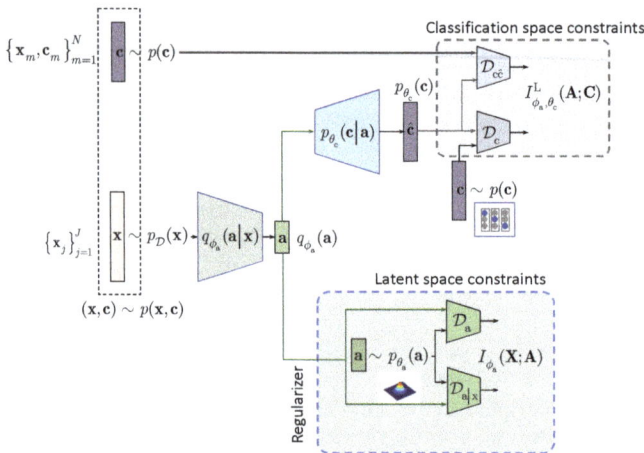

Figure 1. Classification with the hand-crafted latent space regularization.

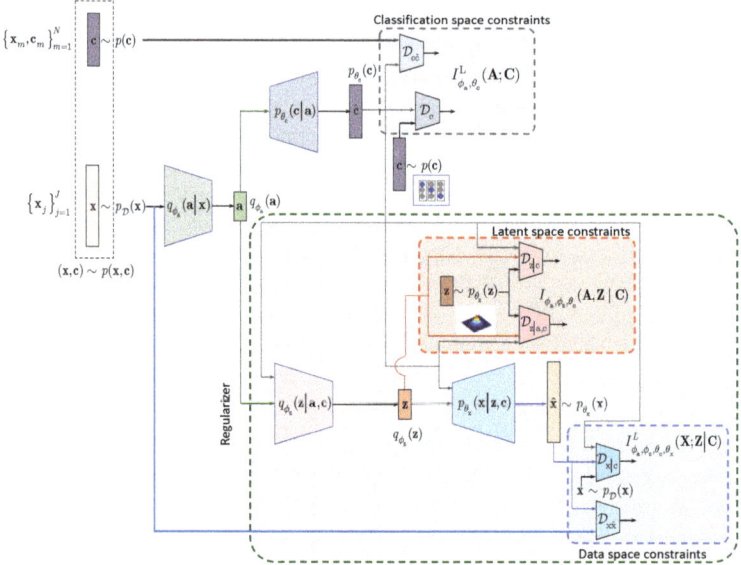

Figure 2. Classification with the learnable latent space regularization.

Summary: The considered methods of semi-supervised learning can be differentiated based on: (i) *the targeted tasks* (auto-encoding, clustering, generation or classification that can be accomplished depending on available labeled data); (ii) *the architecture in terms of the latent space representation* (with a single representation vector or with multiple representation vectors); (iii) *the usage of IB or other underlying frameworks* (methods derived from the IB directly or using regularization techniques); (iv) *the label space regularization* (based on available unlabeled data, augmented labeled data, synthetically generated labeled and unlabeled data, especially designed adversarial examples); (v) *the latent space regularization* (hand-crafted regularizers and priors or learnable priors under the reconstruction and constrastive setups) and (vi) *the reconstruction space regularization in case of reconstruction setup* (based on unlabeled and labeled data, augmented data under certain perturbations, synthetically generated examples).

In this work, our main focus is the latent space regularization for the hand-crafted and learnable priors under the reconstruction setup within the IB framework. Our main task is the semi-supervised classification. We will not consider any augmentation and adversarial techniques besides a simple stochastic encoding based on the addition of data independent noise at the system input or even deterministic encoding without any form of augmentation. The regularization of the label space and reconstruction space is solely based on the terms derived from the IB framework and only includes available labeled and unlabeled data without any form of augmentation. In this way, we want to investigate the role and impact of the latent space regularization as such in the IB-based semi-supervised classification. The usage of the above mentioned techniques of augmentation should be further investigated and will likely provide an additional performance improvement.

3. IB with Hand-Crafted Priors (HCP)

We assume that a semi-supervised classifier has an access to $\{\mathbf{x}_m, \mathbf{c}_m\}_{m=1}^N$ training labeled samples, where $\mathbf{x}_m \in \mathbb{R}^D$ denotes m^{th} data sample and $\mathbf{c}_m$ corresponding encoded class label from the set $\{1, 2, \cdots, M_c\}$, generated from the joint distribution $p(\mathbf{c}, \mathbf{x})$, and non-labeled data samples $\{\mathbf{x}_j\}_{j=1}^J$ with $J \gg N$. To integrate the knowledge about the labeled and non-labeled data at training, one can formulate the IB as:

$$\mathcal{L}^{\text{HCP}}(\boldsymbol{\phi}_a) = I_{\boldsymbol{\phi}_a}(\mathbf{X}; \mathbf{A}) - \beta_c I_{\boldsymbol{\phi}_a}(\mathbf{A}; \mathbf{C}), \qquad (1)$$

where **a** denotes the latent representation, β_c is a Lagrangian multiplier and the IB terms are defined as $I_{\boldsymbol{\phi}_a}(\mathbf{X}; \mathbf{A}) = \mathbb{E}_{q_{\boldsymbol{\phi}_a}(\mathbf{x}, \mathbf{a})} \left[\log \frac{q_{\boldsymbol{\phi}_a}(\mathbf{a}|\mathbf{x})}{q_{\boldsymbol{\phi}_a}(\mathbf{a})} \right]$ and $I_{\boldsymbol{\phi}_a}(\mathbf{A}; \mathbf{C}) = \mathbb{E}_{p(\mathbf{c}, \mathbf{x})} \left[\mathbb{E}_{q_{\boldsymbol{\phi}_a}(\mathbf{a}|\mathbf{x})} \left[\log \frac{q_{\boldsymbol{\phi}_a}(\mathbf{c}|\mathbf{a})}{p(\mathbf{c})} \right] \right]$.

According to the above IB formulation the encoder $q_{\boldsymbol{\phi}_a}(\mathbf{a}|\mathbf{x})$ is trained to minimize the mutual information between **X** and **A** while ensuring that the decoder $q_{\boldsymbol{\phi}_a}(\mathbf{c}|\mathbf{a})$ can reliably decide on labels **C** from the compressed representation **A**. The trade-off between the compression and recognition terms is controlled by β_c. Thus, it is assumed that the information retained in the latent representation **A** represents the sufficient statistics for the class labels **C**.

However, since optimal $q_{\boldsymbol{\phi}_a}(\mathbf{c}|\mathbf{a})$ is unknown, the second term $I_{\boldsymbol{\phi}_a}(\mathbf{A}; \mathbf{C})$ is lower bounded by $I_{\boldsymbol{\phi}_a, \theta_c}(\mathbf{A}; \mathbf{C})$ using a variational approximation $p_{\theta_c}(\mathbf{c}|\mathbf{a})$:

$$\begin{aligned}
I_{\boldsymbol{\phi}_a}(\mathbf{A}; \mathbf{C}) &\triangleq \mathbb{E}_{p(\mathbf{c}, \mathbf{x})} \left[\mathbb{E}_{q_{\boldsymbol{\phi}_a}(\mathbf{a}|\mathbf{x})} \left[\log \frac{q_{\boldsymbol{\phi}_a}(\mathbf{c}|\mathbf{a})}{p(\mathbf{c})} \right] \right] = \mathbb{E}_{p(\mathbf{c}, \mathbf{x})} \left[\mathbb{E}_{q_{\boldsymbol{\phi}_a}(\mathbf{a}|\mathbf{x})} \left[\log \frac{q_{\boldsymbol{\phi}_a}(\mathbf{c}|\mathbf{a})}{p(\mathbf{c})} \frac{p_{\theta_c}(\mathbf{c}|\mathbf{a})}{p_{\theta_c}(\mathbf{c}|\mathbf{a})} \right] \right] \\
&= \mathbb{E}_{p(\mathbf{c}, \mathbf{x})} \left[\mathbb{E}_{q_{\boldsymbol{\phi}_a}(\mathbf{a}|\mathbf{x})} \left[\log \frac{p_{\theta_c}(\mathbf{c}|\mathbf{a})}{p(\mathbf{c})} \right] \right] + \mathbb{E}_{p(\mathbf{c}, \mathbf{x})} \left[\mathbb{E}_{q_{\boldsymbol{\phi}_a}(\mathbf{a}|\mathbf{x})} \left[\log \frac{q_{\boldsymbol{\phi}_a}(\mathbf{c}|\mathbf{a})}{p_{\theta_c}(\mathbf{c}|\mathbf{a})} \right] \right] \\
&= \mathbb{E}_{p(\mathbf{c}, \mathbf{x})} \left[\mathbb{E}_{q_{\boldsymbol{\phi}_a}(\mathbf{a}|\mathbf{x})} \left[\log \frac{p_{\theta_c}(\mathbf{c}|\mathbf{a})}{p(\mathbf{c})} \right] \right] + \mathbb{E}_{p(\mathbf{c}, \mathbf{x})} \left[D_{\text{KL}}(q_{\boldsymbol{\phi}_a}(\mathbf{c}|\mathbf{a}) || p_{\theta_c}(\mathbf{c}|\mathbf{a})) \right] \\
&\geq \mathbb{E}_{p(\mathbf{c}, \mathbf{x})} \left[\mathbb{E}_{q_{\boldsymbol{\phi}_a}(\mathbf{a}|\mathbf{x})} \left[\log \frac{p_{\theta_c}(\mathbf{c}|\mathbf{a})}{p(\mathbf{c})} \right] \right],
\end{aligned} \qquad (2)$$

where $D_{\text{KL}}(q_{\boldsymbol{\phi}_a}(\mathbf{c}|\mathbf{a}) || p_{\theta_c}(\mathbf{c}|\mathbf{a})) = \mathbb{E}_{q_{\boldsymbol{\phi}_a}(\mathbf{a}|\mathbf{x})} \left[\log \frac{q_{\boldsymbol{\phi}_a}(\mathbf{c}|\mathbf{a})}{p_{\theta_c}(\mathbf{c}|\mathbf{a})} \right]$ and the inequality follows from the fact that $D_{\text{KL}}(q_{\boldsymbol{\phi}_a}(\mathbf{c}|\mathbf{a}) || p_{\theta_c}(\mathbf{c}|\mathbf{a})) \geq 0$. We denote the term $I_{\boldsymbol{\phi}_a, \theta_c}(\mathbf{A}; \mathbf{C}) = \mathbb{E}_{p(\mathbf{c}, \mathbf{x})} \left[\mathbb{E}_{q_{\boldsymbol{\phi}_a}(\mathbf{a}|\mathbf{x})} \left[\log \frac{p_{\theta_c}(\mathbf{c}|\mathbf{a})}{p(\mathbf{c})} \right] \right]$. Thus,

$I_{\phi_a}(\mathbf{A};\mathbf{C}) \geq I_{\phi_a,\theta_c}(\mathbf{A};\mathbf{C})$.

Thus, the IB (1) can be reformulated as:

$$\mathcal{L}^{\text{HCP}_L}(\phi_a, \theta_c) = I_{\phi_a}(\mathbf{X};\mathbf{A}) - \beta_c I_{\phi_a,\theta_c}(\mathbf{A};\mathbf{C}). \tag{3}$$

The considered IB is schematically shown in Figure 1 and we will proceed next with the detailed development of each component of the IB formulation.

3.1. Decomposition of the First Term: Hand-Crafted Regularization

The first mutual information term $I_{\phi_a}(\mathbf{X};\mathbf{A})$ in (3) can be decomposed using a factorization by a parametric marginal distribution $p_{\theta_a}(\mathbf{a})$ that represents a prior on the latent representation $\mathbf{a}$:

$$\begin{aligned}
I_{\phi_a}(\mathbf{X};\mathbf{A}) &= \mathbb{E}_{q_{\phi_a}(\mathbf{x},\mathbf{a})}\left[\log \frac{q_{\phi_a}(\mathbf{x},\mathbf{a})}{q_{\phi_a}(\mathbf{a})p_{\mathcal{D}}(\mathbf{x})}\right] = \mathbb{E}_{q_{\phi_a}(\mathbf{x},\mathbf{a})}\left[\log \frac{q_{\phi_a}(\mathbf{a}|\mathbf{x})}{q_{\phi_a}(\mathbf{a})}\frac{p_{\theta_a}(\mathbf{a})}{p_{\theta_a}(\mathbf{a})}\right] \\
&= \mathbb{E}_{p_{\mathcal{D}}(\mathbf{x})}\underbrace{\left[D_{\text{KL}}\left(q_{\phi_a}(\mathbf{a}|\mathbf{X}=\mathbf{x})\|p_{\theta_a}(\mathbf{a})\right)\right]}_{\mathcal{D}_{\mathbf{a}|\mathbf{x}}} - \underbrace{D_{\text{KL}}\left(q_{\phi_a}(\mathbf{a})\|p_{\theta_a}(\mathbf{a})\right)}_{\mathcal{D}_{\mathbf{a}}},
\end{aligned} \tag{4}$$

where the first term denotes the KL-divergence $\mathcal{D}_{\mathbf{a}|\mathbf{x}} \triangleq D_{\text{KL}}\left(q_{\phi_a}(\mathbf{a}|\mathbf{X}=\mathbf{x})\|p_{\theta_a}(\mathbf{a})\right) = \mathbb{E}_{q_{\phi_a}(\mathbf{a}|\mathbf{x})}\left[\log \frac{q_{\phi_a}(\mathbf{a}|\mathbf{x})}{p_{\theta_a}(\mathbf{a})}\right]$ and the term denotes the KL-divergence $\mathcal{D}_{\mathbf{a}} \triangleq D_{\text{KL}}\left(q_{\phi_a}(\mathbf{a})\|p_{\theta_a}(\mathbf{a})\right) = \mathbb{E}_{q_{\phi_a}(\mathbf{a})}\left[\log \frac{q_{\phi_a}(\mathbf{a})}{p_{\theta_a}(\mathbf{a})}\right]$.

It should be pointed out that the encoding $q_{\phi_a}(\mathbf{a}|\mathbf{x})$ can be both stochastic or deterministic. *Stochastic encoding* $q_{\phi_a}(\mathbf{a}|\mathbf{x})$ can be implemented via: (a) *multiplicative encoding* applied to the input $\mathbf{x}$ as $\mathbf{a} = f_{\phi_a}(\mathbf{x} \odot \boldsymbol{\epsilon})$ or in the latent space $\mathbf{a} = f_{\phi_a}(\mathbf{x}) \odot \boldsymbol{\epsilon}$, where $f_{\phi_a}(\mathbf{x})$ is the output of the encoder, $\odot$ denotes the element-wise product and $\boldsymbol{\epsilon}$ follows some data independent or data dependent distribution as in information dropout [7]; (b) *additive encoding* applied to the input $\mathbf{x}$ as $\mathbf{a} = f_{\phi_a}(\mathbf{x} + \boldsymbol{\epsilon})$ with the data independent perturbations, e.g., such as in PixelGAN [19], or in the latent space with generally data-dependent perturbations of form $\mathbf{a} = f_{\phi_a}(\mathbf{x}) + \sigma_{\phi_a}(\mathbf{x}) \odot \boldsymbol{\epsilon}$, where $f_{\phi_a}(\mathbf{x})$ and $\sigma_{\phi_a}(\mathbf{x})$ are outputs of the encoder and $\boldsymbol{\epsilon}$ is assumed to be a zero mean unit variance vector such as in VAE [1] or (c) *concatenative/mixing encoding* $\mathbf{a} = f_{\phi_a}([\mathbf{x}, \boldsymbol{\epsilon}])$ that is generally applied at the input of encoder. Deterministic encoding is based on the mapping $\mathbf{a} = f_{\phi_a}(\mathbf{x})$, i.e., no randomization is introduced, e.g., such as one of encoding modalities of AAE [2].

3.2. Decomposition of the Second Term

In this section, we factorize the second term in (3) to address the semi-supervised training, i.e., to integrate the knowledge of both non-labeled and labeled data available at training:

$$\begin{aligned}
I_{\phi_a,\theta_c}(\mathbf{A};\mathbf{C}) &\triangleq \mathbb{E}_{p(\mathbf{c},\mathbf{x})}\left[\mathbb{E}_{q_{\phi_a}(\mathbf{a}|\mathbf{x})}\left[\log \frac{p_{\theta_c}(\mathbf{c}|\mathbf{a})}{p(\mathbf{c})}\frac{p_{\theta_c}(\mathbf{c})}{p_{\theta_c}(\mathbf{c})}\right]\right] \\
&= -\mathbb{E}_{p(\mathbf{c})}[\log p_{\theta_c}(\mathbf{c})] - \mathbb{E}_{p(\mathbf{c})}\left[\log \frac{p(\mathbf{c})}{p_{\theta_c}(\mathbf{c})}\right] + \mathbb{E}_{p(\mathbf{c},\mathbf{x})}\left[\mathbb{E}_{q_{\phi_a}(\mathbf{a}|\mathbf{x})}[\log p_{\theta_c}(\mathbf{c}|\mathbf{a})]\right] \\
&= H(p(\mathbf{c}); p_{\theta_c}(\mathbf{c})) - D_{\text{KL}}(p(\mathbf{c})\|p_{\theta_c}(\mathbf{c})) - H_{\theta_c,\phi_a}(\mathbf{C}|\mathbf{A}),
\end{aligned} \tag{5}$$

with $H(p(\mathbf{c}); p_\theta(\mathbf{c})) = -\mathbb{E}_{p(\mathbf{c})}[\log p_{\theta_c}(\mathbf{c})]$ denoting a cross-entropy between $p(\mathbf{c})$ and $p_{\theta_c}(\mathbf{c})$, and $\mathcal{D}_c \triangleq D_{\text{KL}}(p(\mathbf{c})\|p_{\theta_c}(\mathbf{c})) = \mathbb{E}_{p(\mathbf{c})}\left[\log \frac{p(\mathbf{c})}{p_{\theta_c}(\mathbf{c})}\right]$ to be a KL-divergence between the prior class label distribution $p(\mathbf{c})$ and the estimated one $p_{\theta_c}(\mathbf{c})$. One can assume different forms of labels' $\mathbf{c}$ encoding but one of the most often used forms is one-hot-label encoding that leads to the categorical distribution $p(\mathbf{c}) = \text{cat}(\mathbf{c})$.

Finally, the conditional entropy is defined as $\mathcal{D}_{c\hat{c}} \triangleq H_{\theta_c,\phi_a}(\mathbf{C}|\mathbf{A}) = -\mathbb{E}_{p(\mathbf{c},\mathbf{x})}\left[\mathbb{E}_{q_{\phi_a}(\mathbf{a}|\mathbf{x})}\left[\log p_{\theta_c}(\mathbf{c}|\mathbf{a})\right]\right]$.

Since $H(p(\mathbf{c}); p_{\theta_c}(\mathbf{c})) \geq 0$, one can lower bound (5) as $I_{\phi_a,\theta_c}(\mathbf{A};\mathbf{C}) \geq I^L_{\phi_a,\theta_c}(\mathbf{A};\mathbf{C})$ where:

$$I^L_{\phi_a,\theta_c}(\mathbf{A};\mathbf{C}) \triangleq -\underbrace{D_{\mathrm{KL}}\left(p(\mathbf{c})\|p_{\theta_c}(\mathbf{c})\right)}_{\mathcal{D}_c} - \underbrace{H_{\theta_c,\phi_a}(\mathbf{C}|\mathbf{A})}_{\mathcal{D}_{c\hat{c}}}. \quad (6)$$

3.3. Supervised and Semi-Supervised Models with/without Hand-Crafted Priors

Summarizing the above variational decomposition of (3) with the terms (4) and (6), we will proceed with four practical scenarios.

Supervised training without latent space regularization (**baseline**): is based on term $\mathcal{D}_{c\hat{c}}$ in (6)

$$\mathcal{L}^{\mathrm{HCP}}_{\mathrm{S-NoReg}}(\theta_c, \phi_a) = \mathcal{D}_{c\hat{c}}. \quad (7)$$

Semi-supervised training without latent space regularization is based on terms $\mathcal{D}_{c\hat{c}}$ and $\mathcal{D}_c$ in (6):

$$\mathcal{L}^{\mathrm{HCP}}_{\mathrm{SS-NoReg}}(\theta_c, \phi_a) = \mathcal{D}_{c\hat{c}} + \mathcal{D}_c. \quad (8)$$

Supervised training with latent space regularization is based on term $\mathcal{D}_{c\hat{c}}$ in (6) and either term $\mathcal{D}_{a|x}$ or $\mathcal{D}_a$ or jointly $\mathcal{D}_{a|x}$ and $\mathcal{D}_a$ in (4):

$$\mathcal{L}^{\mathrm{HCP}}_{\mathrm{S-Reg}}(\theta_c, \phi_a) = \mathbb{E}_{p_\mathcal{D}(\mathbf{x})}\left[\mathcal{D}_{a|x}\right] + \mathcal{D}_a + \beta_c \mathcal{D}_{c\hat{c}}. \quad (9)$$

Semi-supervised training with latent space regularization deploys all terms in (4) and (6):

$$\mathcal{L}^{\mathrm{HCP}}_{\mathrm{SS-Reg}}(\theta_c, \phi_a) = \mathbb{E}_{p_\mathcal{D}(\mathbf{x})}\left[\mathcal{D}_{a|x}\right] + \mathcal{D}_a + \beta_c \mathcal{D}_{c\hat{c}} + \beta_c \mathcal{D}_c. \quad (10)$$

The empirical evaluation of these setups on MNIST dataset is given in Section 5. The same architecture of encoder and decoder was used to establish the impact of each term in a function of available labeled data.

4. IB with Learnable Priors (LP)

In this section, we extend the results obtained for the hand-crafted priors to the learnable priors. Instead of applying the hand-crafted regularization of the latent representation **a** as suggested by the IB (3) and shown in Figure 1, we will assume that the latent representation **a** is regularized by an especially designed AE as shown in Figure 2. The AE-based regularization has two components: (i) the latent space **z** regularization and (ii) the observation space regularization. The design and training of this latent space regularizer in a form of the AE is guided by its own IB. In the general case, all elements of AE, i.e., its encoder-decoder pair, latent and observation space regularizers are conditioned by the learned class label **c**. The resulting Lagrangian with the learnable prior is (formally one should consider $I_{\phi_a,\phi_z,\theta_c}(\mathbf{X};\mathbf{Z}|\mathbf{C})$ for the term A. However, since $I_{\phi_a,\phi_z,\theta_c}(\mathbf{X};\mathbf{Z}|\mathbf{C}) \leq I_{\phi_a,\phi_z,\theta_c}(\mathbf{A};\mathbf{Z}|\mathbf{C})$ due to the Markovianity of considered architecture, we consider the decomposition starting from **A** [20], Data Processing Inequality, Theorem 2.8.1):

$$\mathcal{L}^{\mathrm{LP}}(\phi_a, \phi_z, \theta_c, \theta_x) = \underbrace{I_{\phi_a,\phi_z,\theta_c}(\mathbf{A};\mathbf{Z}|\mathbf{C})}_{A} - \beta_x \underbrace{I_{\phi_a,\phi_z,\theta_c,\theta_x}(\mathbf{X};\mathbf{Z}|\mathbf{C})}_{B} - \beta_c \underbrace{I^L_{\phi_a,\theta_c}(\mathbf{A};\mathbf{C})}_{C}, \quad (11)$$

where β_x is a Lagrangian multiplier controlling the reconstruction of **x** at the decoder and β_c is the same as in (1).

The terms A and B, conditioned by the class c, play a role of the latent space regularizer by imposing the learnable constrains on the vector $\mathbf{a}$. These two terms correspond to the hand-crafted counterpart $I_{\phi_a}(\mathbf{X};\mathbf{A})$ in (3). The term C in the learnable IB formulation corresponds to the classification part of hand-crafted IB in (3) and can be factorized along the same lines as in (6). Therefore, we will proceed with the factorization of terms A and B.

One can also consider the following IB formulation with the learnable priors with no conditioning on c in term A in (11) leading to an unconditional counterpart D below that can be viewed as an IB generalization of semi-supervised AAE [2]:

$$\mathcal{L}_{\text{AAE}}^{\text{LP}}(\phi_a,\phi_z,\theta_c,\theta_x) = \underbrace{I_{\phi_a,\phi_z}(\mathbf{A};\mathbf{Z})}_{D} - \beta_x \underbrace{I_{\phi_a,\phi_z,\theta_c,\theta_x}(\mathbf{X};\mathbf{Z}|\mathbf{C})}_{B} - \beta_c \underbrace{I_{\phi_a,\theta_c}^L(\mathbf{A};\mathbf{C})}_{C}. \tag{12}$$

4.1. Decomposition of Latent Space Regularizer

We will denote $p_{\phi_a,\phi_z,\theta_c}(\mathbf{x},\mathbf{a},\mathbf{c},\mathbf{z}) = p_\mathcal{D}(\mathbf{x})q_{\phi_a}(\mathbf{a}|\mathbf{x})p_{\theta_c}(\mathbf{c}|\mathbf{a})q_{\phi_z}(\mathbf{z}|\mathbf{a},\mathbf{c})$ and decompose the term A in (11) using variational factorization as:

$$\begin{aligned}
I_{\phi_a,\phi_z,\theta_c}(\mathbf{A},\mathbf{Z}|\mathbf{C}) &= \mathbb{E}_{p_{\phi_a,\phi_z,\theta_c}(\mathbf{x},\mathbf{a},\mathbf{c},\mathbf{z})}\left[\log\frac{q_{\phi_z}(\mathbf{z}|\mathbf{a},\mathbf{c})}{q_{\phi_z}(\mathbf{z}|\mathbf{c})}\frac{p_{\theta_z}(\mathbf{z})}{p_{\theta_z}(\mathbf{z})}\right] \\
&= \mathbb{E}_{p_\mathcal{D}(\mathbf{x})}\left[\mathbb{E}_{q_{\phi_a}(\mathbf{a}|\mathbf{x})}\left[\mathbb{E}_{p_{\theta_c}(\mathbf{c}|\mathbf{a})}\underbrace{\left[D_{\text{KL}}\left(q_{\phi_z}(\mathbf{z}|\mathbf{A}=\mathbf{a},\mathbf{C}=\mathbf{c})\|p_{\theta_z}(\mathbf{z})\right)\right]}_{\mathcal{D}_{z|a,c}}\right]\right], \\
&\quad - \mathbb{E}_{p_\mathcal{D}(\mathbf{x})}\left[\mathbb{E}_{q_{\phi_a}(\mathbf{a}|\mathbf{x})}\left[\mathbb{E}_{p_{\theta_c}(\mathbf{c}|\mathbf{a})}\underbrace{\left[D_{\text{KL}}\left(q_{\phi_z}(\mathbf{z}|\mathbf{C}=\mathbf{c})\|p_{\theta_z}(\mathbf{z})\right)\right]}_{\mathcal{D}_{z|c}}\right]\right],
\end{aligned} \tag{13}$$

where $\mathcal{D}_{z|a,c} \triangleq D_{\text{KL}}\left(q_{\phi_z}(\mathbf{z}|\mathbf{a},\mathbf{c})\|p_{\theta_z}(\mathbf{z})\right) = \mathbb{E}_{q_{\phi_z}(\mathbf{z}|\mathbf{a},\mathbf{c})}\left[\log\frac{q_{\phi_z}(\mathbf{z}|\mathbf{a},\mathbf{c})}{p_{\theta_z}(\mathbf{z})}\right]$ and $\mathcal{D}_{z|c} \triangleq D_{\text{KL}}\left(q_{\phi_z}(\mathbf{z}|\mathbf{c})\|p_{\theta_z}(\mathbf{z})\right) = \mathbb{E}_{q_{\phi_z}(\mathbf{z}|\mathbf{c})}\left[\log\frac{q_{\phi_z}(\mathbf{z}|\mathbf{c})}{p_{\theta_z}(\mathbf{z})}\right]$ denote the KL-divergence terms and $q_{\phi_z}(\mathbf{z}|\mathbf{c}) = \mathbb{E}_{p_\mathcal{D}(\mathbf{x})}\left[\mathbb{E}_{q_{\phi_a}(\mathbf{a}|\mathbf{x})}\left[q_{\phi_z}(\mathbf{z}|\mathbf{a},\mathbf{c})\right]\right]$.

4.2. Decomposition of Reconstruction Space Regularizer

Denoting $p_{\phi_a,\phi_z,\theta_c,\theta_x}(\mathbf{x},\mathbf{a},\mathbf{c},\mathbf{z}) = p_\mathcal{D}(\mathbf{x})q_{\phi_a}(\mathbf{a}|\mathbf{x})p_{\theta_c}(\mathbf{c}|\mathbf{a})q_{\phi_z}(\mathbf{z}|\mathbf{a},\mathbf{c})p_{\theta_x}(\mathbf{x}|\mathbf{z},\mathbf{c})$, we decompose the term B in (11) as:

$$\begin{aligned}
I_{\phi_a,\phi_z,\theta_c,\theta_x}(\mathbf{X};\mathbf{Z}|\mathbf{C}) &= \mathbb{E}_{p_{\phi_a,\phi_z,\theta_c,\theta_x}(\mathbf{x},\mathbf{a},\mathbf{c},\mathbf{z})}\left[\log\frac{p_{\theta_x}(\mathbf{x}|\mathbf{z},\mathbf{c})}{p_\mathcal{D}(\mathbf{x}|\mathbf{c})}\frac{p_{\theta_x}(\mathbf{x})}{p_{\theta_x}(\mathbf{x})}\right] \\
&= \mathbb{E}_{p_{\theta_c}(\mathbf{c})}\left[H(p_\mathcal{D}(\mathbf{x}|\mathbf{c});p_{\theta_x}(\mathbf{x}))\right] \\
&\quad - \mathbb{E}_{p_{\theta_c}(\mathbf{c})}\underbrace{\left[D_{\text{KL}}\left(p_\mathcal{D}(\mathbf{x}|\mathbf{C}=\mathbf{c})\|p_{\theta_x}(\mathbf{x})\right)\right]}_{\mathcal{D}_{x|c}} - \underbrace{H_{\phi_a,\phi_z,\theta_c,\theta_x}(\mathbf{X}|\mathbf{Z},\mathbf{C})}_{\mathcal{D}_{x\hat{x}}},
\end{aligned} \tag{14}$$

where $p_{\theta_c}(\mathbf{c}) = \mathbb{E}_{p_\mathcal{D}(\mathbf{x})}\left[\mathbb{E}_{q_{\phi_a}(\mathbf{a}|\mathbf{x})}\left[p_{\theta_c}(\mathbf{c}|\mathbf{a})\right]\right]$. The terms are defined as $H(p_\mathcal{D}(\mathbf{x}|\mathbf{c});p_{\theta_x}(\mathbf{x})) = -\mathbb{E}_{p_\mathcal{D}(\mathbf{x}|\mathbf{c})}\left[\log p_{\theta_x}(\mathbf{x})\right]$, $\mathcal{D}_{x|c} \triangleq D_{\text{KL}}\left(p_\mathcal{D}(\mathbf{x}|\mathbf{C}=\mathbf{c})\|p_{\theta_x}(\mathbf{x})\right) = \mathbb{E}_{p_\mathcal{D}(\mathbf{x}|\mathbf{c})}\left[\log\frac{p_\mathcal{D}(\mathbf{x}|\mathbf{c})}{p_{\theta_x}(\mathbf{x})}\right]$ and $\mathcal{D}_{x\hat{x}} \triangleq H_{\phi_a,\phi_z,\theta_c,\theta_x}(\mathbf{X}|\mathbf{Z},\mathbf{C}) = -\mathbb{E}_{p_\mathcal{D}(\mathbf{x})}\left[\mathbb{E}_{q_{\phi_a}(\mathbf{a}|\mathbf{x})}\left[\mathbb{E}_{p_{\theta_c}(\mathbf{c}|\mathbf{a})}\left[\mathbb{E}_{q_{\phi_z}(\mathbf{z}|\mathbf{a},\mathbf{c})}\left[\log p_{\theta_x}(\mathbf{x}|\mathbf{z},\mathbf{c})\right]\right]\right]\right]$. Since $\mathbb{E}_{p_{\theta_c}(\mathbf{c})}\left[H(p_\mathcal{D}(\mathbf{x}|\mathbf{c});p_{\theta_x}(\mathbf{x}))\right] \geq 0$, we can lower bound $I_{\phi_a,\phi_z,\theta_c,\theta_x}(\mathbf{X};\mathbf{Z}|\mathbf{C}) \geq I_{\phi_a,\phi_z,\theta_c,\theta_x}^L(\mathbf{X};\mathbf{Z}|\mathbf{C}) \triangleq -\mathcal{D}_{x|c} - \mathcal{D}_{x\hat{x}}$.

4.3. Semi-Supervised Models with Learnable Priors

Summarizing the above variational decomposition of (11) with the terms (13) and (14), we will consider semi-supervised training with latent space regularization as:

$$\mathcal{L}^{\text{LP}}_{\text{SS-Reg}}(\theta_c, \theta_x, \phi_a, \phi_z) = \mathbb{E}_{p_{\mathcal{D}}(\mathbf{x})}\left[\mathbb{E}_{q_{\phi_a}(\mathbf{a}|\mathbf{x})}\left[\mathbb{E}_{p_{\theta_c}(\mathbf{c}|\mathbf{a})}\left[\mathcal{D}_{\mathbf{z}|\mathbf{a},\mathbf{c}}\right]\right]\right] + \mathbb{E}_{p_{\mathcal{D}}(\mathbf{x})}\left[\mathbb{E}_{q_{\phi_a}(\mathbf{a}|\mathbf{x})}\left[\mathbb{E}_{p_{\theta_c}(\mathbf{c}|\mathbf{a})}\left[\mathcal{D}_{\mathbf{z}|\mathbf{c}}\right]\right]\right]$$
$$+ \beta_x \mathcal{D}_{\mathbf{x}\hat{\mathbf{x}}} + \beta_x \mathbb{E}_{p_{\theta_c}(\mathbf{c})}\left[\mathcal{D}_{\mathbf{x}|\mathbf{c}}\right] + \beta_c \mathcal{D}_{\mathbf{c}\hat{\mathbf{c}}} + \beta_c \mathcal{D}_{\mathbf{c}}.$$
(15)

To create a link to the semi-supervised AAE [2], we also consider (12), where all latent and reconstruction space regularizers are independent of $\mathbf{c}$, i.e., do not contain conditioning on $\mathbf{c}$. *Semi-supervised training with latent space regularization and MSE reconstruction* based on (12):

$$\mathcal{L}^{\text{LP}}_{\text{SS-AAE}}(\theta_c, \theta_x, \phi_a, \phi_z) = \mathcal{D}_{\mathbf{z}} + \beta_x \mathcal{D}_{\mathbf{x}\hat{\mathbf{x}}} + \beta_c \mathcal{D}_{\mathbf{c}\hat{\mathbf{c}}} + \beta_c \mathcal{D}_{\mathbf{c}},$$
(16)

where $\mathcal{D}_{\mathbf{z}} \triangleq D_{\text{KL}}\left(q_{\phi_z}(\mathbf{z}) \| p_{\theta_z}(\mathbf{z})\right) = \mathbb{E}_{q_{\phi_z}(\mathbf{z})}\left[\log \frac{q_{\phi_z}(\mathbf{z})}{p_{\theta_z}(\mathbf{z})}\right]$.

Semi-supervised training with latent space regularization and with MSE and adversarial reconstruction based on (12) deploys all terms:

$$\mathcal{L}^{\text{LP}}_{\text{SS-AAE}_{\text{complete}}}(\theta_c, \theta_x, \phi_a, \phi_z) = \mathcal{D}_{\mathbf{z}} + \beta_x \mathcal{D}_{\mathbf{x}\hat{\mathbf{x}}} + \beta_x \mathcal{D}_{\mathbf{x}} + \beta_c \mathcal{D}_{\mathbf{c}\hat{\mathbf{c}}} + \beta_c \mathcal{D}_{\mathbf{c}},$$
(17)

where $\mathcal{D}_{\mathbf{x}} \triangleq D_{\text{KL}}\left(p_{\mathcal{D}}(\mathbf{x}) \| p_{\theta_x}(\mathbf{x})\right) = \mathbb{E}_{p_{\mathcal{D}}(\mathbf{x})}\left[\log \frac{p_{\mathcal{D}}(\mathbf{x})}{p_{\theta_x}(\mathbf{x})}\right]$.

4.4. Links to State-Of-The-Art Models

The considered HCP and LP models can be linked with several state-of-the-art unsupervised models such VAE [21,22], β-VAE [23], AAE [2] and BIB-AE [15] and semi-supervised models such as AAE [2], CatGAN [3], VAE (M1 + M2) [1] and SeGMA [18].

4.4.1. Links to Unsupervised Models

The proposed LP model (11) generalizes unsupervised models without the categorical latent representation. In addition, the unsupervised models in a form of the auto-encoder are used as a latent space regularizer in the LP setup. For these reasons, we will briefly consider four models of interest, namely VAE, β-VAE, AAE, and BIB-AE.

Before we proceed with the analysis, we will define an unsupervised IB for these models. We will assume the fused encoders $q_{\phi_a}(\mathbf{a}|\mathbf{x})$ and $q_{\phi_z}(\mathbf{z}|\mathbf{a})$ without conditioning on $\mathbf{c}$ in the inference model according to Figure 2. We also assume no conditionally on $\mathbf{c}$ in the generative model.

The Lagrangian of unsupervised IB is defined according to [15]:

$$\mathcal{L}^{U_L}(\theta_x, \phi_z) = I_{\phi_z}(\mathbf{X}; \mathbf{Z}) - \beta_x I_{\phi_z, \theta_x}(\mathbf{Z}; \mathbf{X}),$$
(18)

where similarly to the supervised counterpart (4), we define the first term as:

$$I_{\phi_z}(\mathbf{X}; \mathbf{Z}) = \mathbb{E}_{q_{\phi_z}(\mathbf{x},\mathbf{z})}\left[\log \frac{q_{\phi_z}(\mathbf{x},\mathbf{z})}{q_{\phi_z}(\mathbf{z})p_{\mathcal{D}}(\mathbf{x})}\right] = \mathbb{E}_{q_{\phi_z}(\mathbf{x},\mathbf{z})}\left[\log \frac{q_{\phi_z}(\mathbf{z}|\mathbf{x})}{q_{\phi_z}(\mathbf{z})}\frac{p_{\theta_z}(\mathbf{z})}{p_{\theta_z}(\mathbf{z})}\right]$$
$$= \mathbb{E}_{p_{\mathcal{D}}(\mathbf{x})}\underbrace{\left[D_{\text{KL}}\left(q_{\phi_z}(\mathbf{z}|\mathbf{X}=\mathbf{x}) \| p_{\theta_z}(\mathbf{z})\right)\right]}_{\mathcal{D}_{\mathbf{z}|\mathbf{x}}} - \underbrace{D_{\text{KL}}\left(q_{\phi_z}(\mathbf{z}) \| p_{\theta_z}(\mathbf{z})\right)}_{\mathcal{D}_{\mathbf{z}}},$$
(19)

and similarly to (14) the second term is defined as:

$$I_{\phi_z,\theta_x}(\mathbf{Z};\mathbf{X}) = \mathbb{E}_{p_{\mathcal{D}}(\mathbf{x})}\left[\mathbb{E}_{q_{\phi_z}(\mathbf{z}|\mathbf{x})}\left[\log\frac{p_{\theta_x}(\mathbf{x}|\mathbf{z})}{p_{\mathcal{D}}(\mathbf{x})}\frac{p_{\theta_x}(\mathbf{x})}{p_{\theta_x}(\mathbf{x})}\right]\right]$$
$$= H(p_{\mathcal{D}}(\mathbf{x}|\mathbf{c}); p_{\theta_x}(\mathbf{x})) - \underbrace{D_{\mathrm{KL}}\left(p_{\mathcal{D}}(\mathbf{x})\|p_{\theta_x}(\mathbf{x})\right)}_{\mathcal{D}_{\mathbf{x}}} - \underbrace{H_{\phi_z,\theta_x}(\mathbf{X}|\mathbf{Z})}_{\mathcal{D}_{\mathbf{x}\hat{\mathbf{x}}}}, \quad (20)$$

where the definition of all terms should follow from the above equations. Since $H(p_{\mathcal{D}}(\mathbf{x}|\mathbf{c}); p_{\theta_x}(\mathbf{x})) \geq 0$, we can lower bound $I_{\phi_z,\theta_x}(\mathbf{Z};\mathbf{X}) \geq -\mathcal{D}_{\mathbf{x}} - \mathcal{D}_{\mathbf{x}\hat{\mathbf{x}}}$.

Having defined the unsupervised IB variational bounded decomposition, we can proceed with an analysis of the related state-of-the-art methods along the lines of analysis introduced in Summary part of Section 2.

VAE [21,22] and β-VAE [23]:

1. *The targeted tasks*: auto-encoding and generation.
2. *The architecture in terms of the latent space representation*: the encoder outputs two vectors representing the mean and standard deviation vectors that control a new latent representation $\mathbf{z} = f_{\phi_z}(\mathbf{x}) + \sigma_{\phi_z}(\mathbf{x}) \odot \epsilon$, where $f_{\phi_z}(\mathbf{x})$ and $\sigma_{\phi_z}(\mathbf{x})$ are outputs of the encoder and ϵ is assumed to be a zero mean unit variance Gaussian vector.
3. *The usage of IB or other underlying frameworks*: both VAE and β-VAE use evidence lower bound (ELBO) and are not derived from the IB framework. However, it can be shown [15] that the Lagrangian (18) can be reformulated for VAE and β−VAE as:

$$\mathcal{L}_{\beta-\mathrm{VAE}}(\theta_x, \phi_z) = \mathbb{E}_{p_{\mathcal{D}}(\mathbf{x})}\left[\mathcal{D}_{\mathbf{z}|\mathbf{x}}\right] + \beta_{\mathbf{x}}\mathcal{D}_{\mathbf{x}\hat{\mathbf{x}}}, \quad (21)$$

where $\beta_{\mathbf{x}} = 1$ for VAE. It can be noted that the VAE and β-VAE are based on an upper bound on the mutual information term $I_{\phi_z}(\mathbf{X};\mathbf{Z}) \leq \mathbb{E}_{p_{\mathcal{D}}(\mathbf{x})}\left[\mathcal{D}_{\mathbf{z}|\mathbf{x}}\right]$, since $D_{\mathrm{KL}}\left(q_{\phi_z}(\mathbf{z})\|p_{\theta_z}(\mathbf{z})\right) \geq 0$. Similar considerations apply to the second term since $D_{\mathrm{KL}}\left(p_{\mathcal{D}}(\mathbf{x})\|p_{\theta_x}(\mathbf{x})\right) \geq 0$.

4. *The label space regularization*: does not apply here due to the unsupervised setting.
5. *The latent space regularization*: is based on the hand-crafted prior with Gaussian pdf.
6. *The reconstruction space regularization in case of reconstruction loss*: is based on the mean square error (MSE) counterpart of $\mathcal{D}_{\mathbf{x}\hat{\mathbf{x}}}$ that corresponds to the Guassian likelihood assumption.

Unsupervised AAE [2]:

1. *The targeted tasks*: auto-encoding and generation.
2. *The architecture in terms of the latent space representation*: the encoder outputs one vector in stochastic or deterministic way as $\mathbf{z} = f_{\phi_z}(\mathbf{x})$.
3. *The usage of IB or other underlying frameworks*: AAE is not derived from the IB framework. As shown in [15], the AAE equivalent Lagrangian (18) can be linked with the IB formulation and defined as:

$$\mathcal{L}_{\mathrm{AAE}}(\theta_x, \phi_z) = \mathcal{D}_{\mathbf{z}} + \beta_{\mathbf{x}}\mathcal{D}_{\mathbf{x}\hat{\mathbf{x}}}, \quad (22)$$

where $\beta_{\mathbf{x}} = 1$ in the original AAE formulation. It should be pointed out that the IB formulation of AAE contains the term $\mathcal{D}_{\mathbf{x}\hat{\mathbf{x}}}$, whose origin can be explained in the same way as for the VAE. Despite the fact that the term $\mathcal{D}_{\mathbf{z}}$ indeed appears in (22) with the opposite sign, it cannot be interpreted either as an upper bound on $I_{\phi_z}(\mathbf{X};\mathbf{Z})$ similarly to the VAE or as a lower bound. The goal of AAE is to minimize the reconstruction loss or to maximize the log-likelihood by ensuring that the latent space marginal distribution $q_{\phi_z}(\mathbf{z})$ matches the prior $p_{\theta_z}(\mathbf{z})$. The latter corresponds to the minimization of $D_{\mathrm{KL}}\left(q_{\phi_z}(\mathbf{z})\|p_{\theta_z}(\mathbf{z})\right)$, i.e., $\mathcal{D}_{\mathbf{z}}$ term.

4. *The label space regularization*: does not apply here due to the unsupervised setting.
5. *The latent space regularization*: is based on the hand-crafted prior with zero mean unit variance Gaussian pdf for each dimension.

6. *The reconstruction space regularization in case of reconstruction loss*: is based on the MSE.

BIB-AE [15]:

1. *The targeted tasks*: auto-encoding and generation.
2. *The architecture in terms of the latent space representation*: the encoder outputs one vector using any form of stochastic or deterministic encoding.
3. *The usage of IB or other underlying frameworks*: the BIB-AE is derived from the unsupervised IB (18) and its Lagrangian is defined as:

$$\mathcal{L}_{\text{BIB-AE}}(\boldsymbol{\theta}_x, \boldsymbol{\phi}_z) = \mathbb{E}_{p_\mathcal{D}(x)}\left[\mathcal{D}_{z|x}\right] - \mathcal{D}_z + \beta_x \mathcal{D}_x + \beta_x \mathcal{D}_{x\hat{x}}. \tag{23}$$

4. *The label space regularization*: does not apply here due to the unsupervised setting.
5. *The latent space regularization*: is based on the hand-crafted prior with Gaussian pdf applied to both conditional and unconditional terms. In fact, the prior for $\mathcal{D}_z$ can be any but $\mathcal{D}_{z|x}$ requires analytical parametrisation.
6. *The reconstruction space regularization in case of reconstruction loss*: is based on the MSE counterpart of $\mathcal{D}_{x\hat{x}}$ and the discriminator $\mathcal{D}_x$. This is a disctintive feature in comparison to VAE and AAE.

In summary, BIB-AE includes VAE and AAE as two particular cases. In turns, it should be clear that the regularizer of semi-supervised model considered in this paper resembles the BIB-AE model and extends it to the conditional case that will be considered below.

4.4.2. Links to Semi-Supervised Models

The proposed LP model (11) is also related to several state-of-the-art semi-supervised models used for the classification. As pointed out in the introduction, we only consider available labeled and unlabeled samples in our analysis. The extension to the augmented samples, i.e., permutations, syntehtically generated samples, i.e., fakes, and the adversarial examples for both latent space and label space regularizations can be performed along the line of analysis but it goes beyond the scope and focus of this paper.

Semi-supervised AAE [2]:

1. *The targeted tasks*: auto-encoding, clustering, (conditional) generation and classification.
2. *The architecture in terms of the latent space representation*: the encoder outputs two vectors representing the discrete class and continuous type of style. The class distribution is assumed to follow categorical distribution and style Gaussian one. Both constraints on the prior distributions are ensured using adversarial framework with two corresponding discriminators. In its original setting, AAE does not use any augmented samples or adversarial examples.

   Remark: It should be pointed out that in our architecture we consider the latent space to be represented by the vector **a**, which is fed to the classifier and regularizer that gives a natural consideration of IB and corresponding regularization and priors. In the case of semi-supervised AAE, the latent space is considered by the class and style representations directly. Therefore, to make it coherent with our case, one should assume that the class vector of semi-supervised AAE corresponds to the vector **c** and the style vector to the vector **z**.

3. *The usage of IB or other underlying frameworks*: AAE is not derived from the IB framework. However, as shown in our analysis the semi-supervised AAE represents the learnable prior case in part of latent space regularization. The corresponding Lagrangian of semi-supervised AAE is given by (16) and considered in Section 4.3.
4. *The label space regularization*: is based on the adversarial discriminator in assumption that the class labels follow categorical distribution. This is applied to both labeled and unlabeled samples.
5. *The latent space regularization*: is based on the learnable prior with Gaussian pdf of AE.

6. *The reconstruction space regularization in case of reconstruction loss*: is only based on the MSE.

CatGAN [3]: is based on an extension of classical GAN binary discriminator designed to distinguish between the original images and fake images generated from the latent space distribution to a multi-class discriminator. The author assumes the one-hot-vector encoding of class labels. The system is considered for the unsupervised and semi-supervised modes. For both modes the one-hot-vector encoding is used to encoded class labels. For the unsupervised mode, the system has an access only to the unlabeled data and the output of the classifier is considered to be a clustering to a predefined number of clusters/classes. The main idea behind the unsupervised training consists of a training of the discriminator that any sample from the set of original images is assigned to one of the classes with high fidelity whereas any fake or adversarial sample is assigned to all classes almost equiprobably. This corresponds to the fake samples and the regularization in the label space is based on the considered and extended framework of entropy minimization-based regularization. In the case of absence of fakes, this regularization coincides with the semi-supervised AAE label space regularization under the categorical distribution and adversarial discriminator that is equivalent to enforcing the minimum entropy of label space. However, the encoding of fake samples is equivalent to a sort of rejection option expressed via the activation of classes that have maximum entropy or uniform distribution over the classes. Equivalently, the above types of encoding can be considered to be the maximization of mutual information between the original data and encoded class labels and minimization of mutual information between the fakes/adversarial samples and the class labels. Semi-supevised CatGAN model adds a cross-entropy term computed for the true labeled samples.

Therefore, in summary:

1. *The targeted tasks*: auto-encoding, clustering, generation and classification.
2. *The architecture in terms of the latent space representation*: there is no encoder as such and instead the system has a generator/decoder that generates samples from a random latent space **a** following some hand-crafted prior. The second element of architecture is a classifier with the min/max entropy optimization for the original and fake samples. The encoding of classes is assumed to be a one-hot-vector encoding.
3. *The usage of IB or other underlying frameworks*: CatGAN is not derived from the IB framework. However, as shown in [15], one can apply the IB formulation to the adversarial generative models as in the case of CatGAN assuming that the term $I_{\phi_a}(\mathbf{X};\mathbf{A}) = 0$ in (3) due to the absence of encoder as such. The minimization problem (3) reduces to the maximization of the second term $I_{\phi_a,\theta_c}(\mathbf{A};\mathbf{C})$ expressed via its lower bound of variational decomposition (6). The first term $\mathcal{D}_c$ enforces that the class labels of unlabeled samples follow the defined prior distribution $p(\mathbf{c})$ with the above property of entropy minimization under one-hot-vector encoding whereas the second term $\mathcal{D}_{c\hat{c}}$ reflects the supervised part for labeled samples. In the original CatGAN formulation, the author does not use the expression for the mutual information for the decoder/generator training as it is shown above but instead uses the decomposition of mutual information via the difference of corresponding entropies (see, first two terms in (9) in [3]). As we have pointed out, we do not include in our analysis the term corresponding to the fake samples as in original CatGAN. However, we do believe that this form of regularization does play an important role for the semi-supervised classification. The impact of this terms requires additional studies.
4. *The label space regularization*: is based on the above assumptions for labeled samples, which are included into the cross-entropy term, unlabeled samples included into the entropy minimization term and fake samples included into the entropy maximization term in the original CatGAN method.
5. *The latent space regularization*: is based on the hand-crafted prior.
6. *The reconstruction space regularization in case of reconstruction loss*: is based on the adversarial discriminator only.

SeGMA [18]: is a semi-supervised clustering and generative system with a single latent vector representation auto-encoder similar in spirit to the unsupervised version of AAE that can be also used for the classification. The latent space of SeGMA is assumed to follow a mixture of Gaussians. Using a small labeled data set, classes are assigned to components of this mixture of Gaussians by minimizing the cross-entropy loss induced by the class posterior distribution of a simple Gaussian classifier. The resulting mixture describes the distribution of the whole data, and representatives of individual classes are generated by sampling from its components. In the classification setup, SeGMA uses the latent space clustering scheme for the classification.

Therefore, in summary:

1. *The targeted tasks*: auto-encoding, clustering, generation and classification.
2. *The architecture in terms of the latent space representation*: a single vector representation following mixture of Gaussians distribution.
3. *The usage of IB or other underlying frameworks*: SeGMA is not derived from the IB framework but a link to the regularized ELBO an other related auto-encoders with interpretable latent space is demonstrated. However, as in previous methods it can be linked to the considered IB interpretation of the semi-supervised methods with hand-crafted priors (16). An equivalent Lagrangian of SeGMA is:

$$\mathcal{L}_{\text{SeGMA}}(\theta_c, \theta_x, \phi_z) = \mathcal{D}_z + \beta_x \mathcal{D}_{x\hat{x}} + \beta_c \mathcal{D}_{c\hat{c}}, \tag{24}$$

where the latent space discriminator $\mathcal{D}_z$ is assumed to be the maximum mean discrepancy (MMD) penalty that is analytically defined for the mixture of Gaussians pdf, $\mathcal{D}_{x\hat{x}}$ is represented by the MSE and $\mathcal{D}_{c\hat{c}}$ represents the cross-entropy for the labeled data defined over class labels deduced from the latent space representation.

4. *The label space regularization*: is based on the above assumptions for labeled samples, which are included into the cross-entropy term as discussed above.
5. *The latent space regularization*: is based on the hand-crafted mixture of Gaussians pdf.
6. *The reconstruction space regularization in case of reconstruction loss*: is based on the MSE.

VAE (M1 + M2) [1]: is based on the combination of several models. The model M1 represents a vanilla VAE considered in Section 4.4.1. Therefore, model M1 is a particular case of considered unsupervised IB. The model M2 is a combination of encoder producing a continuous latent representation and following Gaussian distribution and a classifier that takes as an input original data in parallel to the model M1. The class labels are encoded using the one-hot-vector representations and follow categorical distribution with a hyper-parameter following the symmetric Dirichlet distribution. The decoder of model M2 takes as an input the continuous latent representation and output of classifier. The decoder is trained under the MSE distortion metric. It is important to point out that the classifier works with the input data directly but not with the common latent space such as in the considered LP model. For this reason, it is an obvious analogy with the considered LP model (11) under the assumption that $\mathbf{a} = \mathbf{x}$ and all performed IB analysis directly applies to. However, as pointed by the authors, the performance of model M2 in the semi-supervised classification for the limited number of labeled samples is relatively poor. That is why the third hybrid model M1 + M2 is considered when the models M1 and M2 and used in a stacked way. At the first stage, the model M1 is learned as the usual VAE. Then the latent space of model M1 is used as an input to the model M2 trained in a semi-supervised way. Such a two-stage approach closely resembles the learnable prior architecture presented in Figure 2. However, our model is end-to-end trained with the explainable common latent space and IB origin, while the model M1 + M2 is trained in two stages with the use of regularized ELBO for the derivation of model M2.

1. *The targeted tasks*: auto-encoding, clustering, (conditional) generation and classification.

2. *The architecture in terms of the latent space representation*: the stacked combination of models M1 and M2 is used as discussed above.
3. *The usage of IB or other underlying frameworks*: VAE M1 + M2 is not derived from the IB framework but it is linked to the regularized ELBO with the cross-entropy for the labeled samples. The corresponding IB Lagrangian of semi-supervised VAE M1 + M2 under the assumption of end-to-end training can be defined as:

$$\mathcal{L}_{SS-VAE\,M1+M2}^{LP}(\theta_c, \theta_x, \phi_a, \phi_z) = \mathbb{E}_{p_\mathcal{D}(\mathbf{x})}\left[\mathcal{D}_{z|x}\right] + \beta_x \mathcal{D}_{x\hat{x}} + \beta_c \mathcal{D}_{c\hat{c}} + \beta_c \mathcal{D}_c. \qquad (25)$$

4. *The label space regularization*: is based on the assumption of categorical distribution of labels.
5. *The reconstruction space regularization in case of reconstruction loss*: is only based on the MSE.

5. Experimental Results

5.1. Experimental Setup

The tested system is based on (i) the deterministic encoder and decoder, (ii) the stochastic encoder of type $\mathbf{a} = f_{\phi_a}(\mathbf{x} + \epsilon)$ with the data independent perturbations ϵ and deterministic decoder. The density ratio estimator [24] is used to measure all KL-divergences. The results of semi-supervised classification on the MNIST dataset are reported in Table 1, where symbol D indicates the deterministic setup (i) and symbol S corresponds to the stochastic one (ii). To choose the optimal parameters of systems, e.g., the Lagrangian multipliers in the considered models, we used 3-run cross-validation with the randomly chosen labeled examples as shown in Appendices B–G. Once the model parameters were chosen, we run 10 time cross-validation and the average results are shown in Table 1.

Additionally, we performed a 10-run cross-validation on the SVHN dataset [25]. We used the same architecture as for MNIST with the same encoders, decoders and discriminators. In contrast to VAE M1 + M2, we used normalized raw data without any pre-processing. Additionally, in contrast to AAE, where an extra set of 531,131 unlabeled images was used for the semi-supervised training, in our experiments only a train set of 73,257 images was used for training. Moreover, the experiments were performed: (i) for the optimal parameters chosen after 3-run cross-validation for the MNIST dataset with no special adaption to SVHN dataset and (ii) under the network architectures with exactly the same number of used filters as given in Appendices B–G for the MNIST dataset. In summary, our goal is to test the generalization capacity of the proposed approach but not just to achieve the best performance by fine-tuning of network parameters. The obtained results are represented in Table 1.

We compare the considered architectures with several state-of-the-art semi-supervised methods such as AAE [2], CatGAN [3], VAE (M1 + M2) [1], IB multiview [5], MV-InfoMax [5] and InfoMax [3] with 100, 1000 and 60,000 training labeled samples. The expected training times for the considered models are given in Table 2. The source code is available at https://github.com/taranO/IB-semi-supervised-classification. The analysis of the latent space of trained models for the MNIST dataset is given in Appendix A.

5.2. Discussion MNIST

The deterministic and stochastic systems based on the learnable priors clearly demonstrate the state-of-the-art performance in comparison to the considered semi-supervised counterparts.

Baseline Neural Network (NN): the obtained results allow concluding that, if the amount of labeled training data is large, as shown in "all" column (Table 1), the latent space regularization has no practically significant impact on the classification performance for both hand crafted and learnable priors. The deep classifier is capable of learning a latent representation retaining only sufficient statistics in the latent space solely based on the cross-entropy component of IB classification term decomposition as shown in Table A1, row $\mathcal{D}_{c\hat{c}}$ and column "all". The classes appear to be well separable under this form of visualization. At the same time, the decrease of number of labeled samples leads to the degradation of classification accuracy as show in Table 1 for columns "1000" and "100". This degradation is also clearly observed in Table A1, row $\mathcal{D}_{c\hat{c}}$ and column "100", where there is larger overlap between the classes compared to the column "all". The stochastic encoding via the addition of noise to the input samples does not enhance the performance with respect to the deterministic decoding for the small amount of labeled examples. One can assume that the presence of additive noise is not typical for the considered data, whereas the samples clearly differ in the geometrical appearance. Therefore, we can only assume that random geometrical permutations would be a more interesting alternative to the additive noise permutations/encoding.

Table 1. Semi-supervised classification performance (percentage error) for the optimal parameters (Appendices B–G) defined on the MNIST (D—deterministic; S—stochastic).

		MNIST (100)	MIST (1000)	MNIST (all)	SVHN (1000)
NN Baseline ($\mathcal{D}_{c\hat{c}}$)	[D]	26.31 ($\pm$0.91)	7.50 ($\pm$0.19)	0.68 ($\pm$0.05)	36.16 ($\pm$0.77)
	[S]	26.78 ($\pm$1.66)	7.54 ($\pm$0.25)	0.70 ($\pm$0.05)	36.28 ($\pm$0.93)
InfoMax [3]	[S]	33.41	21.5	15.86	-
VAE [5]	[S]	14.26	8.71	5.02	-
MV-InfoMax [5]	[S]	13.22	7.39	6.07	-
IB multiview [5]	[S]	3.03	2.34	2.22	-
VAE (M1 + M2) [5]	[S]	3.33 ($\pm$0.14)	2.40 ($\pm$0.02)	0.96	36.02 ($\pm$0.10)
CatGAN	[S]	1.91 ($\pm$0.10)	1.73 ($\pm$0.18)	0.91	-
AAE	[D]	1.90 ($\pm$0.10)	1.60 ($\pm$0.08)	0.85 ($\pm$0.02)	17.70 ($\pm$0.30)
No priors on latent space					
$\mathcal{D}_{c\hat{c}} + \mathcal{D}_c$	[D]	20.72 ($\pm$1.58)	4.99 ($\pm$0.28)	0.69 ($\pm$0.04)	25.78 ($\pm$0.90)
	[S]	19.60 ($\pm$1.37)	4.49 ($\pm$0.25)	0.67 ($\pm$0.05)	26.34 ($\pm$0.80)
Hand crafted latent space priors					
$\beta_c \mathcal{D}_{c\hat{c}} + \mathcal{D}_a$	[D]	27.44 ($\pm$1.40)	6.77 ($\pm$0.34)	0.91 ($\pm$0.05)	35.94 ($\pm$1.08)
	[S]	27.48 ($\pm$1.07)	6.91 ($\pm$0.45)	0.88 ($\pm$0.05)	35.80 ($\pm$1.21)
$\beta_c \mathcal{D}_{c\hat{c}} + \mathcal{D}_a + \beta_c \mathcal{D}_c$	[D]	12.04 ($\pm$4.46)	2.43 ($\pm$0.12)	0.81 ($\pm$0.05)	24.70 ($\pm$0.46)
	[S]	11.80 ($\pm$3.82)	2.40 ($\pm$0.10)	0.82 ($\pm$0.04)	24.62 ($\pm$0.54)
Learnable latent space priors					
$\beta_c \mathcal{D}_{c\hat{c}} + \beta_c \mathcal{D}_c + \mathcal{D}_z + \beta_x \mathcal{D}_{x\hat{x}}$	[D]	1.55 ($\pm$0.21)	1.25 ($\pm$0.10)	0.74 ($\pm$0.04)	20.07 ($\pm$0.36)
	[S]	1.49 ($\pm$0.18)	1.43 ($\pm$0.06)	0.78 ($\pm$0.04)	20.00 ($\pm$0.31)
$\beta_c \mathcal{D}_{c\hat{c}} + \beta_c \mathcal{D}_c + \mathcal{D}_z + \beta_x \mathcal{D}_{x\hat{x}} + \beta_x \mathcal{D}_x$	[D]	1.38 ($\pm$0.09)	1.21 ($\pm$0.10)	0.77 ($\pm$0.06)	19.75 ($\pm$0.52)
	[S]	1.42 ($\pm$0.10)	1.16 ($\pm$0.09)	0.79 ($\pm$0.02)	19.71 ($\pm$0.26)

No priors on latent space: to investigate the impact of unlabeled data, we add the adversarial regularizer $\mathcal{D}_c$ to the baseline classifier based on $\mathcal{D}_{c\hat{c}}$. The term $\mathcal{D}_c$ enforces the distribution of class labels for the unlabeled samples to follow the categorical distribution. At this stage, no regularization of latent space is applied. The addition of the adversarial regularizer $\mathcal{D}_c$, see "100" column (Table 1), allows reducing the classification error in comparison to the baseline classifier. Moreover, the stochastic encoder slightly outperforms the deterministic one for all numbers of labeled samples. However, the achieved classification error is far away from the performance of baseline classifier trained on the whole labeled data set. Thus, the cross-entropy and adversarial classification terms alone can hardly cope with the lack of labeled data, and proper regularization of the latent space is the main mechanism capable of retaining the most relevant representation.

Hand crafted latent space priors: along this line we investigate the impact of hand-crafted regularization in the form of the added discriminator $\mathcal{D}_a$ imposing Gaussian prior on the latent representation **a**. The sole regularization of latent space with the hand-crafted prior on the Gaussianity does not reflect the complex nature of latent space of real data. As a result the performance of the regularized classifier $\beta_c \mathcal{D}_{c\hat{c}} + \mathcal{D}_a$ does not lead to a remarkable improvement in comparison to the non-regularized counterpart $\mathcal{D}_{c\hat{c}}$ for both stochastic and deterministic types of encoding. When in addition the label space regularization $\mathcal{D}_c$ is added to the final classifier $\beta_c \mathcal{D}_{c\hat{c}} + \mathcal{D}_a + \beta_c \mathcal{D}_c$, it leads to the factor of 2 classification error reduction over the cross-entropy baseline classifier but it is still far away from the fully supervised baseline classifier trained on the fully labeled data set. At the same time, there is no significant difference between the stochastic and deterministic types of encoding.

Learnable latent space priors: along this line we will investigate the impact of learnable priors by adding the corresponding regularizations of the latent space of auto-encoder and data reconstruction. We investigate the role of reconstruction space regularization based on the MSE expressed via $\mathcal{D}_{x\hat{x}}$ and joint $\mathcal{D}_{x\hat{x}}$ and $\mathcal{D}_x$. The addition of discriminator $\mathcal{D}_x$ slightly enhances the classification but requires almost doubled training time as shown in Table 2. The stochastic encoding does not show any obvious advantage over the deterministic one in this setup. The separability of classes shown in Table A1, row $\beta_c \mathcal{D}_{c\hat{c}} + \beta_c \mathcal{D}_c + \mathcal{D}_z + \beta_x \mathcal{D}_{x\hat{x}} + \beta_x \mathcal{D}_x$ and column "100", is very close to those of column "all" and row $\mathcal{D}_{c\hat{c}}$, i.e., the semi-supervised system with 100 labeled examples is capable of closely approximating the fully supervised one. We show the t-sne only for this setup since it practically coincides with $\beta_c \mathcal{D}_{c\hat{c}} + \beta_c \mathcal{D}_c + \mathcal{D}_z + \beta_x \mathcal{D}_{x\hat{x}}$. However, it should be pointed out that the learnable priors ensures the reconstruction of data from the compressed latent space and the learned representation is the sufficient statistics for the data reconstruction task but not for the classification one. Since the entropy of the classification task is significantly lower to those of reconstruction, such a learned representation contains more information than actually needed for the classification task. A fraction of retained information is irrelevant to the classification problem and might be a potential source of classification errors. This likely explains a gap in performance between the considered semi-supervised system and fully supervised one.

Table 2. Execution time (hours) per 100 epochs on one NVIDIA GPU. For the SVHN the models with the learnable latent space priors were trained with a learning rate 0.0001 that explains the longer time but without optimization of Lagrangians, i.e., the Lagrangians were re-used from pre-trained MNIST model. All the others models were trained with a learning rate 0.001.

	MNIST	SVHN
NN Baseline ($\mathcal{D}_{c\hat{c}}$)	0.47–0.65	0.85–0.92
No priors on latent space		
$\mathcal{D}_{c\hat{c}} + \mathcal{D}_c$	0.47–0.65	0.85–0.92
Hand crafted latent space priors		
$\beta_c \mathcal{D}_{c\hat{c}} + \mathcal{D}_a$	0.47–0.65	1–1.05
$\beta_c \mathcal{D}_{c\hat{c}} + \mathcal{D}_a + \beta_c \mathcal{D}_c$	0.97–1.18	1.5–1.6
Learnable latent space priors		
$\beta_c \mathcal{D}_{c\hat{c}} + \beta_c \mathcal{D}_c + \mathcal{D}_z + \beta_x \mathcal{D}_{x\hat{x}}$	1.23–1.6	2.25–2.3
$\beta_c \mathcal{D}_{c\hat{c}} + \beta_c \mathcal{D}_c + \mathcal{D}_z + \beta_x \mathcal{D}_{x\hat{x}} + \beta_x \mathcal{D}_x$	1.98–2.42	3.5–3.55

5.3. Discussion SVHN

In the SVHN test, we did not try to optimize the Lagrangian coefficients as it was done for MNIST. However, to compensate for a potential non-optimality, we perform the model training with the reduced learning rate as indicated in Table 2. As a result, the training time on the SVHN dataset is longer. Therefore, 10-run validation of the proposed framework on the SVHN dataset was done with the optimal Lagrangian multipliers determined on the MNIST dataset. In this respect, one might observe a small degradation of the obtained results compared to the state-of-the-art. Additionally, we did not apply any pre-processing such as PCA that was used in VAE M1 + M2 and we did not use the extended unlabeled dataset as it was done in case of AAE. One can clearly observe the same behavior of semi-supervised classifiers as for MNIST data set discussed in Section 5.2. Therefore, we can clearly confirm the role of learnable priors in the overall performance observed for both datasets.

6. Conclusions and Future Work

We have introduced a novel formulation of variational information bottleneck for semi-supervised classification. To overcome the problem of original bottleneck and to compensate the lack of labeled data in the semi-supervised setting, we considered two models of latent space regularization via hand-crafted and learnable priors. On a toy example of MNIST dataset we investigated how the parameters of proposed framework influence the performance of classifier. By end-to-end training, we demonstrate how the proposed framework compares to the state-of-the-art methods and approaches the performance of fully supervised classifier.

The envisioned future work is along the lines of providing a stronger compression yet preserving only classification task relevant information since retaining more task irrelevant information does not provide distinguishable classification features, i.e., it only ensures reliable data reconstruction. In this work, we have considered IB for the predictive latent space model. We think that the contrastive multi-view IB formulation would be an interesting candidate for the regularization of latent space. Additionally, we did not use the adversarially generated examples to impose the constraint on the minimization of mutual information between them and class labels or equivalently to maximize the entropy of class label distribution for these adversarial examples according to the framework of entropy minimization. This line of "adversarial" regularization seems to be a very interesting complement to the considered variational bottleneck. In this work, we considered a particular form of stochastic encoding by the addition of data independent noise to the input with the preservation of the same class labels. This also corresponds to the consistency regularization when samples can be more generally permuted including the geometrical transformations. It is also interesting to point out that the same form of generic permutations is used in the unsupervised contrastive loss-based multi-view formulations for the continual latent space representation as opposed to the categorical one in the consistency regularization. Finally, the conditional generation can be an interesting line of research considering the generation from discrete labels and continuous latent space of the autoencoder.

Author Contributions: Conceptualization, S.V. and O.T.; methodology, O.T., M.K., T.H. and D.R.; software, O.T.; validation, O.T.; formal analysis, M.K., T.H. and D.R.; investigation, O.T.; writing—original draft preparation, S.V. and O.T., writing—review and editing, ALL; visualization, S.V. and O.T.; supervision, S.V.; project administration, S.V., All authors have read and agreed to the published version of the manuscript.

Funding: This research was funded by the Swiss National Science Foundation SNF No. 200021_182063.

Conflicts of Interest: The authors declare no conflict of interest.

Abbreviations

The following abbreviations are used in this manuscript:

IB	Information bottleneck
VAE	Variational autoencode
AAE	Adversarial autoencoder
CatGAN	Categorical generative adversarial networks
KL-divergences	Kullback–Leibler divergences
MSE	Mean squared error
HCP	IB with hand-crafted priors
LP	IB with learnable priors
NN	Neural Network
SS	Semi-supervised

Appendix A. Latent Space of Trained Models

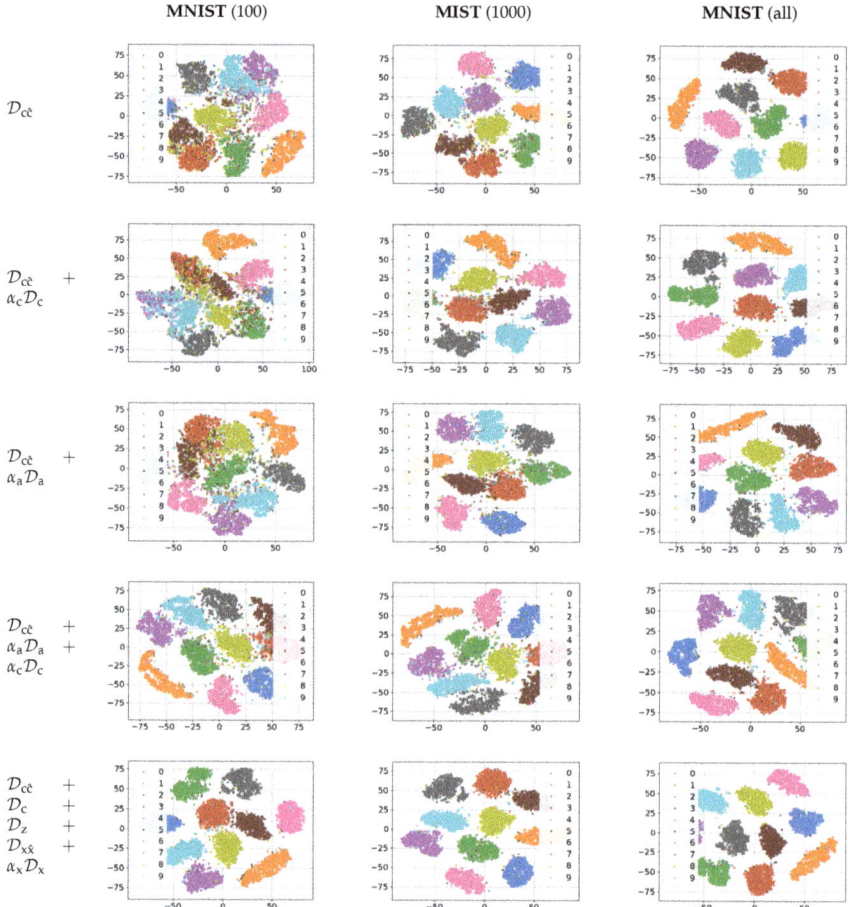

Figure A1. Latent space **a** (of size 1024) of classifier.

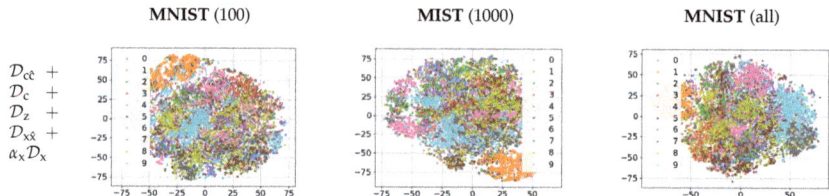

Figure A2. Latent space **z** (of size 20) of auto-encoder.

In this section, we consider the properties of classifier's latent space for both the hand-crafted and learnable priors under different amount of training samples. Figures A1 and A2 show t-sne plots for the perplexity 30 for 100, 1000 and 60,000 ("all") training labels of the MNIST dataset.

The first raw of Figure A1 with the label "$\mathcal{D}_{c\hat{c}}$" corresponds to the classifier considered in Appendix B. The latent space **a** of the classifier with "all" labels demonstrates the perfect separability of classes. The classes are far away from each other and there are practically no outliers leading to the misclassification. The decrease of the number of labels in the supervised setup, see the columns 1000 and 100, leads to a visible degradation of separability between the classes.

The regularization of class label space by the regularizer $\mathcal{D}_c$ or by the hand-crafted latent space regularizer $\mathcal{D}_a$ shown in raws "$\mathcal{D}_{c\hat{c}} + \alpha_c \mathcal{D}_c$" considered in Appendix C and "$\mathcal{D}_{c\hat{c}} + \alpha_a \mathcal{D}_a$" considered in Appendix D for the small number of training samples equal 100 does not significantly enhance the class separability with respect to "$\mathcal{D}_{c\hat{c}}$".

At the same time, the joint usage of the above regularizers according to the model "$\mathcal{D}_{c\hat{c}} + \alpha_c \mathcal{D}_c + \alpha_a \mathcal{D}_a$" according to the model in Appendix E leads to the better separability of classes for 100 labels in comparison with the previous cases. At the same time, the addition of these regularizers does not have any impact on the latent space for "all" label case.

The introduction of learnable regularization of latent space along with the class label regularization according to the model "$\mathcal{D}_{c\hat{c}} + \mathcal{D}_c + \mathcal{D}_z + \mathcal{D}_{x\hat{x}} + \alpha_x \mathcal{D}_x$" considered in Appendix G enhances the class separability in the latent space of classifier for 100 label case that is also very close to the fully supervised case.

For the comparison reasons, we also visualize the latent space of the auto-encoder z for the above model in Figure A2.

Appendix B. Supervised Training without Latent Space Regularization (Baseline)

The baseline architecture is based on the cross-entropy term $\mathcal{D}_{c\hat{c}}$ (7) in the main part of paper and depicted in Figure A3:

$$\mathcal{L}_{\text{S-NoReg}}^{\text{HCP}}(\boldsymbol{\theta}_c, \boldsymbol{\phi}_a) = \mathcal{D}_{c\hat{c}}. \tag{A1}$$

The parameters of encoder and decoder are shown in Table A1. The performance of baseline supervised classifier with and without batch normalization corresponds to the parameter $\alpha_c = 0$ in Table A3 (deterministic scenario) and Table A4 (stochastic scenario).

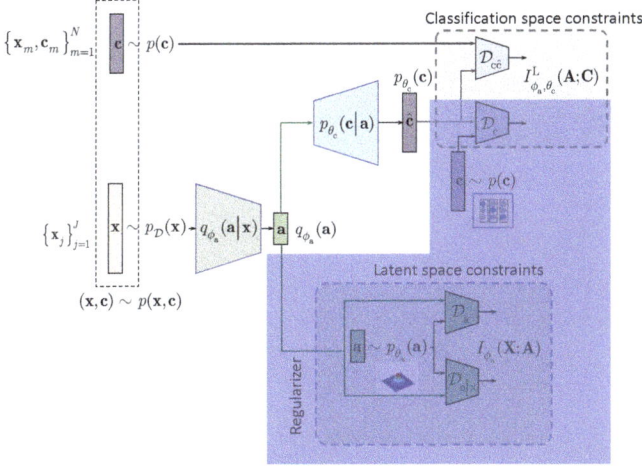

Figure A3. Baseline classifier based on $\mathcal{D}_{c\hat{c}}$. The blue shadowed regions are not used.

Table A1. The network parameters of baseline classifier trained on $\mathcal{D}_{c\hat{c}}$. The encoder is trained with and without batch normalization (BN) after Conv2D layers.

Encoder	
Size	Layer
28 × 28 × 1	Input
14 × 14 × 32	Conv2D, LeakyReLU
7 × 7 × 64	Conv2D, LeakyReLU
4 × 4 × 128	Conv2D, LeakyReLU
2048	Flatten
1024	FC, ReLU
Decoder	
Size	Layer
1024	Input
500	FC, ReLU
10	FC, Softmax

Appendix C. Semi-Supervised Training without Latent Space Regularization and with Class Label Regularizer

This model is based on terms $\mathcal{D}_{c\hat{c}}$ and $\mathcal{D}_c$ in (8) in the main part of paper and schematically shown in Figure A4:

$$\mathcal{L}^{\text{HCP}}_{\text{SS-NoReg}}(\boldsymbol{\theta}_c, \boldsymbol{\phi}_a) = \mathcal{D}_{c\hat{c}} + \alpha_c \mathcal{D}_c. \tag{A2}$$

The parameters of encoder, decoder and discriminator are shown in Table A2. The KL-divergence term $\mathcal{D}_c$ is implemented in a form of density ratio estimator (DRE). In the considered practical implementation, the parameter α_c controls the trade-off between the cross-entropy and class discriminator terms. The discriminator $\mathcal{D}_c$ is trained in an adversarial way based on samples generated by the decoder and from targeted distribution.

The performance of semi-supervised classifier with and without batch normalization is shown in Table A3 (deterministic scenario) and Table A4 (stochastic scenario).

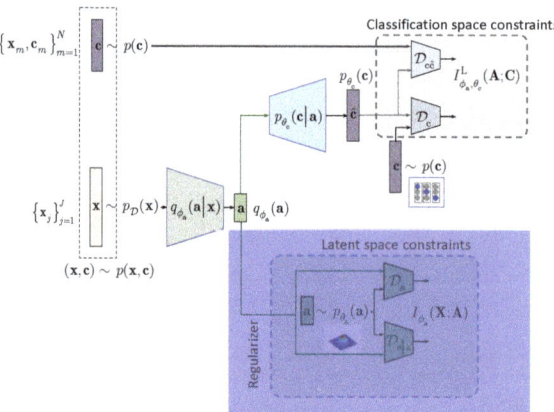

Figure A4. Semi-supervised classifier based on the cross-entropy $\mathcal{D}_{c\hat{c}}$ and categorical class discriminator $\mathcal{D}_c$. No latent space regularization is applied. The blue shadowed regions are not used.

Table A2. The network parameters of semi-supervised classifier trained on $\mathcal{D}_{c\hat{c}}$ and $\mathcal{D}_c$. The encoder is trained with and without batch normalization (BN) after Conv2D layers.

Encoder	
Size	Layer
$28 \times 28 \times 1$	Input
$14 \times 14 \times 32$	Conv2D, LeakyReLU
$7 \times 7 \times 64$	Conv2D, LeakyReLU
$4 \times 4 \times 128$	Conv2D, LeakyReLU
2048	Flatten
1024	FC, ReLU
Decoder	
Size	Layer
1024	Input
500	FC, ReLU
10	FC, Softmax
$\mathcal{D}_c$	
Size	Layer
10	Input
500	FC, ReLU
500	FC, ReLU
1	FC, Sigmoid

Table A3. The performance (percentage error) of **deterministic** classifier based on $\mathcal{D}_{c\hat{c}} + \alpha_c \mathcal{D}_c$ for the encoder with and without batch normalization as a function of Lagrangian multiplier α_c and the number of labelled examples.

Encoder Model	α_c	Runs			Mean	std
		1	2	3		
MNIST 100						
without BN	0	26.56	26.24	28.04	26.95	0.96
	0.005	20.44	21.93	18.98	20.45	1.48
	0.0005	18.55	20.43	20.59	**19.86**	1.14
	1	19.23	22.42	20.57	20.74	1.60
with BN	0	29.37	29.27	30.62	29.75	0.75
	0.005	27.97	28.02	26.27	27.42	1.00
	0.0005	25.99	23.70	24.47	**24.72**	1.17
	1	27.78	31.98	35.88	31.88	4.05
MNIST 1000						
without BN	0	7.74	6.99	6.97	7.23	0.44
	0.005	5.62	6.06	5.60	**5.76**	0.26
	0.0005	6.30	6.12	6.02	6.15	0.14
	1	5.99	6.27	6.28	6.18	0.16
with BN	0	7.45	6.95	7.52	7.31	0.31
	0.005	5.57	5.08	5.22	**5.29**	0.25
	0.0005	5.60	6.05	6.22	5.96	0.32
	1	6.05	6.41	5.82	6.09	0.30

Table A3. Cont.

Encoder Model	α_c	Runs			Mean	std
		1	2	3		
MNIST all						
without BN	0	0.83	0.83	0.74	**0.80**	0.05
	0.005	0.83	0.82	0.88	0.84	0.03
	0.0005	0.86	0.92	0.82	0.87	0.05
	1	0.72	0.85	0.87	0.81	0.08
with BN	0	0.73	0.67	0.79	0.73	0.06
	0.005	0.72	0.73	0.70	0.72	0.02
	0.0005	0.75	0.77	0.72	0.75	0.03
	1	0.67	0.68	0.73	**0.69**	0.03

Table A4. The performance (percentage error) of **stochastic** classifier with supervised noisy data (noise std = 0.1, # noise realisation = 3) based on $\mathcal{D}_{c\hat{c}} + \alpha_c \mathcal{D}_c$ for the encoder with and without batch normalization as a function of Lagrangian multiplier α_c and the number of labelled examples.

Encoder Model	α_c	Runs			Mean	std
		1	2	3		
MNIST 100						
without BN	0	25.75	26.61	26.59	26.32	0.49
	0.005	23.34	21.38	24.37	23.03	1.52
	0.0005	19.92	15.83	16.03	**17.26**	2.31
	1	22.51	20.48	21.28	21.42	1.02
with BN	0	30.26	31.24	29.3	30.27	0.97
	0.005	21.17	24.41	24.75	**23.44**	1.98
	0.0005	22.97	26.38	24.44	24.60	1.71
	1	26.62	30.43	28.44	28.50	1.91
MNIST 1000						
without BN	0	7.68	7.30	7.23	7.4	0.24
	0.005	5.59	5.16	5.80	5.52	0.33
	0.0005	5.59	6	5.84	5.81	0.21
	1	6.66	6.8	7.62	7.03	0.52
with BN	0	6.97	7.06	7.66	7.23	0.38
	0.005	4.42	4.54	4.08	**4.35**	0.24
	0.0005	5.28	5.56	5.14	5.33	0.21
	1	5.77	5.88	5.72	5.79	0.08
MNIST all						
without BN	0	0.8	0.91	0.87	0.86	0.06
	0.005	0.77	0.82	0.88	0.82	0.06
	0.0005	0.86	0.81	0.87	0.85	0.03
	1	0.93	0.85	0.92	0.90	0.04
with BN	0	0.65	0.67	0.71	0.68	0.03
	0.005	0.69	0.77	0.68	0.71	0.05
	0.0005	0.78	0.71	0.74	0.74	0.04
	1	0.71	0.64	0.62	**0.66**	0.05

Appendix D. Supervised Training with Hand Crafted Latent Space Regularization

This model is based on the cross-entropy term $\mathcal{D}_{c\hat{c}}$ and either term $\mathcal{D}_{a|x}$ or $\mathcal{D}_a$ or jointly $\mathcal{D}_{a|x}$ and $\mathcal{D}_a$ as defined by (9) in the main part of paper. In our implementation, we consider the regularization based on the adversarial term $\mathcal{D}_a$ similar to AAE due to the flexibility of imposing different priors on the latent space distribution. The implemented system is shown in Figure A5 and the training is based on:

$$\mathcal{L}_{\text{S-Reg}}^{\text{HCP}}(\theta_c, \phi_a) = \mathcal{D}_{c\hat{c}} + \alpha_a \mathcal{D}_a, \tag{A3}$$

where α_a is a regularization parameter controlling a trade-off between the cross-entropy term and latent space regularization term. We have replaced the Lagrangians above with respect to (9) in the main part of paper and used it in front of $\mathcal{D}_a$ in contrast to the original formulation (9). It is done to keep the term $\mathcal{D}_{c\hat{c}}$ without a multiplier as the reference to the baseline classifier.

The parameters of encoder, decoder and discriminator are summarized in Table A5. The performance of this classifier without and with batch normalization is shown in Table A6 (deterministic scenario) and Table A7 (stochastic scenario).

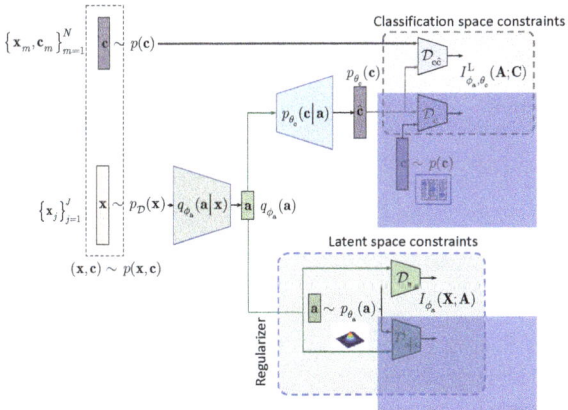

Figure A5. Supervised classifier based on the cross-entropy $\mathcal{D}_{c\hat{c}}$ and hand crafted latent space regularization $\mathcal{D}_a$. The blue shadowed parts are not used.

Table A5. The network parameters of supervised classifier trained on $\mathcal{D}_{c\hat{c}}$ and $\mathcal{D}_a$. The encoder is trained with and without batch normalization (BN) after Conv2D layers. $\mathcal{D}_a$ is trained in the adversarial way.

Encoder	
Size	Layer
$28 \times 28 \times 1$	Input
$14 \times 14 \times 32$	Conv2D, LeakyReLU
$7 \times 7 \times 64$	Conv2D, LeakyReLU
$4 \times 4 \times 128$	Conv2D, LeakyReLU
2048	Flatten
1024	FC
Decoder	
Size	Layer
1024	Input
500	FC, ReLU
10	FC, Softmax
$\mathcal{D}_a$	
Size	Layer
1024	Input
500	FC, ReLU
500	FC, ReLU
1	FC, Sigmoid

Table A6. The performance (percentage error) of **deterministic** classifier based on $\mathcal{D}_{c\hat{c}} + \alpha_a \mathcal{D}_a$ for the encoder with and without batch normalization as a function of Lagrangian multiplier.

Encoder Model	α_a	Runs 1	Runs 2	Runs 3	Mean	std
MNIST 100						
without BN	0	26.79	27.26	27.39	27.15	0.32
	0.005	28.05	25.95	30.72	28.24	2.39
	0.0005	26.67	27.69	28.46	**27.61**	0.89
	1	33.42	33.05	34.81	33.76	0.92
with BN	0	30.37	29.32	29.82	29.83	0.52
	0.005	28.02	31.49	30.80	**30.11**	1.84
	0.0005	34.54	31.92	29.82	31.09	2.36
	1	34.43	44.35	44.25	41.01	5.70
MNIST 1000						
without BN	0	7.16	8.12	7.55	7.61	0.48
	0.005	7.02	6.34	6.59	6.65	0.34
	0.0005	6.73	6.34	6.82	**6.63**	0.26
	1	9.49	9.93	10.56	9.99	0.54
with BN	0	7.39	7.83	7.92	**7.72**	0.28
	0.005	7.94	7.15	8.53	7.88	0.69
	0.0005	8.00	9.62	9.51	9.05	0.91
	1	15.79	14.88	13.71	14.79	1.04
MNIST all						
without BN	0	0.76	0.70	0.81	**0.76**	0.06
	0.005	1.07	1.03	1.13	1.08	0.05
	0.0005	0.84	0.78	0.89	0.84	0.06
	1	4.78	7.24	4.71	5.58	1.44
with BN	0	0.68	0.68	0.69	**0.68**	0.01
	0.005	0.90	0.81	1.12	0.94	0.16
	0.0005	0.87	0.80	0.89	0.85	0.05
	1	2.37	3.61	4.35	3.44	1.00

Table A7. The performance (percentage error) of **stochastic** classifier with supervised noisy data (noise std = 0.1, # noise realisation = 3) based on $\mathcal{D}_{c\hat{c}} + \alpha_a \mathcal{D}_a$ for the encoder with and without batch normalization as a function of Lagrangian multiplier.

Encoder Model	α_a	Runs 1	Runs 2	Runs 3	Mean	std
MNIST 100						
without BN	0.005	28.13	25.16	29.9	27.73	2.40
	0.0005	28.05	30.03	28.11	28.73	1.13
	1	32.33	34.09	33.73	33.38	0.93
with BN	0.005	32.25	33.47	26.01	30.58	4.00
	0.0005	33.37	36.15	35.65	35.06	1.48
	1	33.37	42.37	32.46	36.07	5.48
MNIST 1000						
without BN	0.005	7.37	7.17	6.65	7.06	0.37
	0.0005	7.48	6.68	6.67	**6.94**	0.46
	1	9.48	9.94	11.61	10.34	1.12
with BN	0.005	7.82	7.97	7.81	7.87	0.09
	0.0005	9.5	8.68	9.37	9.18	0.44
	1	12.99	10.52	9.98	11.16	1.60
MNIST all						
without BN	0.005	1.19	1.09	1.06	1.11	0.07
	0.0005	0.79	0.88	0.82	0.83	0.05
	1	6.22	4.81	5	5.34	0.77
with BN	0.005	0.94	1.07	1.04	1.02	0.07
	0.0005	0.78	0.81	0.78	**0.79**	0.02
	1	4.49	3.35	2.18	3.34	1.16

Appendix E. Semi-Supervised Training with Hand Crafted Latent Space and Class Label Regularizations

This model is based on the cross-entropy term $\mathcal{D}_{c\hat{c}}$ and either term $\mathcal{D}_{a|x}$ or $\mathcal{D}_a$ or jointly $\mathcal{D}_{a|x}$ and $\mathcal{D}_a$ and the label class regularizer $\mathcal{D}_c$ as defined by (10) in the main part of paper. In our implementation, we consider the regularization based on the adversarial term $\mathcal{D}_a$ only as shown in Figure A6. The training is based on:

$$\mathcal{L}_{S-Reg}^{HCP}(\theta_c, \phi_a) = \mathcal{D}_{c\hat{c}} + \alpha_c \mathcal{D}_c + \alpha_a \mathcal{D}_a. \tag{A4}$$

The parameters of encoder, decoder and both discriminators are shown in Table A8. The performance of this classifier without and with batch normalization is shown in Table A9 (deterministic scenario) and Table A10 (stochastic scenario).

Table A8. The network parameters of semi-supervised classifier trained on $\mathcal{D}_{c\hat{c}}$, $\mathcal{D}_a$ and $\mathcal{D}_c$. The encoder is trained with and without batch normalization (BN) after Conv2D layers. $\mathcal{D}_a$ and $\mathcal{D}_c$ are trained in the adversarial way.

Encoder	
Size	Layer
$28 \times 28 \times 1$	Input
$14 \times 14 \times 32$	Conv2D, LeakyReLU
$7 \times 7 \times 64$	Conv2D, LeakyReLU
$4 \times 4 \times 128$	Conv2D, LeakyReLU
2048	Flatten
1024	FC
Decoder	
Size	Layer
1024	Input
500	FC, ReLU
10	FC, Softmax
$\mathcal{D}_c$	
Size	Layer
10	Input
500	FC, ReLU
500	FC, ReLU
1	FC, Sigmoid
$\mathcal{D}_a$	
Size	Layer
1024	Input
500	FC, ReLU
500	FC, ReLU
1	FC, Sigmoid

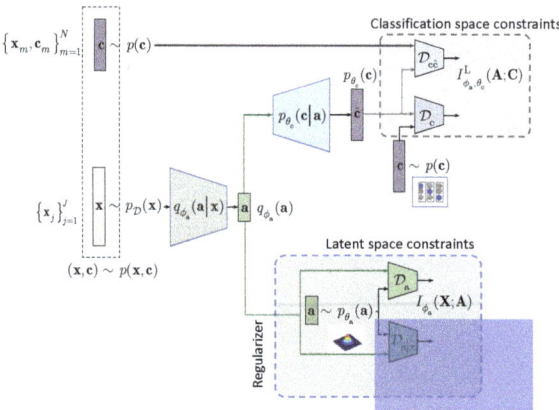

Figure A6. Semi-supervised classifier based on the cross-entropy $\mathcal{D}_{c\hat{c}}$ and hand crafted latent space regularization $\mathcal{D}_a$. The blue shadowed parts are not used.

Table A9. The performance (percentage error) of **deterministic** classifier based on $\mathcal{D}_{c\hat{c}} + \alpha_a \mathcal{D}_a + \alpha_c \mathcal{D}_c$ for the encoder with and without batch normalization.

Encoder Model	α_a	α_c	Runs 1	2	3	Mean	std
MNIST 100							
without BN	0.005	0.005	21.39	18.12	18.34	19.28	1.83
	0.0005	0.0005	15.33	22.36	13.80	17.16	4.56
	0.005	0.0005	25.66	26.25	28.81	26.91	1.67
	0.0005	0.005	9.82	13.44	13.06	**12.11**	1.99
with BN	0.005	0.005	23.45	21.19	28.87	24.50	3.94
	0.0005	0.0005	28.57	19.06	26.37	24.67	4.98
	0.005	0.0005	26.18	26.18	25.49	25.95	0.40
	0.0005	0.005	8.96	13.82	14.76	**12.52**	3.11
MNIST 1000							
without BN	0.005	0.005	3.91	4.21	3.70	3.94	0.26
	0.0005	0.0005	3.54	3.72	3.54	3.60	0.10
	0.005	0.0005	6.19	5.80	7.31	6.43	0.78
	0.0005	0.005	2.80	2.82	2.83	**2.82**	0.02
with BN	0.005	0.005	3.30	2.94	2.93	3.06	0.21
	0.0005	0.0005	2.80	2.53	2.50	2.61	0.17
	0.005	0.0005	3.51	3.75	4.12	3.79	0.31
	0.0005	0.005	2.58	2.27	2.24	**2.37**	0.19
MNIST all							
without BN	0.005	0.005	1.04	1.07	1.07	1.06	0.02
	0.0005	0.0005	0.86	0.90	0.88	0.88	0.02
	0.005	0.0005	1.08	0.92	1.09	1.03	0.10
	0.0005	0.005	0.85	0.93	0.93	0.90	0.05
with BN	0.005	0.005	1.10	1.01	0.93	1.01	0.09
	0.0005	0.0005	0.84	0.88	0.83	0.85	0.03
	0.005	0.0005	1.10	1.12	0.93	1.05	0.10
	0.0005	0.005	0.76	0.82	0.79	**0.79**	0.03

Table A10. The performance (percentage error) of **stochastic** classifier with supervised noisy data (noise std = 0.1, # noise realisation = 3) based on $\mathcal{D}_{c\hat{c}} + \alpha_a \mathcal{D}_a + \alpha_c \mathcal{D}_c$ for the encoder with and without batch normalization.

Encoder Model	α_a	α_c	Runs			Mean	std
			1	2	3		
MNIST 100							
without BN	0.005	0.005	12.4	18.05	16.73	15.73	2.96
	0.0005	0.0005	15.01	11.16	14.74	13.64	2.15
	0.005	0.0005	23.31	26.61	25.41	25.11	1.67
	0.0005	0.005	9.21	9.02	10.12	**9.45**	0.59
with BN	0.005	0.005	13.55	22.48	14.72	16.92	4.85
	0.0005	0.0005	8.37	15.01	26.92	16.77	9.40
	0.005	0.0005	32.12	30.27	31.44	31.28	0.94
	0.0005	0.005	5.46	17	11.54	11.33	5.77
MNIST 1000							
without BN	0.005	0.005	3.9	4.25	4.02	4.06	0.18
	0.0005	0.0005	3.64	3.82	4.11	3.86	0.24
	0.005	0.0005	6.68	5.34	6.36	6.13	0.70
	0.0005	0.005	3.03	2.88	2.66	2.86	0.19
with BN	0.005	0.005	2.96	3.37	2.98	3.10	0.23
	0.0005	0.0005	2.87	3.10	2.73	2.90	0.19
	0.005	0.0005	3.72	3.8	4.14	3.89	0.22
	0.0005	0.005	2.57	2.39	2.28	**2.41**	0.15
MNIST all							
without BN	0.005	0.005	1.05	1.09	1.1	1.08	0.33
	0.0005	0.0005	0.94	0.96	0.9	0.93	0.03
	0.005	0.0005	1.16	1.14	1.13	1.14	0.02
	0.0005	0.005	0.88	0.92	0.91	0.90	0.02
with BN	0.005	0.005	0.98	0.84	0.94	0.92	0.07
	0.0005	0.0005	0.79	0.96	0.82	0.86	0.09
	0.005	0.0005	1.04	1.05	1.03	1.04	0.01
	0.0005	0.005	0.74	0.78	0.84	**0.79**	0.05

Appendix F. Semi-Supervised Training with Learnable Latent Space Regularization

This model is based on the cross-entropy term $\mathcal{D}_{c\hat{c}}$, the MSE term representing $\mathcal{D}_{x\hat{x}}$, the label class regularizer $\mathcal{D}_c$ and either term $\mathcal{D}_{z|x}$ or $\mathcal{D}_z$ or jointly $\mathcal{D}_{z|x}$ and $\mathcal{D}_z$ as defined by (16) in the main part of paper. In our implementation, we consider the regularization of the latent space based on the adversarial term $\mathcal{D}_z$ only to compare it with the vanila AAE as shown in Figure A7. The encoder is also not conditioned on **c** as in the original semi-supervised AAE. Thus, the tested system is based on:

$$\mathcal{L}_{\text{SS-AAE}}^{\text{LP}}(\theta_c, \theta_x, \phi_a, \phi_z) = \beta_c \mathcal{D}_{c\hat{c}} + \beta_c \mathcal{D}_c + \mathcal{D}_z + \beta_x \mathcal{D}_{x\hat{x}}. \tag{A5}$$

We set the parameters $\beta_x = \beta_c = 1$ to compare our system with the vanila AAE. However, these parameters can be also optimized in practice.

The parameters of encoder and decoder are shown in Table A11. The performance of this classifier without and with batch normalization is shown in Table A12 (deterministic scenario) and Table A13 (stochastic scenario).

Table A11. The encoder and decoder of semi-supervised classifier trained based on $\mathcal{D}_{c\hat{c}}$, $\mathcal{D}_c$ and $\mathcal{D}_z$. The encoder is trained with and without batch normalization (BN) after Conv2D layers. $\mathcal{D}_c$ and $\mathcal{D}_z$ are trained in the adversarial way.

Encoder	
Size	Layer
$28 \times 28 \times 1$ *	Input
$14 \times 14 \times 32$	Conv2D, LeakyReLU
$7 \times 7 \times 64$	Conv2D, LeakyReLU
$4 \times 4 \times 128$	Conv2D, LeakyReLU
2048	Flatten
1024	FC, ReLU
10 10	FC, Softmax FC

Decoder	
Size	Layer
10 + 10	Input
$7 \times 7 \times 128$	FC, Reshape, BN, ReLU
$14 \times 14 \times 128$	Conv2DTrans, BN, ReLU
$28 \times 28 \times 128$	Conv2DTrans, BN, ReLU
$28 \times 28 \times 64$	Conv2DTrans, BN, ReLU
$28 \times 28 \times 1$	Conv2DTrans, Sigmoid

Dz	
Size	Layer
10	Input
500	FC, ReLU
500	FC, ReLU
1	FC, Sigmoid

Dc	
Size	Layer
10	Input
500	FC, ReLU
500	FC, ReLU
1	FC, Sigmoid

Table A12. The performance (percentage error) of **deterministic** classifier based on $\mathcal{D}_{c\hat{c}} + \mathcal{D}_c + \mathcal{D}_z + \mathcal{D}_{x\hat{x}}$ for the encoder with and without batch normalization.

Encoder Model	Runs			Mean	std
	1	2	3		
MNIST 100					
without BN	2.15	2.05	1.78	1.99	0.19
with BN	1.57	1.56	1.92	**1.68**	0.21
MNIST 1000					
without BN	1.55	1.47	1.53	1.52	0.04
with BN	1.37	1.34	1.73	**1.48**	0.22
MNIST all					
without BN	0.78	0.7	0.82	0.77	0.06
with BN	0.79	0.77	0.76	**0.77**	0.02

Table A13. The performance (percentage error) of **stochastic** classifier with supervised noisy data (noise std = 0.1, # noise realisation = 3) based on $\mathcal{D}_{c\hat{c}} + \mathcal{D}_c + \mathcal{D}_z + \mathcal{D}_{x\hat{x}}$ for the encoder with and without batch normalization.

Encoder Model	Runs			Mean	std
	1	2	3		
MNIST 100					
without BN	1.55	3.19	2.11	2.28	0.83
with BN	1.4	1.33	1.72	**1.48**	0.21
MNIST 1000					
without BN	1.73	1.53	1.6	1.62	0.10
with BN	1.28	1.43	1.2	**1.30**	0.12
MNIST all					
without BN	0.94	0.86	0.86	0.89	0.05
with BN	0.77	0.65	0.84	**0.75**	0.10

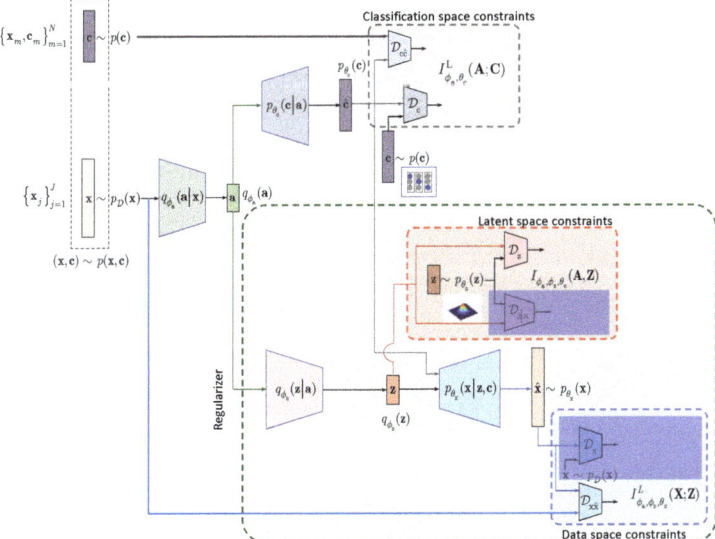

Figure A7. Semi-supervised classifier with learnable priors: the cross-entropy $\mathcal{D}_{c\hat{c}}$, MSE $\mathcal{D}_{x\hat{x}}$, class label $\mathcal{D}_c$ and latent space regularization $\mathcal{D}_a$. The blue shadowed parts are not used.

Appendix G. Semi-Supervised Training with Learnable Latent Space Regularization and Adversarial Reconstruction

This model is similar to the previously considered model but in addition to the MSE reconstruction term representing $\mathcal{D}_{x\hat{x}}$ it also contains the adversarial reconstruction term $\mathcal{D}_x$ as defined by (17) in the main part of paper. In our implementation, we consider the regularization of the latent space based on the adversarial term $\mathcal{D}_z$ as shown in Figure A8. The training is based on:

$$\mathcal{L}_{SS-AAE}^{LP}(\theta_c, \theta_x, \phi_a, \phi_z) = \mathcal{D}_z + \mathcal{D}_{x\hat{x}} + \mathcal{D}_{c\hat{c}} + \mathcal{D}_c + \alpha_x \mathcal{D}_x. \tag{A6}$$

The parameters of encoder and decoder are shown in Table A14. The performance of this classifier without and with batch normalization is shown in Table A15 (deterministic scenario) and Table A16 (stochastic scenario).

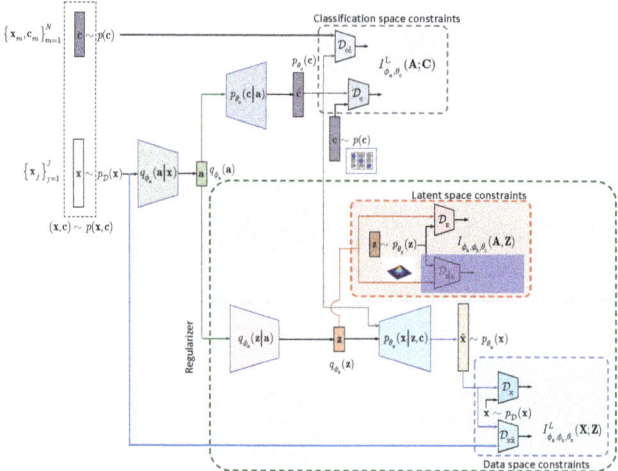

Figure A8. Semi-supervised classifier with learnable priors: the cross-entropy $\mathcal{D}_{c\hat{c}}$, MSE $\mathcal{D}_{x\hat{x}}$, adversarial reconstruction $\mathcal{D}_x$, class label $\mathcal{D}_c$ and latent space regularizer $\mathcal{D}_z$. The blue shadowed parts are not used.

Table A14. The network parameters of semi-supervised classifier trained based on $\mathcal{D}_{c\hat{c}}$, $\mathcal{D}_c$ and $\mathcal{D}_z$. The encoder is trained with and without batch normalization (BN) after Conv2D layers. $\mathcal{D}_c$ and $\mathcal{D}_z$ are trained in the adversarial way.

Encoder		
Size	Layer	
$28 \times 28 \times 1$	Input	
$14 \times 14 \times 32$	Conv2D, LeakyReLU	
$7 \times 7 \times 64$	Conv2D, LeakyReLU	
$4 \times 4 \times 128$	Conv2D, LeakyReLU	
2048	Flatten	
1024	FC, ReLU	
10 10	FC, Softmax	FC
Dz		
Size	Layer	
10	Input	
500	FC, ReLU	
500	FC, ReLU	
1	FC, Sigmoid	
Dc		
Size	Layer	
10	Input	
500	FC, ReLU	
500	FC, ReLU	
1	FC, Sigmoid	
Decoder		
Size	Layer	
10 + 10	Input	
$7 \times 7 \times 128$	FC, Reshape, BN, ReLU	
$14 \times 14 \times 128$	Conv2DTrans, BN, ReLU	
$28 \times 28 \times 128$	Conv2DTrans, BN, ReLU	
$28 \times 28 \times 64$	Conv2DTrans, BN, ReLU	
$28 \times 28 \times 1$	Conv2DTrans, Sigmoid	

Table A14. *Cont.*

Dx	
Size	Layer
$28 \times 28 \times 1$	Input
$14 \times 14 \times 64$	Conv2D, LeakyReLU
$7 \times 7 \times 64$	Conv2D, LeakyReLU
$4 \times 4 \times 128$	Conv2D, LeakyReLU
$4 \times 4 \times 256$	Conv2D, LeakyReLU
4096	Flatten
1	FC, Sigmoid

Table A15. The performance (percentage error) of **deterministic** classifier based on $\mathcal{D}_{c\hat{c}} + \mathcal{D}_c + \mathcal{D}_z + \mathcal{D}_{x\hat{x}} + \alpha_x \mathcal{D}_x$ for the encoder with and without batch normalization.

Encoder Model	α_x	Runs 1	Runs 2	Runs 3	Mean	std
MNIST 100						
without BN	0.005	2.85	3.36	2.77	2.99	0.32
	0.0005	2.58	2.49	3.08	2.72	0.32
	1	19.62	19.96	15.97	18.52	2.21
with BN	0.005	1.56	1.33	1.35	1.41	0.13
	0.0005	1.68	1.66	2.02	1.79	0.20
	1	20.85	13.6	21.67	18.71	4.44
MNIST 1000						
without BN	0.005	2.29	2.35	2.11	2.25	0.12
	0.0005	1.69	1.88	2.24	1.94	0.28
	1	3.47	3.30	4.12	3.63	0.43
with BN	0.005	1.18	1.21	1.09	1.16	0.06
	0.0005	1.44	1.28	1.29	1.34	0.09
	1	4.14	2.94	2.48	3.19	0.86
MNIST all						
without BN	0.005	0.97	1.01	1.04	1.01	0.04
	0.0005	0.88	0.85	0.93	0.89	0.04
	1	1.31	1.28	1.47	1.35	0.10
with BN	0.005	0.81	0.83	0.75	0.80	0.04
	0.0005	0.73	0.78	0.75	**0.75**	0.03
	1	0.88	0.86	1.27	1.00	0.23

Table A16. The performance (percentage error) of **stochastic** classifier with supervised noisy data (noise std = 0.1, # noise realisation = 3) based on $\mathcal{D}_{c\hat{c}} + \mathcal{D}_c + \mathcal{D}_z + \mathcal{D}_{x\hat{x}} + \alpha_x \mathcal{D}_x$ for the encoder with and without batch normalization.

Encoder Model	α_x	Runs 1	Runs 2	Runs 3	Mean	std
MNIST 100						
without BN	0.005	2.45	3.04	2.67	2.72	0.30
	0.0005	2.63	2.3	2.45	2.46	0.17
with BN	0.005	1.34	1.21	6.4	2.98	2.96
	0.0005	1.35	1.51	1.93	**1.60**	0.30
MNIST 1000						
without BN	0.005	2.31	2.26	2.2	2.26	0.06
	0.0005	1.71	2.16	1.86	1.91	0.23
with BN	0.005	1.23	1.31	1.10	**1.21**	0.11
	0.0005	1.42	1.62	1.37	1.47	0.13
MNIST all						
without BN	0.005	0.93	1.01	1.05	1.00	0.06
	0.0005	0.92	0.83	0.88	0.88	0.05
with BN	0.005	0.88	0.86	0.91	0.88	0.03
	0.0005	0.77	0.80	0.80	**0.79**	0.02

References

1. Kingma, D.P.; Mohamed, S.; Rezende, D.J.; Welling, M. Semi-supervised learning with deep generative models. In *Advances in Neural Information Processing Systems*; MIT Press: Montreal, QC, Canada, 2014; pp. 3581–3589.
2. Makhzani, A.; Shlens, J.; Jaitly, N.; Goodfellow, I.; Frey, B. Adversarial autoencoders. *arXiv* **2015**, arXiv:1511.05644.
3. Springenberg, J.T. Unsupervised and semi-supervised learning with categorical generative adversarial networks. *arXiv* **2015**, arXiv:1511.06390.
4. Chen, T.; Kornblith, S.; Norouzi, M.; Hinton, G. A simple framework for contrastive learning of visual representations. *arXiv* **2020**, arXiv:2002.05709.
5. Federici, M.; Dutta, A.; Forré, P.; Kushman, N.; Akata, Z. Learning Robust Representations via Multi-View Information Bottleneck. *arXiv* **2020**, arXiv:2002.07017.
6. Tishby, N.; Zaslavsky, N. Deep learning and the information bottleneck principle. In Proceedings of the 2015 IEEE Information Theory Workshop (ITW), Jerusalem, Israel, 26 April–1 May 2015; pp. 1–5.
7. Achille, A.; Soatto, S. Information dropout: Learning optimal representations through noisy computation. *IEEE Trans. Pattern Anal. Mach. Intell.* **2018**, *40*, 2897–2905. [CrossRef] [PubMed]
8. Berthelot, D.; Carlini, N.; Goodfellow, I.; Papernot, N.; Oliver, A.; Raffel, C.A. Mixmatch: A holistic approach to semi-supervised learning. In *Advances in Neural Information Processing Systems*; MIT Press: Vancouver, BC, Canada, 2019; pp. 5049–5059.
9. Grandvalet, Y.; Bengio, Y. Semi-supervised learning by entropy minimization. In *Advances in Neural Information Processing Systems*; MIT Press: Vancouver, BC, Canada, 2004; pp. 529–536.
10. Lee, D.H. Pseudo-label: The simple and efficient semi-supervised learning method for deep neural networks. In *ICML Workshop: Challenges in Representation Learning (WREPL)*; ICML: Atlanta, GR, USA, **2013**; Volume 3.
11. Cirecsan, D.C.; Meier, U.; Gambardella, L.M.; Schmidhuber, J. Deep, big, simple neural nets for handwritten digit recognition. *Neural Comput.* **2010**, *22*, 3207–3220.
12. Cubuk, E.D.; Zoph, B.; Mane, D.; Vasudevan, V.; Le, Q.V. Autoaugment: Learning augmentation policies from data. *arXiv* **2018**, arXiv:1805.09501.
13. Amjad, R.A.; Geiger, B.C. Learning representations for neural network-based classification using the information bottleneck principle. *IEEE Trans. Pattern Anal. Mach. Intell.* **2019**, *42*, 2225–2239. [CrossRef] [PubMed]
14. Alemi, A.A.; Fischer, I.; Dillon, J.V.; Murphy, K. Deep variational information bottleneck. *arXiv* **2016**, arXiv:1612.00410.
15. Voloshynovskiy, S.; Kondah, M.; Rezaeifar, S.; Taran, O.; Hotolyak, T.; Rezende, D.J. Information bottleneck through variational glasses. In *NeurIPS Workshop on Bayesian Deep Learning*; Vancouver Convention Center: Vancouver, BC, Canada, 2019.
16. Uğur, Y.; Zaidi, A. Variational Information Bottleneck for Unsupervised Clustering: Deep Gaussian Mixture Embedding. *Entropy* **2020**, *22*, 213. [CrossRef]
17. Maaløe, L.; Sønderby, C.K.; Sønderby, S.K.; Winther, O. Auxiliary deep generative models. *arXiv* **2016**, arXiv:1602.05473.
18. Śmieja, M.; Wołczyk, M.; Tabor, J.; Geiger, B.C. SeGMA: Semi-Supervised Gaussian Mixture Auto-Encoder. *arXiv* **2019**, arXiv:1906.09333.
19. Makhzani, A.; Frey, B.J. Pixelgan autoencoders. In *Advances in Neural Information Processing Systems*; MIT Press: Long Beach, CA, USA, **2017**; pp. 1975–1985.
20. Cover, T.M.; Thomas, J.A. *Elements of Information Theory*; John Wiley & Sons: Hoboken, NJ, USA, 2012.
21. Kingma, D.; Welling, M. Auto-Encoding Variational Bayes. *arXiv* **2014**, arXiv:1312.6114.
22. Rezende, D.J.; Mohamed, S.; Wierstra, D. Stochastic backpropagation and approximate inference in deep generative models. *arXiv* **2014**, arXiv:1401.4082.
23. Higgins, I.; Matthey, L.; Pal, A.; Burgess, C.; Glorot, X.; Botvinick, M.; Mohamed, S.; Lerchner, A. beta-VAE: Learning Basic Visual Concepts with a Constrained Variational Framework. In Proceedings of the International Conference on Learning Representations (ICLR), Toulon, France, 24–26 April 2017.

24. Goodfellow, I.; Pouget-Abadie, J.; Mirza, M.; Xu, B.; Warde-Farley, D.; Ozair, S.; Courville, A.; Bengio, Y. Generative adversarial nets. In *Advances in Neural Information Processing Systems*; MIT Press: Montreal, QC, Canada, 2014; pp. 2672–2680.
25. Netzer, Y.; Wang, T.; Coates, A.; Bissacco, A.; Wu, B.; Ng, A.Y. Reading Digits in Natural Images with Unsupervised Feature Learning. In *NIPS Workshop on Deep Learning and Unsupervised Feature Learning*; NIPS Workshop: Granada, Spain, 2011; Volume 2011, p. 5.

 © 2020 by the authors. Licensee MDPI, Basel, Switzerland. This article is an open access article distributed under the terms and conditions of the Creative Commons Attribution (CC BY) license (http://creativecommons.org/licenses/by/4.0/).

Article

The Conditional Entropy Bottleneck

Ian Fischer

Google Research, Mountain View, CA 94043, USA; iansf@google.com

Received: 30 July 2020; Accepted: 28 August 2020; Published: 8 September 2020

Abstract: Much of the field of Machine Learning exhibits a prominent set of failure modes, including vulnerability to adversarial examples, poor out-of-distribution (OoD) detection, miscalibration, and willingness to memorize random labelings of datasets. We characterize these as failures of *robust generalization*, which extends the traditional measure of generalization as accuracy or related metrics on a held-out set. We hypothesize that these failures to robustly generalize are due to the learning systems retaining *too much* information about the training data. To test this hypothesis, we propose the *Minimum Necessary Information* (MNI) criterion for evaluating the quality of a model. In order to train models that perform well with respect to the MNI criterion, we present a new objective function, the *Conditional Entropy Bottleneck* (CEB), which is closely related to the *Information Bottleneck* (IB). We experimentally test our hypothesis by comparing the performance of CEB models with deterministic models and Variational Information Bottleneck (VIB) models on a variety of different datasets and robustness challenges. We find strong empirical evidence supporting our hypothesis that MNI models improve on these problems of robust generalization.

Keywords: information theory; information bottleneck; machine learning

1. Introduction

Despite excellent progress in classical generalization (e.g., accuracy on a held-out set), the field of Machine Learning continues to struggle with the following issues:

- **Vulnerability to adversarial examples.** Most machine-learned systems are vulnerable to adversarial examples. Many defenses have been proposed, but few have demonstrated robustness against a powerful, general-purpose adversary. Many proposed defenses are ad-hoc and fail in the presence of a concerted attacker [1,2].
- **Poor out-of-distribution detection.** Most models do a poor job of signaling that they have received data that is substantially different from the data they were trained on. Even generative models can report that an entirely different dataset has higher likelihood than the dataset they were trained on [3]. Ideally, a trained model would give less confident predictions for data that was far from the training distribution (as well as for adversarial examples). Barring that, there would be a clear, principled statistic that could be extracted from the model to tell whether the model *should* have made a low-confidence prediction. Many different approaches to providing such a statistic have been proposed [4–9], but most seem to do poorly on what humans intuitively view as obviously different data.
- **Miscalibrated predictions.** Related to the issues above, classifiers tend to be overconfident in their predictions [4]. Miscalibration reduces confidence that a model's output is fair and trustworthy.
- **Overfitting to the training data.** Zhang et al. [10] demonstrated that classifiers can memorize fixed random labelings of training data, which means that it is possible to learn a classifier with perfect *inability* to generalize. This critical observation makes it clear that a fundamental test of generalization is that the model should *fail* to learn when given what we call *information-free* datasets.

We consider these to be problems of *robust generalization*, which we define and discuss in Section 2.1. In this work, we hypothesize that these problems of robust generalization all have a common cause: models retain *too much* information about the training data. We formalize this by introducing the *Minimum Necessary Information* (MNI) criterion for evaluating a learned representation (Section 2.2). We then introduce an objective function that directly optimizes the MNI, the *Conditional Entropy Bottleneck* (CEB) (Section 2.3) and compare it with the closely-related *Information Bottleneck* (IB) objective [11] in Section 2.5. In Section 2.6, we describe practical ways to optimize CEB in a variety of settings.

Finally, we give empirical evidence for the following claims:

- **Better classification accuracy.** MNI models can achieve superior accuracy on classification tasks than models that capture either more or less information than the minimum necessary information (Sections 3.1.1 and 3.1.6).
- **Improved robustness to adversarial examples.** Retaining excessive information about the training data results in vulnerability to a variety of whitebox and transfer adversarial examples. MNI models are substantially more robust to these attacks (Sections 3.1.2 and 3.1.6).
- **Strong out-of-distribution detection.** The CEB objective provides a useful metric for out-of-distribution (OoD) detection, and CEB models can detect OoD examples as well or better than non-MNI models (Section 3.1.3).
- **Better calibration.** MNI models are better calibrated than non-MNI models (Section 3.1.4).
- **No memorization of information-free datasets.** MNI models fail to learn in information-free settings, which we view as a minimum bar for demonstrating robust generalization (Section 3.1.5).

2. Materials and Methods

2.1. Robust Generalization

In classical generalization, we are interested in a model's performance on held-out data on some task of interest, such as classification accuracy. In *robust generalization*, we want: **(RG1)** *to maintain the model's performance in the classical generalization setting*; **(RG2)** *to ensure the model's performance in the presence of an adversary (unknown at training time)*; and **(RG3)** *to detect adversarial and non-adversarial data that strongly differ from the training distribution.*

Adversarial training approaches considered in the literature so far [12–14] violate **(RG1)**, as they typically result in substantial decreases in accuracy. Similarly, provable robustness approaches (e.g., Cohen et al. [15], Wong et al. [16]) provide guarantees for a particular adversary known at training time, also at a cost to test accuracy. To our knowledge, neither approaches provide any mechanism to satisfy **(RG3)**. On the other hand, approaches for detecting adversarial and non-adversarial out-of-distribution (OoD) examples [4–9] are either known to be vulnerable to adversarial attack [1,2], or do not demonstrate that the approach provides robustness against unknown adveraries, both of which violate **(RG2)**.

Training on information-free datasets [10] provides an additional way to check if a learning system is compatible with **(RG1)**, as memorization of such datasets necessarily results in maximally poor performance on any test set. Model calibration is not obviously a necessary condition for robust generalization, but if a model is well-calibrated on a held-out set, its confidence may provide some signal for distinguishing OoD examples, so we mention it as a relevant metric for **(RG3)**.

To our knowledge, the only works to date that have demonstrated progress on robust generalization for modern machine learning datasets are the *Variational Information Bottleneck* [17,18] (VIB), and *Information Dropout* [19]. Alemi et al. [17] presented preliminary results that VIB improves adversarial robustness on image classification tasks while maintaining high classification accuracy (**(RG1)** and **(RG2)**). Alemi et al. [18] showed that VIB models provide a useful signal, the *Rate*, R, for detecting OoD examples (**(RG3)**). Achille and Soatto [19] also showed preliminary results on adversarial robustness and demonstrated failure to train on information-free datasets.

In this work, we do not claim to "solve" robust generalization, but we do show notable improvement on all three conditions simply by changing the training objective. This evidence supports our core hypothesis that problems of robust generalization are caused in part by retaining too much information about the training data.

2.2. The Minimum Necessary Information

We define the *Minimum Necessary Information* (MNI) criterion for a learned representation in three parts:

- **Information.** We would like a representation Z that captures useful information about a dataset (X, Y). Entropy is the unique measure of information [20], so the criterion prefers information-theoretic approaches. (We assume familiarity with the mutual information and its relationships to entropy and conditional entropy: $I(X;Y) = H(X) - H(X|Y) = H(Y) - H(Y|X)$ [21] (p. 20).)
- **Necessity.** The semantic value of information is given by a *task*, which is specified by the set of variables in the dataset. Here we will assume that the task of interest is to predict Y given X, as in any supervised learning dataset. The information we capture in our representation Z must be necessary to solve this task. As a variable X may have *redundant* information that is useful for predicting Y, a representation Z that captures the necessary information may not be minimal or unique (the MNI criterion does not require uniqueness of Z).
- **Minimality.** Given all representations $\mathcal{Z}$ that can solve the task, we require one that retains the smallest amount of information about the task: $\inf_{Z \in \mathcal{Z}} I(Z; X, Y)$.

Necessity can be defined as $I(X;Y) \leq I(Y;Z)$. Any less information than that would prevent Z from solving the task of predicting Y from X. *Minimality* can be defined as $I(X;Y) \geq I(X;Z)$. Any more information than that would result in Z capturing information from X that is either redundant or irrelevant for predicting Y. Since the information captured by Z is constrained from above and below, we have the following necessary and sufficient conditions for perfectly achieving the Minimum Necessary Information, which we call the *MNI Point*:

$$I(X;Y) = I(X;Z) = I(Y;Z) \tag{1}$$

The MNI point defines a unique point in the information plane. The geometry of the information plane can be seen in Figure 1. The MNI criterion does not make any Markov assumptions on the models or algorithms that learn the representations. However, the algorithms we discuss here all do rely on the standard Markov chain $Z \leftarrow X \leftrightarrow Y$. See Fischer [22] for an example of an objective that doesn't rely on a Markov chain during training.

A closely related concept to Necessity is called *sufficiency* by Achille and Soatto [19] and other authors. We avoid the term due to potential confusion with minimum sufficient statistics, which maintain the mutual information between a model and the data it generates [21] (p. 35). The primary difference between necessity and sufficiency is the reliance on the Markov constraint to define sufficiency. Ref. [19] also does not identify the MNI point as an idealized target, instead defining the optimization problem: minimize $I(X;Z)$ s.t. $H(Y|Z) = H(Y|X)$.

In general it may not be possible to satisfy Equation (1). As discussed in Anantharam et al. [23–25], for any given dataset (X, Y), there is some maximum value for any possible representation Z:

$$1 \geq \eta_{KL} = \sup_{Z \leftarrow X \rightarrow Y} \frac{I(Y;Z)}{I(X;Z)} \tag{2}$$

with equality only when $X \rightarrow Y$ is a *deterministic* map. Training datasets are often deterministic in one direction or the other. e.g., common image datasets map each distinct image to a single label.

Thus, in practice, we can often get very close to the MNI on the training set given a sufficiently powerful model.

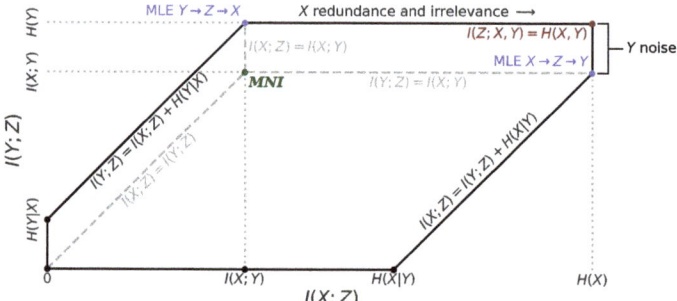

Figure 1. Geometry of the feasible regions in the $(I(X;Z), I(Y;Z))$ information plane for for any algorithm, with key points and edges labeled. The **black** edges bound the feasible region for an (X, Y) pair where $H(X|Y) > I(X;Y) > H(Y|X)$, which would generally be the case in an image classification task, for example. The gray dashed lines bound the feasible regions when the underlying model depends on a Markov chain. The $I(X;Z) = I(Y;Z)$ and $I(Y;Z) = I(X;Y)$ lines are the upper bound for $Z \leftarrow X \leftrightarrow Y$. The $I(X;Z) = I(Y;Z)$ and $I(X;Z) = I(X;Y)$ lines are the right bound for $Z \leftarrow Y \leftrightarrow X$. The blue points correspond to the best possible Maximum Likelihood Estimates (MLE) for the corresponding Markov chain models. The red point corresponds to the maximum information Z could ever capture about (X, Y). The Minimum Necessary Information (MNI) point is green. As $I(X;Z)$ increases, Z captures more information that is either redundant or irrelevant with respect to predicting Y. Similarly, any variation in Y that remains once we know X is just noise as far as the task is concerned. The MNI point is the unique point that has no redundant or irrelevant information from X, and everything but the noise from Y.

2.2.1. MNI and Robust Generalization

To satisfy **(RG1)** (classical generalization), a model must have $I(X;Z) \geq I(X;Y) = I(Y;Z)$ on the *test* dataset. Shamir et al. [26] show that $|I(X;Z) - \hat{I}(X;Z)| \approx O\left(\frac{2^{\hat{I}(X;Z)}}{\sqrt{N}}\right)$, where $\hat{I}(\cdot)$ indicates the training set information and N is the size of the training set. More recently, Bassily et al. [27] gave a similar result in a PAC setting. Both results indicate that models that are *compressed on the training data* should do *better at generalizing* to similar test data.

Less clear is how an MNI model might improve on **(RG2)** (adversarial robustness). In this work, we treat it as a hypothesis that we investigate empirically rather than theoretically. The intuition behind the hypothesis can be described in terms of the idea of *robust* and *non-robust features* from Ilyas et al. [28]: non-robust features in X should be compressed as much as possible when we learn Z, whereas robust features should be retained as much as is necessary. If Equation (1) is satisfied, Z must have "scaled" the importance of the the features in X according to their importance for predicting Y. Consequently, an attacker that tries to take advantage of a non-robust feature will have to change it much more in order to confuse the model, possibly exceeding the constraints of the attack before it succeeds.

For **(RG3)** (detection), the MNI criterion does not directly apply, as that will be a property of specific modeling choices. However, if the model provides an accurate way to measure $I(X = x; Z = z)$ for a particular pair (x, z), Alemi et al. [18] suggests that can be a valuable signal for OoD detection.

2.3. The Conditional Entropy Bottleneck

We would like to learn a representation Z of X that will be useful for predicting Y. We can represent this problem setting with the Markov chain $Z \leftarrow X \leftrightarrow Y$. We would like Z to satisfy Equation (1).

Given the conditional independence $Z \perp\!\!\!\perp Y|X$ in our Markov chain, $I(Y;Z) \leq I(X;Y)$, by the data processing inequality. Thus, maximizing $I(Y;Z)$ is consistent with the MNI criterion.

However, $I(X;Z)$ does not clearly have a constraint that targets $I(X;Y)$, as $0 \leq I(X;Z) \leq H(X)$. Instead, we can notice the following identities at the MNI point:

$$I(X;Y|Z) = I(X;Z|Y) = I(Y;Z|X) = 0 \tag{3}$$

The conditional mutual information is always non-negative, so learning a compressed representation Z of X is equivalent to minimizing $I(X;Z|Y)$. Using our Markov chain and the chain rule of mutual information [21]:

$$I(X;Z|Y) = I(X,Y;Z) - I(Y;Z) = I(X;Z) - I(Y;Z) \tag{4}$$

This leads us to the general *Conditional Entropy Bottleneck*:

$$\text{CEB} \equiv \min_Z I(X;Z|Y) - \gamma I(Y;Z) \tag{5}$$

$$= \min_Z H(Z) - H(Z|X) - H(Z) + H(Z|Y)$$

$$\quad - \gamma(H(Y) + H(Y|Z)) \tag{6}$$

$$\Leftrightarrow \min_Z -H(Z|X) + H(Z|Y) + \gamma H(Y|Z) \tag{7}$$

In line 7, we can optionally drop $H(Y)$ because it is constant with respect to Z. Here, any $\gamma > 0$ is valid, but for deterministic datasets (Section 2.2), $\gamma = 1$ will achieve the MNI for a sufficiently powerful model. Further, we should expect $\gamma = 1$ to yield *consistent* models and other values of γ not to: since $I(Y;Z)$ shows up in two forms in the objective, weighing them differently forces the optimization procedure to count bits of $I(Y;Z)$ in two different ways, potentially leading to a situation where $H(Z) - H(Z|Y) \neq H(Y) - H(Y|Z)$ at convergence. Given knowledge of those four entropies, we can define a consistency metric for Z:

$$C_{I(Y;Z)}(Z) \equiv |H(Z) - H(Z|Y) - H(Y) + H(Y|Z)| \tag{8}$$

2.4. Variational Bound on CEB

We will variationally upper bound the first term of Equation (5) and lower bound the second term using three distributions: $e(z|x)$, the *encoder* which defines the joint distribution we will use for sampling, $p(x,y,z) \equiv p(x,y)e(z|x)$; $b(z|y)$, the *backward encoder*, an approximation of $p(z|y)$; and $c(y|z)$, the *classifier*, an approximation of $p(y|z)$ (the name is arbitrary, as Y may not be labels). All of $e(\cdot)$, $b(\cdot)$, and $c(\cdot)$ may have learned parameters, just like the encoder and decoder of a VAE [29], or the encoder, classifier, and marginal in VIB.

In the following, we write expectations $\langle \log e(z|x) \rangle$. They are always with respect to the joint distribution; here, that is $p(x,y,z) \equiv p(x,y)e(z|x)$. The first term of Equation (5):

$$I(X;Z|Y) = -H(Z|X) + H(Z|Y) = \langle \log e(z|x) \rangle - \langle \log p(z|y) \rangle \tag{9}$$

$$= \langle \log e(z|x) \rangle - \langle \log b(z|y) \rangle - \langle \text{KL}[p(z|y)||b(z|y)] \rangle \tag{10}$$

$$\leq \langle \log e(z|x) \rangle - \langle \log b(z|y) \rangle \tag{11}$$

The second term of Equation (5):

$$I(Y;Z) = H(Y) - H(Y|Z) \propto -H(Y|Z) = \langle \log p(y|z) \rangle \tag{12}$$

$$= \langle \log c(y|z) \rangle + \langle \text{KL}[p(y|z)||c(y|z)] \rangle \tag{13}$$

$$\geq \langle \log c(y|z) \rangle \tag{14}$$

These variational bounds give us a tractable objective function for amortized inference, the *Variational Conditional Entropy Bottleneck* (VCEB):

$$\text{VCEB} \equiv \min_{e,b,c} \langle \log e(z|x) \rangle - \langle \log b(z|y) \rangle - \gamma \langle \log c(y|z) \rangle \qquad (15)$$

There are a number of other ways to optimize Equation (5). We describe a few of them in Section 2.6 and Appendices B and C.

2.5. Comparison to the Information Bottleneck

The Information Bottleneck (IB) [11] learns a representation Z from X subject to a soft constraint:

$$IB \equiv \min_Z I(X;Z) - \beta I(Y;Z) \qquad (16)$$

where β^{-1} controls the strength of the constraint. As $\beta \to \infty$, IB recovers the standard cross-entropy loss.

In Figure 2 we show information diagrams comparing which regions IB and CEB maximize and minimize. See Yeung [30] for a theoretical explanation of information diagrams. CEB avoids trying to both minimize and maximize the central region at the same time. In Figure 3 we show the feasible regions for CEB and IB, labeling the MNI point on both. CEB's rectification of the information plane means that we can always measure in absolute terms how much more we could compress our representation *at the same predictive performance*: $I(X;Z|Y) \geq 0$. For IB, it is not possible to tell *a priori* how far we are from optimal compression.

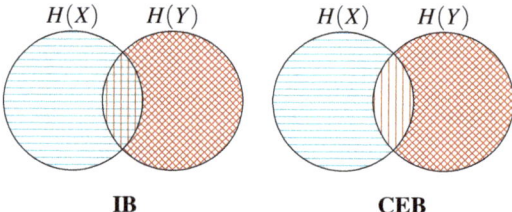

Figure 2. Information diagrams showing how IB and CEB maximize and minimize different regions. ⊗ regions inaccessible to the objective due to the Markov chain $Z \leftarrow X \leftrightarrow Y$. ⊕ regions being maximized by the objective ($I(Y;Z)$ in both cases). ⊖ regions being minimized by the objective. **IB** minimizes the intersection between Z and both $H(X|Y)$ and $I(X;Y)$. **CEB** only minimizes the intersection between Z and $H(X|Y)$.

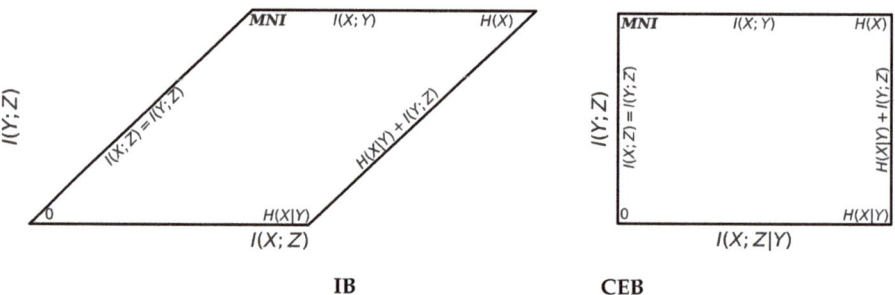

Figure 3. Geometry of the feasible regions for IB and CEB, with all points labeled. CEB rectifies IB's parallelogram by subtracting $I(Y;Z)$ at every point. Everything outside of the black lines is unattainable by any model on any dataset. Compare the IB feasible region to the dashed region in Figure 1.

From Equations (4), (5) and (16), it is clear that CEB and IB are equivalent for $\gamma = \beta - 1$. To simplify comparison of the two objectives, we can parameterize them with:

$$\rho = \log \gamma = \log(\beta - 1) \tag{17}$$

Under this parameterization, for deterministic datasets, sufficiently powerful models will target the MNI point at $\rho = 0$. As ρ increases, more information is captured by the model. $\rho < 0$ *may* capture less than the MNI. $\rho > 0$ *may* capture more than the MNI.

2.5.1. Amortized IB

As described in Tishby et al. [11], IB is a tabular method, so it is not usable for amortized inference. The tabular optimization procedure used for IB trivially applies to CEB, just by setting $\beta = \gamma + 1$. Two recent works have extended IB for amortized inference. Achille and Soatto [19] presents *InfoDropout*, which uses IB to motivate a variation on Dropout [31]. Alemi et al. [17] presents the *Variational Information Bottleneck* (VIB):

$$VIB \equiv \langle \log e(z|x) \rangle - \langle \log m(z) \rangle - \beta \langle \log c(y|z) \rangle \tag{18}$$

Instead of the backward encoder, VIB has a *marginal posterior*, $m(z)$, which is a variational approximation to $e(z) = \int dx\, p(x) e(z|x)$.

Following Alemi et al. [32], we define the *Rate* (R):

$$R \equiv \langle \log e(z|x) \rangle - \langle \log m(z) \rangle \geq I(X;Z) \tag{19}$$

We similarly define the *Residual Information* (Re_X):

$$Re_X \equiv \langle \log e(z|x) \rangle - \langle \log b(z|y) \rangle \geq I(X;Z|Y) \tag{20}$$

During optimization, observing R does not tell us how tightly we are adhering to the MNI. However, observing Re_X tells us exactly how many bits we are from the MNI point, assuming that our current classifier is optimal.

For convenience, define $CEB_x \equiv CEB_{\rho=x}$, and likewise for VIB. We can compare variational CEB with VIB by taking their difference at $\rho = 0$:

$$CEB_0 - VIB_0 = \langle \log b(z|y) \rangle - \langle \log m(z) \rangle \tag{21}$$
$$- \langle \log c(y|z) \rangle + \langle \log p(y) \rangle \tag{22}$$

Solving for $m(z)$ when that difference is 0:

$$m(z) = \frac{b(z|y) p(y)}{c(y|z)} \tag{23}$$

Since the optimal $m^*(z)$ is the marginalization of $e(z|x)$, at convergence we must have:

$$m^*(z) = \int dx\, p(x) e(z|x) = \frac{p(z|y) p(y)}{p(y|z)} \tag{24}$$

This solution may be difficult to find, as $m(z)$ only gets information about y indirectly through $e(z|x)$. For otherwise equivalent models, we may expect VIB_0 to converge to a looser approximation of $I(X;Z) = I(Y;Z) = I(X;Y)$ than CEB. Since VIB optimizes an upper bound on $I(X;Z)$, VIB_0 will report R converging to $I(X;Y)$, but may capture less than the MNI. In contrast, if Re_X converges to 0, the variational tightness of $b(z|y)$ to the optimal $p(z|y)$ depends only on the tightness of $c(y|z)$ to the optimal $p(y|z)$.

2.6. Model Variants

We introduce some variants on the basic variational CEB classification model that we will use in Section 3.1.6.

2.6.1. Bidirectional CEB

We can learn a shared representation Z that can be used to predict both X and Y with the following bidirectional CEB model: $Z_X \leftarrow X \leftrightarrow Y \rightarrow Z_Y$. This corresponds to the following joint: $p(x,y,z_X,z_Y) \equiv p(x,y)e(z_X|x)b(z_Y|y)$. The main CEB objective can then be applied in both directions:

$$\text{CEB}_{\text{bidir}} \equiv \min -H(Z_X|X) + H(Z_X|Y) + \gamma_X H(Y|Z_X) \\ - H(Z_Y|Y) + H(Z_Y|X) + \gamma_Y H(X|Z_Y) \tag{25}$$

For the two latent representations to be useful, we want them to be consistent with each other (minimally, they must have the same parametric form). Fortunately, that consistency is trivial to encourage by making the natural variational substitutions: $p(z_Y|x) \rightarrow e(z_Y|x)$ and $p(z_X|y) \rightarrow b(z_X|y)$. This gives variational $\text{CEB}_{\text{bidir}}$:

$$\min \langle \log e(z_X|x) \rangle - \langle \log b(z_X|y) \rangle - \gamma_X \langle \log c(y|z_X) \rangle \\ + \langle \log b(z_Y|y) \rangle - \langle \log e(z_Y|x) \rangle - \gamma_Y \langle \log d(x|z_Y) \rangle \tag{26}$$

where $d(x|z)$ is a *decoder* distribution. At convergence, we learn a unified Z that is consistent with both Z_X and Z_Y, permitting generation of either output given either input in the trained model, in the same spirit as Vedantam et al. [33], but without needing to train a joint encoder $q(z|x,y)$.

2.6.2. Consistent Classifier

We can reuse the backwards encoder as a classifier: $c(y|z) \propto b(z|y)p(y)$. We refer to this as the *Consistent Classifier*: $c(y|z) \equiv \text{softmax } b(z|y)p(y)$. If the labels are uniformly distributed, the $p(y)$ factor can be dropped; otherwise, it suffices to use the empirical $p(y)$. Using the consistent classifier for classification problems results in a model that only needs parameters for the two encoders, $e(z|x)$ and $b(z|y)$. This classifier differs from the more common *maximum a posteriori* (MAP) classifier because $b(z|y)$ is not the sampling distribution of either Z or Y.

2.6.3. CatGen Decoder

We can further generalize the idea of the consistent classifier to arbitrary prediction tasks by relaxing the requirement that we perfectly marginalize Y in the softmax. Instead, we can marginalize Y over any minibatch of size K we see at training time, under an assumption of a uniform distribution over the training examples we sampled:

$$p(y|z) = \frac{p(z|y)p(y)}{\int dy'\, p(z|y')p(y')} \tag{27}$$

$$\approx \frac{p(z|y)\frac{1}{K}}{\sum_{k=1}^{K} p(z|y_k)\frac{1}{K}} = \frac{p(z|y)}{\sum_{k=1}^{K} p(z|y_k)} \tag{28}$$

$$\approx \frac{b(z|y)}{\sum_{k=1}^{K} b(z|y_k)} \equiv c(y|z) \tag{29}$$

We can immediately see that this definition of $c(y|z)$ gives a valid distribution, as it is just a softmax over the minibatch. That means it can be directly used in the original objective without violating the variational bound. We call this decoder *CatGen*, for *Categorical Generative Model* because it can trivially "generate" Y: the softmax defines a categorical distribution over the batch; sampling from it gives indices of $Y = y_j$ that most closely correspond to $Z = z_i$.

Maximizing $I(Y; Z)$ in this manner is a universal task, in that it can be applied to any paired data X, Y. This includes images and labels – the CatGen model may be used in place of both $c(y|z_X)$ and $d(x|z_Y)$ in the $\text{CEB}_{\text{bidir}}$ model (using $e(z|x)$ for $d(x|z_Y)$). This avoids a common concern when dealing with multivariate predictions: if predicting X is disproportionately harder than predicting Y, it can be difficult to balance the model [33,34]. For CatGen models, predicting X is never any harder than predicting Y, since in both cases we are just trying to choose the correct example out of K possibilities.

It turns out that CatGen is mathematically equivalent to *Contrastive Predictive Coding* (CPC) [35] after an offset of $\log K$. We can see this using the proof from Poole et al. [36], and substituting $\log b(z|y)$ for $f(y, z)$:

$$I(Y; Z) \geq \frac{1}{K} \sum_{k=1}^{K} \mathbb{E}_{\Pi_j y_k, z \sim p(y_j) p(x_k|y_k) e(z|x_k)} \left[\log \frac{e^{f(y_k, z)}}{\frac{1}{K} \sum_{i=1}^{K} e^{f(y_i, z)}} \right] \qquad (30)$$

$$= \frac{1}{K} \sum_{k=1}^{K} \mathbb{E}_{\Pi_j y_k, z \sim p(y_j) p(x_k|y_k) e(z|x_k)} \left[\log \frac{b(z|y_k)}{\frac{1}{K} \sum_{i=1}^{K} b(z|y_i)} \right] \qquad (31)$$

The advantage of the CatGen approach over CPC in the CEB setting is that we already have parameterized the forward and backward encoders to compute $I(X; Z|Y)$, so we don't need to introduce any new parameters when using CatGen to maximize the $I(Y; Z)$ term.

As with CPC, the CatGen bound is constrained by $\log K$, but when targeting the MNI, it is more likely that we can train with $\log K \geq I(X; Y)$. This is trivially the case for the datasets we explore here, where $I(X; Y) \leq \log 10$. It is also practical for larger datasets like ImageNet, where models are routinely trained with batch sizes in the thousands (e.g., Goyal et al. [37]), and $I(X; Y) \leq \log 1000$.

3. Results

We evaluate deterministic, VIB, and CEB models on Fashion MNIST [38] and CIFAR10 [39]. Our experiments focus on comparing the performance of otherwise *identical* models when we change only the objective function and vary ρ. Thus, we are interested in relative differences in performance that can be directly attributed to the difference in objective and ρ. These experiments cover the three aspects of *Robust Generalization* (Section 2.1): **(RG1)** (classical generalization) in Sections 3.1 and 3.1.6; **(RG2)** (adversarial robustness) in Sections 3.1 and 3.1.6; and **(RG3)** (detection) in Section 3.1.

3.1. (RG1), (RG2), and (RG3): Fashion MNIST

Fashion MNIST [38] is an interesting dataset in that it is visually complex and challenging, but small enough to train in a reasonable amount of time. We trained 60 different models on Fashion MNIST, four each for the following 15 types: a deterministic model (*Determ*); seven VIB models (VIB_{-1}, ..., VIB_5); and seven CEB models (CEB_{-1}, ..., CEB_5). Subscripts indicate ρ. All 60 models share the same inference architecture and are trained with otherwise identical hyperparameters. See Appendix A for details.

3.1.1. (RG1): Accuracy and Compression

In Figure 4 we see that both VIB and CEB have improved accuracy over the deterministic baseline, consistent with compressed representations generalizing better. Also, CEB outperforms VIB at every ρ, which we can attribute to the tighter variational bound given by minimizing Re_X rather than R. In the case of a simple classification problem with a uniform distribution over classes in the training set (like Fashion MNIST), we can directly compute $I(X; Y) = \log C$, where C is the number of classes. In order to compare the relative complexity of the learned representations for the VIB and CEB models, in the second panel of Figure 4 we show the maximum *rate lower bound* seen during training:

$R_X \equiv \left\langle \log \frac{e(z|x)}{\frac{1}{K}\sum_k^K e(z|x_k)} \right\rangle \leq I(X;Z)$ using the encoder's minibatch marginal for both VIB and CEB. This lower bound on $I(X;Z)$ is the "InfoNCE with a tractable encoder" bound from Poole et al. [36]. The two sets of models show nearly the same R_X at each value of ρ. Both models converge to exactly $I(X;Y) = \log 10 \approx 2.3$ nats at $\rho = 0$, as predicted by the derivation of CEB.

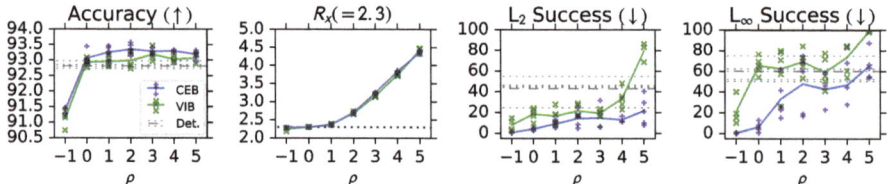

Figure 4. Test accuracy, maximum rate lower bound $R_X \leq I(Z;X)$ seen during training, and robustness to targeted PGD L_2 and L_∞ attacks on CEB, VIB, and Deterministic models trained on Fashion MNIST. At every ρ the CEB models outperform the VIB models on both accuracy and robustness, while having essentially identical maximum rates. *None of these models is adversarially trained.*

3.1.2. (RG2): Adversarial Robustness

The bottom two panels of Figure 4 show robustness to targeted *Projected Gradient Descent* (PGD) L_2 and L_∞ attacks [14]. All of the attacks are targeting the *trouser* class of Fashion MNIST, as that is the most distinctive class. Targeting a less distinctive class, such as one of the shirt classes, would confuse the difficulty of classifying the different shirts and the robustness of the model to adversaries. To measure robustness to the targeted attacks, we count the number of predictions that changed from a correct prediction on the clean image to an incorrect prediction of the target class on the adversarial image, and divide by the original number of correct predictions. Consistent with testing **(RG2)**, these adversaries are completely unknown to the models at training time – none of these models see any adversarial examples during training. CEB again outperforms VIB at every ρ, and the deterministic baseline at all but the least-compressed model ($\rho = 5$). We also see for both models that as ρ decreases, the robustness to both attacks increases, indicating that more compressed models are more robust.

Consistent with the MNI hypothesis, at $\rho = 0$ we end up with CEB models that have hit exactly 2.3 nats for the rate lower bound, have maintained high accuracy, and have strong robustness to both attacks. Moving to $\rho = -1$ gives only a small improvement to robustness, at the cost of a large decrease in accuracy.

3.1.3. (RG3): Out-of-Distribution Detection

We compare the ability of Determ, CEB_0, VIB_0, and VIB_4 to detect four different out-of-distribution (OoD) detection datasets. $U(0,1)$ is uniform noise in the image domain. MNIST uses the MNIST test set. Vertical Flip is the most challenging, using vertically flipped Fashion MNIST test images, as originally proposed in Alemi et al. [18]. CW is the Carlini-Wagner L_2 attack [40] at the default settings found in Papernot et al. [41], and additionally includes the adversarial attack success rate against each model.

We use two different metrics for thresholding, proposed in Alemi et al. [18]. H is the classifier entropy. R is the rate, defined in Section 2.5. These two threshold scores are used with the standard suite of proper scoring rules [42]: *False Positive Rate at 95% True Positive Rate* (FPR 95% TPR), *Area Under the ROC Curve* (AUROC), and *Area Under the Precision-Recall Curve* (AUPR).

Table 1 shows that using R to detect OoD examples can be much more effective than using classifier-based approaches. The deterministic baseline model is far weaker at detection using H than either of the high-performing stochastic models (CEB_0 and VIB_4). Those models both saturate detection performance, providing reliable signals for all four OoD datasets. However, as VIB_0 demonstrates, simply having R available as a signal does not guarantee good detection. As we saw above, the VIB_0 models had noticeably worse classification performance, indicating that they had not achieved the MNI point: $I(Y;Z) < I(X;Z)$ for those models. These results indicate that for detection, violating the MNI criterion by having $I(X;Z) > I(X;Y)$ may not be harmful, but violating the criterion in the opposite direction *is* harmful.

Table 1. Results for out-of-distribution detection (*OoD*). Thrsh. is the threshold score used: H is the entropy of the classifier; R is the rate. Determ cannot compute R, so only H is shown. For VIB and CEB models, H is always inferior to R, similar to findings in Alemi et al. [18], so we omit it. *Adv. Success* is attack success of the CW adversary (bottom four rows). Arrows denote whether higher or lower scores are better. **Bold** indicates the best score in that column for that OoD dataset.

OoD	Model	Thrsh.	FPR @ 95% TPR ↓	AUROC ↑	AUPR In ↑	Adv. Success ↓
U(0,1)	Determ	H	35.8	93.5	97.1	N/A
	VIB_4	R	**0.0**	**100.0**	**100.0**	N/A
	VIB_0	R	80.6	57.1	51.4	N/A
	CEB_0	R	**0.0**	**100.0**	**100.0**	N/A
MNIST	Determ	H	59.0	88.4	90.0	N/A
	VIB_4	R	**0.0**	**100.0**	**100.0**	N/A
	VIB_0	R	12.3	66.7	91.1	N/A
	CEB_0	R	0.1	94.4	99.9	N/A
Vertical Flip	Determ	H	66.8	88.6	90.2	N/A
	VIB_4	R	**0.0**	**100.0**	**100.0**	N/A
	VIB_0	R	17.3	52.7	91.3	N/A
	CEB_0	R	**0.0**	90.7	**100.0**	N/A
CW	Determ	H	15.4	90.7	86.0	100.0%
	VIB_4	R	**0.0**	**100.0**	**100.0**	55.2%
	VIB_0	R	**0.0**	98.7	**100.0**	**35.8%**
	CEB_0	R	**0.0**	99.7	**100.0**	**35.8%**

3.1.4. (RG3): Calibration

A *well-calibrated* model is correct half of the time it gives a confidence of 50% for its prediction. In Figure 5, we show calibration plots at various points during training for four models. Calibration curves help analyze whether models are underconfident or overconfident. Each point in the plots corresponds to a 5% confidence bin. Accuracy is averaged for each bin. All four networks move from under- to overconfidence during training. However, CEB_0 and VIB_0 end up only slightly overconfident, while $\rho = 2$ is already sufficient to make VIB and CEB (not shown) nearly as overconfident as the deterministic model.

3.1.5. (RG1): Overfitting Experiments

We replicate the basic experiment from Zhang et al. [10] by using the images from Fashion MNIST, but replacing the training labels with fixed random labels. This dataset is *information-free* because $I(X;Y) = 0$. We use that dataset to train multiple deterministic models, as well as CEB and VIB models at ρ from 0 through 7. We find that the CEB and VIB models with $\rho < 6$ *never* learn, even after 100 epochs of training, but the deterministic models *always* learn. After about 40 epochs of training they begin to memorize the random labels, indicating severe overfitting and a perfect *failure* to generalize.

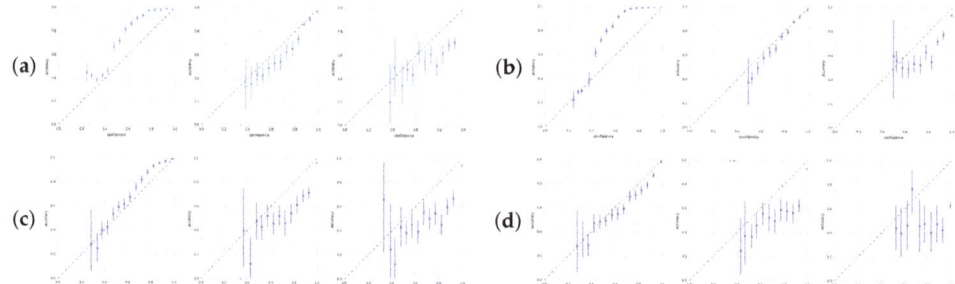

Figure 5. Calibration plots with 90% confidence intervals for four of the models after 2000 steps, 20,000 steps, and 40,000 steps (left, center, and right of each trio): (**a**) is CEB_0; (**b**) is VIB_0; (**c**) is VIB_2; (**d**) is Determ. *Perfect calibration* corresponds to the dashed diagonal lines. *Underconfidence* occurs when the points are above the diagonal. *Overconfidence* is below the diagonal. The $\rho = 0$ models are nearly perfectly calibrated still at 20,000 steps, but even at $\rho = 2$, the VIB model is almost as overconfident as Determ.

3.1.6. (RG1) and (RG2): CIFAR10 Experiments

For CIFAR10 [39] we trained the largest Wide ResNet [43] we could fit on a single GPU with a batch size of 250. This was a 62×7 model trained using AutoAugment [44]. We trained 3 CatGen CEB_{bidir} models each of CEB_0 and CEB_5 and then selected the two models with the highest test accuracy for the adversarial robustness experiments. We evaluated the CatGen models using the consistent classifier, since CatGen models only train $e(z|x)$ and $b(z|y)$. CEB_0 reached **97.51%** accuracy. This result is better than the 28×10 Wide ResNet from AutoAugment by 0.19 percentage points, although it is still worse than the Shake-Drop model from that paper. We additionally tested the model on the CIFAR-10.1 test set [45], getting accuracy of 93.6%. This is a gap of only **3.9** percentage points, which is better than all of the results reported in that paper, and substantially better than the Wide ResNet results (but still inferior to the Shake-Drop AutoAugment results). The CEB_5 model reached 97.06% accuracy on the normal test set and 91.9% on the CIFAR-10.1 test set, showing that increased ρ gave substantially worse generalization.

To test robustness of these models, we swept ϵ for both PGD attacks (Figure 6). The CEB_0 model not only has substantially higher accuracy than the adversarially-trained Wide ResNet from Madry et al. [14] (*Madry*), it also beats the Madry model on both the L_2 and the L_∞ attacks at almost all values of ϵ. We also show that this model is even more robust to two transfer attacks, where we used the CEB_5 model and the Madry model to generate PGD attacks, and then test them on the CEB_0 model. This result indicates that these models are not doing "gradient masking", a failure mode of some attempts at adversarial defense [2], since these are black-box attacks that do not rely on taking gradients through the target model.

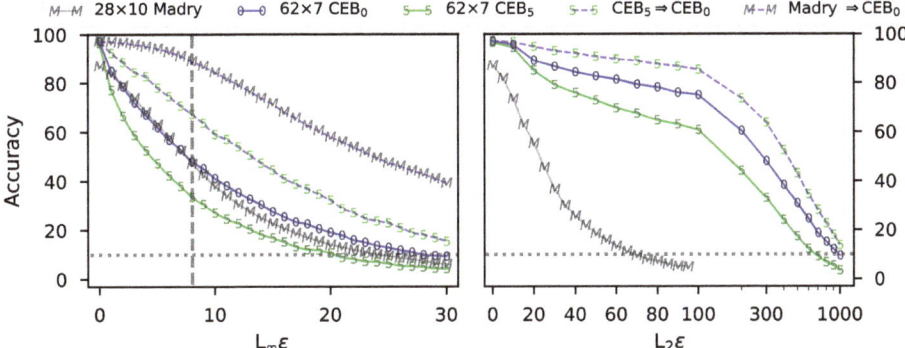

Figure 6. Left: Accuracy on untargeted L_∞ attacks at different values of ε for all 10,000 CIFAR10 test set examples. CEB_0 is the model with the highest accuracy (97.51%) trained at $\rho = 0$. CEB_5 is the model with the highest accuracy (97.06%) trained at $\rho = 5$. Madry is the best adversarially-trained model from Madry et al. [14] with 87.3% accuracy (values provided by Aleksander Madry). $CEB_5 \Rightarrow CEB_0$ is transfer attacks from the CEB_5 model to the CEB_0 model. Madry $\Rightarrow CEB_0$ is transfer attacks from the Madry model to the CEB_0 model. Madry was trained with 7 steps of PGD at $\varepsilon = 8$ (grey dashed line). Chance is 10% (grey dotted line). **Right:** Accuracy on untargeted L_2 attacks at different values of ε. All values are collected at 7 steps of PGD. CEB_0 outperforms Madry everywhere except the region of $L_\infty \varepsilon \in [2, 7]$. Madry appears to have overfit to L_∞, given its poor performance on L_2 attacks relative to either CEB model. *None of the CEB models are adversarially trained.*

4. Conclusions

We have presented the Conditional Entropy Bottleneck (CEB), motivated by the Minimum Necessary Information (MNI) criterion and the hypothesis that failures of *robust generalization* are due in part to learning models that retain *too much* information about the training data. We have shown empirically that simply by switching to CEB, models may substantially improve their robust generalization, including **(RG1)** higher accuracy, **(RG2)** better adversarial robustness, and **(RG3)** stronger OoD detection. We believe that the MNI criterion and CEB offer a promising path forward for many tasks in machine learning by permitting fast amortized inference in an easy-to-implement framework that improves robust generalization.

Funding: This research received no external funding.

Acknowledgments: We would like to thank Alex Alemi and Kevin Murphy for valuable discussions in the preparation of this work.

Conflicts of Interest: The authors declare no conflict of interest.

Appendix A. Model Details

Here we collect a number of results that are not critical to the core of the paper, but may be of interest to particular audiences.

Appendix A.1. Fashion MNIST

All of the models in our Fashion MNIST experiments have the same core architecture: A 7×2 Wide Resnet [43] for the encoder, with a final layer of $D = 4$ dimensions for the latent representation, followed by a two layer MLP classifier using ELU [46] activations with a final categorical distribution over the 10 classes.

The stochastic models parameterize the mean and variance of a $D = 4$ fully covariate multivariate Normal distribution with the output of the encoder. Samples from that distribution are passed into the classifier MLP. Apart from that difference, the stochastic models don't differ from Determ during evaluation. None of the five models uses any form of regularization (e.g., L_1, L_2, DropOut [31], BatchNorm [47]).

The VIB models have an additional learned marginal, $m(z)$, which is a mixture of 240 $D = 4$ fully covariate multivariate Normal distributions. The CEB model instead has the backward encoder, $b(z|y)$ which is a $D = 4$ fully covariate multivariate Normal distribution parameterized by a 1 layer MLP mapping the label, $Y = y$, to the mean and variance. In order to simplify comparisons, for CEB we additionally train a marginal $m(z)$ identical in form to that used by the VIB models. However, for CEB, $m(z)$ is trained using a separate optimizer so that it doesn't impact training of the CEB objective in any way. Having $m(z)$ for both CEB and VIB allows us to compare the rate, R, of each model except Determ.

Appendix A.2. CIFAR-10

For the 62×7 CEB CIFAR-10 models, we used the AutoAugment policies for CIFAR-10. We trained the models for 800 epochs, lowering the learning rate by a factor of 10 at 400 and 600 epochs. We trained all of the models using Adam [48] at a base learning rate of 10^{-3}.

Appendix A.3. Distributional Families

Any distributional family may be used for the encoder. Reparameterizable distributions [29,49] are convenient, but it is also possible to use the score function trick [50] to get a high-variance estimate of the gradient for distributions that have no explicit or implicit reparameterization. In general, a good choice for $b(z|y)$ is the same distributional family as $e(z|x)$, or a mixture thereof. These are modeling choices that need to be made by the practitioner, as they depend on the dataset. In this work, we chose normal distributions because they are easy to work with and will be the common choice for many problems, particularly when parameterized with neural networks, but that choice is incidental rather than fundamental.

Appendix B. Mutual Information Optimization

As an objective function, CEB is independent of the methods used to optimize it. Here we focus on variational objectives because they are simple, tractable, and well-understood, but any approach to optimize mutual information terms can work, so long as they respect the side of the bounds required by the objective. For example, both Oord et al. [35], Hjelm et al. [51] could be used to maximize $I(Y;Z)$.

Appendix B.1. Finiteness of the Mutual Information

The conditions for infinite mutual information given in Amjad and Geiger [52] do not apply to either CEB or VIB, as they both use stochastic encoders $e(z|x)$. In our experiments using continuous representations, we did not encounter mutual information terms that diverged to infinity, although it is possible to make modeling and data choices that make it more likely that there will be numerical instabilities. This is not a flaw specific to CEB or VIB, however, and we found numerical instability to be almost non-existent across a wide variety of modeling and architectural choices for both variational objectives.

Appendix C. Additional CEB Objectives

Here we describe a few more variants of the CEB objective.

Appendix C.1. Hierarchical CEB

Thus far, we have focused on learning a single latent representation (possibly composed of multiple latent variables at the same level). Here, we consider one way to learn a hierarchical model with CEB.

Consider the graphical model $Z_2 \leftarrow Z_1 \leftarrow X \leftrightarrow Y$. This is the simplest hierarchical supervised representation learning model. The general form of its information diagram is given in Figure A1.

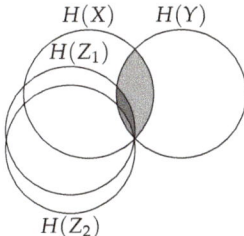

Figure A1. Information diagram for the basic hierarchical CEB model, $Z_2 \leftarrow Z_1 \leftarrow X \leftrightarrow Y$.

The key observation for generalizing CEB to hierarchical models is that the target mutual information doesn't change. By this, we mean that all of the Z_i in the hierarchy should cover $I(X;Y)$ at convergence, which means maximizing $I(Y;Z_i)$. It is reasonable to ask why we would want to train such a model, given that the final set of representations are presumably all effectively identical in terms of information content. Doing so allows us to train deep models in a principled manner such that all layers of the network are consistent with each other and with the data. We need to be more careful when considering the residual information terms, though – it is not the case that we want to minimize $I(X;Z_i|Y)$, which is not consistent with the graphical model. Instead, we want to minimize $I(Z_{i-1};Z_i|Y)$, defining $Z_0 = X$.

This gives the following simple *Hierarchical CEB* objective:

$$CEB_{\text{hier}} \equiv \min \sum_i I(Z_{i-1};Z_i|Y) - I(Y;Z_i) \tag{A1}$$

$$\Leftrightarrow \min \sum_i -H(Z_i|Z_{i-1}) + H(Z_i|Y) + H(Y|Z_i) \tag{A2}$$

Because all of the Z_i are targetting Y, this objective is as stable as regular CEB.

Appendix C.2. Sequence Learning

Many of the richest problems in machine learning vary over time. In Bialek et al. [53], the authors define the *Predictive Information*:

$$I(X_{past}, X_{future}) = \left\langle \log \frac{p(x_{past}, x_{future})}{p(x_{past})p(x_{future})} \right\rangle$$

This is of course just the mutual information between the past and the future. However, under an assumption of temporal invariance (any time of fixed length is expected to have the same entropy), they are able to characterize the predictive information, and show that it is a subextensive quantity: $\lim_{T \to \infty} I(T)/T \to 0$, where $I(T)$ is the predictive information over a time window of length $2T$ (T steps of the past predicting T steps into the future). This concise statement tells us that past observations contain vanishingly small information about the future as the time window increases.

The application of CEB to extracting the predictive information is straightforward. Given the Markov chain $X_{<t} \to X_{\geq t}$, we learn a representation Z_t that optimally covers $I(X_{<t}, X_{\geq t})$ in *Predictive CEB*:

$$CEB_{\text{pred}} \equiv \min I(X_{<t};Z_t|X_{\geq t}) - I(X_{\geq t},Z_t) \tag{A3}$$

$$\Rightarrow \min -H(Z_t|X_{<t}) + H(Z_t|X_{\geq t}) + H(X_{\geq t}|Z_t) \tag{A4}$$

Given a dataset of sequences, CEB_{pred} may be extended to a bidirectional model. In this case, two representations are learned, $Z_{<t}$ and $Z_{\geq t}$. Both representations are for timestep t, the first representing the observations before t, and the second representing the observations from t onwards.

As in the normal bidirectional model, using the same encoder and backwards encoder for both parts of the bidirectional CEB objective ties the two representations together.

Appendix C.2.1. Modeling and Architectural Choices

As with all of the variants of CEB, whatever entropy remains in the data after capturing the entropy of the mutual information in the representation must be modeled by the decoder. In this case, a natural modeling choice would be a probalistic RNN with powerful decoders per time-step to be predicted. However, it is worth noting that such a decoder would need to sample at each future step to decode the subsequent step. An alternative, if the prediction horizon is short or the predicted data are small, is to decode the entire sequence from Z_t in a single, feed-forward network (possibly as a single autoregression over all outputs in some natural sequence). Given the subextensivity of the predictive information, that may be a reasonable choice in stochastic environments, as the useful prediction window may be small.

Likely a better alternative, however, is to use the CatGen decoder, as no generation of the long future sequences is required in that case.

Appendix C.2.2. Multi-Scale Sequence Learning

As in WaveNet [54], it is natural to consider sequence learning at multiple different temporal scales. Combining an architecture like time-dilated WaveNet with CEB is as simple as combining CEB_{pred} with CEB_{hier} (Appendix C.1). In this case, each of the Z_i would represent a wider time dilation conditioned on the aggregate Z_{i-1}.

Appendix C.3. Unsupervised CEB

For unsupervised learning, it seems challenging to put the decision about what information should be kept into objective function hyperparameters, as in the β VAE and penalty VAE [32] objectives. That work showed that it is possible to constrain the amount of information in the learned representation, but it is unclear how those objective functions keep only the "correct" bits of information for the downstream tasks you might care about. This is in contrast to supervised learning while targeting the MNI point, where the task clearly defines the both the correct amount of information and which bits are likely to be important.

Our perspective on the importance of defining a task in order to constrain the information in the representation suggests that we can turn the problem into a data modeling problem in which the practitioner who selects the dataset also "models" the likely form of the useful bits in the dataset for the downstream task of interest.

In particular, given a dataset X, we propose selecting a function $f(X) \to X'$ that transforms X into a new random variable X'. This defines a paired dataset, $P(X, X')$, on which we can use CEB as normal. Note that choosing the identity function for f results in maximal mutual information between X and X' ($H(X)$ nats), which will result in a representation that is far from the MNI for normal downstream tasks.

It may seem that we have not proposed anything useful, as the selection of $f(.)$ is unconstrained, and seems much more daunting than selecting β in a β VAE or σ in a penalty VAE. However, there is a very powerful class of functions that makes this problem much simpler, and that also make it clear using CEB will *only* select bits from X that are useful. That class of functions is the noise functions.

Appendix C.3.1. Denoising CEB Autoencoder

Given a dataset X without labels or other targets, and some set of tasks in mind to be solved by a learned representation, we may select a random noise variable U, and function $X' = f(X, U)$ that we believe will destroy the irrelevant information in X. We may then add representation variables $Z_X, Z_{X'}$ to the model, giving the joint distribution $p(x, x', u, z_X, z_{X'}) \equiv p(x)p(u)p(x'|f(x,u))e(z_X|x)b(z_{X'}|x')$. This joint distribution is represented in Figure A2.

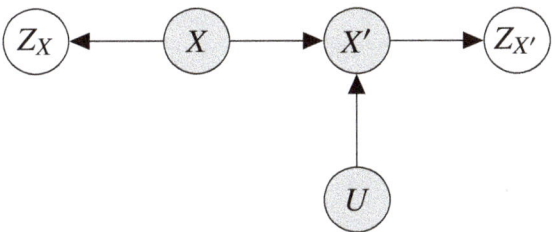

Figure A2. Graphical model for the Denoising CEB Autoencoder.

Denoising Autoencoders were originally proposed in Vincent et al. [55]. In that work, the authors argue informally that reconstruction of corrupted inputs is a desirable property of learned representations. In this paper's notation, we could describe their proposed objective as min $H(X|Z_{X'})$, or equivalently min $\langle \log d(x|z_{X'} = f(x,\eta)) \rangle_{x,\eta \sim p(x)p(\theta)}$.

We also note that, practically speaking, we would like to learn a representation that is consistent with uncorrupted inputs as well. Consequently, we are going to use a bidirectional model.

$$CEB_{\text{denoise}} \equiv \min I(X; Z_X | X') - I(X'; Z_X) \quad \text{(A5)}$$
$$+ I(X'; Z_{X'} | X) - I(X; Z_{X'}) \quad \text{(A6)}$$
$$\Rightarrow \min -H(Z_X|X) + H(Z_X|X') + H(X'|Z_X) \quad \text{(A7)}$$
$$- H(Z_{X'}|X') + H(Z_{X'}|X) + H(X|Z_{X'}) \quad \text{(A8)}$$

This requires two encoders and two decoders, which may seem expensive, but it permits a consistent learned representation that can be used cleanly for downstream tasks. Using a single encoder/decoder pair would result in either an encoder that does not work well with uncorrupted inputs, or a decoder that only generates noisy outputs.

If you are only interested in the learned representation and not in generating good reconstructions, the objective simplifies to the first three terms. In that case, the objective is properly called a *Noising CEB Autoencoder*, as the model predicts the noisy X' from X:

$$CEB_{\text{noise}} \equiv \min I(X; Z_X|X') - I(X'; Z_X) \quad \text{(A9)}$$
$$\Rightarrow \min -H(Z_X|X) + H(Z_X|X') + H(X'|Z_X) \quad \text{(A10)}$$

In these models, the noise function, $X' = f(X, U)$ must encode the practitioner's assumptions about the structure of information in the data. This obviously will vary per type of data, and even per desired downstream task.

However, we don't need to work too hard to find the perfect noise function initially. A reasonable choice for f is:

$$f(x, \eta) = \text{clip}(x + \eta, \mathcal{D}) \quad \text{(A11)}$$
$$\eta \sim \lambda U(-1, 1) * \mathcal{D} \quad \text{(A12)}$$
$$\mathcal{D} = \text{domain}(X) \quad \text{(A13)}$$

In other words, add uniform noise scaled to the domain of X and by a hyperparameter λ, and clip the result to the domain of X. When $\lambda = 1$, X' is indistinguishable from uniform noise. As $\lambda \to 0$, this maintains more and more of the original information from X in X'. For some value of $\lambda > 0$, most of the irrelevant information is destroyed and most of the relevant information is maintained, if we assume that higher frequency content in the domain of X is less likely to contain the desired information. That information is what will be retained in the learned representation.

Theoretical Optimality of Noise Functions

Above we claimed that this learning procedure will only select bits that are useful for the downstream task, given that we select the proper noise function. Here we prove that claim constructively. Imagine an oracle that knows which bits of information should be destroyed, and which retained in order to solve the future task of interest. Further imagine, for simplicity, that the task of interest is classification. What noise function must that oracle implement in order to ensure that $CEB_{denoise}$ can only learn exactly the bits needed for classification? The answer is simple: for every $X = x_i$, select $X' = x'_i$ uniformly at random from among all of the $X = x_j$ that should have the same class label as $X = x_i$. Now, the only way for CEB to maximize $I(X; Z_{X'})$ and minimize $I(X'; Z_{X'})$ is by learning a representation that is isomorphic to classification, and that encodes exactly $I(X; Y)$ nats of information, even though it was only trained "unsupervisedly" on X, X' pairs. Thus, if we can choose the correct noise function that destroys only the bits we don't care about, $CEB_{denoise}$ will learn the desired representation and nothing else (caveated by model, architecture, and optimizer selection, as usual).

References

1. Carlini, N.; Wagner, D. Adversarial examples are not easily detected: Bypassing ten detection methods. In Proceedings of the 10th ACM Workshop on Artificial Intelligence and Security, Dallas, TX, USA, 10–17 August 2017; pp. 3–14.
2. Athalye, A.; Carlini, N.; Wagner, D. Obfuscated Gradients Give a False Sense of Security: Circumventing Defenses to Adversarial Examples. In Proceedings of the 35th International Conference on Machine Learning (ICML), Stockholm, Sweden, 10–15 July 2018.
3. Nalisnick, E.; Matsukawa, A.; Teh, Y.W.; Gorur, D.; Lakshminarayanan, B. Do deep generative models know what they don't know? In Proceedings of the International Conference on Learning Representations, New Orleans, LA, USA, 6–9 May 2019.
4. Guo, C.; Pleiss, G.; Sun, Y.; Weinberger, K.Q. On Calibration of Modern Neural Networks. In Proceedings of the 34th International Conference on Machine Learning, Sydney, Australia, 6–11 August 2017.
5. Lakshminarayanan, B.; Pritzel, A.; Blundell, C. Simple and Scalable Predictive Uncertainty Estimation using Deep Ensembles. In *Advances in Neural Information Processing Systems 30*; Guyon, I., Luxburg, U.V., Bengio, S., Wallach, H., Fergus, R., Vishwanathan, S.; Garnett, R., Eds.; MIT Press: Cambridge, MA, USA, 2017; pp. 6402–6413; Available online: https://papers.nips.cc/paper/7219-simple-and-scalable-predictive-uncertainty-estimation-using-deep-ensembles (accessed on 7 September 2020).
6. Hendrycks, D.; Gimpel, K. A Baseline for Detecting Misclassified and Out-of-Distribution Examples in Neural Networks. In Proceedings of the International Conference on Learning Representations, Toulon, France, 24–26 April 2017.
7. Liang, S.; Li, Y.; Srikant, R. Enhancing The Reliability of Out-of-distribution Image Detection in Neural Networks. In Proceedings of the International Conference on Learning Representations, Vancouver, BC, Canada, 30 April–3 May 2018.
8. Lee, K.; Lee, H.; Lee, K.; Shin, J. Training Confidence-calibrated Classifiers for Detecting Out-of-Distribution Samples. In Proceedings of the International Conference on Learning Representations, Vancouver, BC, Canada, 30 April–3 May 2018.
9. Devries, T.; Taylor, G.W. Learning Confidence for Out-of-Distribution Detection in Neural Networks. *arXiv* **2018**, arXiv:1802.04865.
10. Zhang, C.; Bengio, S.; Hardt, M.; Recht, B.; Vinyals, O. Understanding deep learning requires rethinking generalization. In Proceedings of the International Conference on Learning Representations, Toulon, France, 24–26 April 2017.
11. Tishby, N.; Pereira, F.C.; Bialek, W. The information bottleneck method. In Proceedings of the 37th annual Allerton Conference on Communication, Control, and Computing, Allerton, IL, USA, 22–24 September 1999; pp. 368–377.
12. Goodfellow, I.J.; Shlens, J.; Szegedy, C. Explaining and Harnessing Adversarial Examples. *arXiv* **2014**, arXiv:1412.6572. Available online: https://arxiv.org/abs/1412.6572 (accessed on 7 September 2020).

13. Kurakin, A.; Goodfellow, I.; Bengio, S. Adversarial machine learning at scale. In Proceedings of the International Conference on Learning Representations, Toulon, France, 24–26 April 2017.
14. Madry, A.; Makelov, A.; Schmidt, L.; Tsipras, D.; Vladu, A. Towards Deep Learning Models Resistant to Adversarial Attacks. In Proceedings of the International Conference on Learning Representations, Vancouver, BC, Canada, 30 April–3 May 2018.
15. Cohen, J.M.; Rosenfeld, E.; Kolter, J.Z. Certified adversarial robustness via randomized smoothing. *arXiv* **2019**, arXiv:1902.02918.
16. Wong, E.; Schmidt, F.; Metzen, J.H.; Kolter, J.Z. Scaling provable adversarial defenses. In *Advances in Neural Information Processing Systems*; NIPS: La Jolla, CA, USA, 2018; Available online: https://papers.nips.cc/paper/8060-scaling-provable-adversarial-defenses (accessed on 7 September 2020).
17. Alemi, A.A.; Fischer, I.; Dillon, J.V.; Murphy, K. Deep Variational Information Bottleneck. In Proceedings of the International Conference on Learning Representations, Toulon, France, 24–26 April 2017.
18. Alemi, A.A.; Fischer, I.; Dillon, J.V. Uncertainty in the Variational Information Bottleneck. *arXiv* **2018**, arXiv:1807.00906. Available online: https://arxiv.org/abs/1807.00906 (accessed on 7 September 2020).
19. Achille, A.; Soatto, S. Information dropout: Learning optimal representations through noisy computation. *IEEE Trans. Pattern Anal. Mach. Intell.* **2018**, *40*, 2897–2905. [CrossRef] [PubMed]
20. Shannon, C.E. A Mathematical Theory of Communication. *Bell Syst. Tech. J.* **1948**, *27*, 379–423. [CrossRef]
21. Cover, T.M.; Thomas, J.A. *Elements of Information Theory*, 2nd ed.; John Wiley & Sons: Hoboken, NJ, USA, 2006.
22. Fischer, I. Bounding the Multivariate Mutual Information. Information Theory and Machine Learning Workshop. 2019. Available online: https://drive.google.com/file/d/17lJiJ4v_6h0p-ist_jCrr-o1ODi7yELx/view (accessed on 7 September 2020).
23. Anantharam, V.; Gohari, A.; Kamath, S.; Nair, C. On hypercontractivity and a data processing inequality. In Proceedings of the 2014 IEEE International Symposium on Information Theory, Honolulu, HI, USA, 29 June–4 July 2014; pp. 3022–3026.
24. Polyanskiy, Y.; Wu, Y. Strong data-processing inequalities for channels and Bayesian networks. In *Convexity and Concentration*; Springer: New York, NY, USA, 2017; pp. 211–249.
25. Wu, T.; Fischer, I.; Chuang, I.L.; Tegmark, M. Learnability for the Information Bottleneck. *Entropy* **2019**, *21*, 924.10.3390/e21100924. [CrossRef]
26. Shamir, O.; Sabato, S.; Tishby, N. Learning and generalization with the information bottleneck. *Theor. Comput. Sci.* **2010**, *411*, 2696–2711. [CrossRef]
27. Bassily, R.; Moran, S.; Nachum, I.; Shafer, J.; Yehudayoff, A. Learners that Use Little Information. In Proceedings of the Machine Learning Research, New York, NY, USA, 23–24 February 2018; Janoos, F., Mohri, M., Sridharan, K., Eds.; 2018; Volume 83, pp. 25–55; Available online: http://proceedings.mlr.press/v83/bassily18a.html (accessed on 7 September 2020)
28. Ilyas, A.; Santurkar, S.; Tsipras, D.; Engstrom, L.; Tran, B.; Madry, A. Adversarial examples are not bugs, they are features. In *Advances in Neural Information Processing Systems*; NIPS: La Jolla, CA, USA, 2019; pp. 125–136; Available online: https://papers.nips.cc/paper/8307-adversarial-examples-are-not-bugs-they-are-features (accessed on 7 September 2020).
29. Kingma, D.P.; Welling, M. Auto-encoding variational Bayes. In Proceedings of the International Conference on Learning Representations, Banff, AB, Canada, 14–16 April 2014.
30. Yeung, R.W. A new outlook on Shannon's information measures. *IEEE Trans. Inf. Theory* **1991**, *37*, 466–474. [CrossRef]
31. Srivastava, N.; Hinton, G.; Krizhevsky, A.; Sutskever, I.; Salakhutdinov, R. Dropout: A simple way to prevent neural networks from overfitting. *J. Mach. Learn. Res.* **2014**, *15*, 1929–1958.
32. Alemi, A.A.; Poole, B.; Fischer, I.; Dillon, J.V.; Saurous, R.A.; Murphy, K. Fixing a Broken ELBO. In *ICML2018*. 2018. Available online: https://icml.cc/Conferences/2018/ScheduleMultitrack?event=2442 (accessed on 7 September 2020).
33. Vedantam, R.; Fischer, I.; Huang, J.; Murphy, K. Generative Models of Visually Grounded Imagination. In Proceedings of the International Conference on Learning Representations, Vancouver, BC, Canada, 30 April–3 May 2018.
34. Higgins, I.; Sonnerat, N.; Matthey, L.; Pal, A.; Burgess, C.P.; Bošnjak, M.; Shanahan, M.; Botvinick, M.; Hassabis, D.; Lerchner, A. SCAN: Learning Hierarchical Compositional Visual Concepts. *arXiv* **2018**, arXiv:1707.03389.

35. Oord, A.V.D.; Li, Y.; Vinyals, O. Representation learning with contrastive predictive coding. *arXiv* **2018**, arXiv:1807.03748.
36. Poole, B.; Ozair, S.; van den Oord, A.; Alemi, A.A.; Tucker, G. On Variational Bounds of Mutual Information. In Proceedings of the ICML2019, Long Beach, CA, USA, 9–15 June 2019.
37. Goyal, P.; Dollár, P.; Girshick, R.; Noordhuis, P.; Wesolowski, L.; Kyrola, A.; Tulloch, A.; Jia, Y.; He, K. Accurate, large minibatch SGD: Training imagenet in 1 hour. *arXiv* **2017**, arXiv:1706.02677.
38. Xiao, H.; Rasul, K.; Vollgraf, R. Fashion-MNIST: A Novel Image Dataset for Benchmarking Machine Learning Algorithms. *arXiv* **2017**, arXiv:1708.07747.
39. Krizhevsky, A.; Hinton, G. *Learning Multiple Layers of Features From Tiny Images*; Technical Report; University of Toronto: Toronto, ON, USA, 2009; Available online: https://www.cs.toronto.edu/~kriz/learning-features-2009-TR.pdf (accessed on 7 September 2020).
40. Carlini, N.; Wagner, D. Towards evaluating the robustness of neural networks. In Proceedings of the 2017 IEEE Symposium on Security and Privacy (SP), San Jose, CA, USA, 22–26 May 2017; pp. 39–57.
41. Papernot, N.; Faghri, F.; Carlini, N.; Goodfellow, I.; Feinman, R.; Kurakin, A.; Xie, C.; Sharma, Y.; Brown, T.; Roy, A.; et al. Technical Report on the CleverHans v2.1.0 Adversarial Examples Library. *arXiv* **2018**, arXiv:1610.00768.
42. Lee, K.; Lee, K.; Lee, H.; Shin, J. A Simple Unified Framework for Detecting Out-of-Distribution Samples and Adversarial Attacks. In *Advances in Neural Information Processing Systems 31*; Bengio, S., Wallach, H., Larochelle, H., Grauman, K., Cesa-Bianchi, N., Garnett, R., Eds.; NIPS: La Jolla, CA, USA, 2018; pp. 7167–7177; Available online: https://papers.nips.cc/paper/7947-a-simple-unified-framework-for-detecting-out-of-distribution-samples-and-adversarial-attacks (accessed on 7 September 2020).
43. Zagoruyko, S.; Komodakis, N. Wide Residual Networks. In *Proceedings of the British Machine Vision Conference (BMVC), York, UK, 9–12 September 2019*; Richard, C., Wilson, E.R.H., Smith, W.A.P., Eds.; BMVA Press: London, UK, 2016; pp. 87.1–87.12; Available online: http://www.bmva.org/bmvc/2016/papers/paper087/ (accessed on 7 September 2020).
44. Cubuk, E.D.; Zoph, B.; Mane, D.; Vasudevan, V.; Le, Q.V. AutoAugment: Learning Augmentation Strategies From Data. In Proceedings of the IEEE/CVF Conference on Computer Vision and Pattern Recognition (CVPR), Long Beach, CA, USA, 16–20 June 2019.
45. Recht, B.; Roelofs, R.; Schmidt, L.; Shankar, V. Do CIFAR-10 classifiers generalize to CIFAR-10? *arXiv* **2018**, arXiv:1806.00451.
46. Clevert, D.A.; Unterthiner, T.; Hochreiter, S. Fast and accurate deep network learning by exponential linear units (elus). In Proceedings of the International Conference on Learning Representations Workshop, San Juan, Puerto Rico, 2–4 May 2016.
47. Ioffe, S.; Szegedy, C. Batch Normalization: Accelerating Deep Network Training by Reducing Internal Covariate Shift. In *Proceedings of the Machine Learning Research, Lille, France, 7–9 July 2015*; Bach, F., Blei, D., Eds.; PMLR: Lille, France, 2015; Volume 37, pp. 448–456; Available online: http://proceedings.mlr.press/v37/ioffe15 (accessed on 7 September 2020).
48. Kingma, D.; Ba, J. Adam: A method for stochastic optimization. In Proceedings of the International Conference on Learning Representations, San Diego, CA, USA, 4–8 May 2015.
49. Figurnov, M.; Mohamed, S.; Mnih, A. Implicit Reparameterization Gradients. In *Advances in Neural Information Processing Systems 31*; Bengio, S., Wallach, H., Larochelle, H., Grauman, K., Cesa-Bianchi, N., Garnett, R., Eds.; NIPS: La Jolla, CA, USA, 2018; pp. 441–452; Available online: https://papers.nips.cc/paper/7326-implicit-reparameterization-gradients (accessed on 7 September 2020).
50. Williams, R.J. Simple statistical gradient-following algorithms for connectionist reinforcement learning. *Mach. Learn.* **1992**, *8*, 229–256. [CrossRef]
51. Hjelm, R.D.; Fedorov, A.; Lavoie-Marchildon, S.; Grewal, K.; Bachman, P.; Trischler, A.; Bengio, Y. Learning deep representations by mutual information estimation and maximization. In Proceedings of the International Conference on Learning Representations, New Orleans, LA, USA, 6–9 May 2019.
52. Amjad, R.A.; Geiger, B.C. Learning Representations for Neural Network-Based Classification Using the Information Bottleneck Principle. *IEEE Trans. Pattern Anal. Mach. Intell.* **2020**, *42*, 2225–2239. [CrossRef] [PubMed]
53. Bialek, W.; Nemenman, I.; Tishby, N. Predictability, complexity, and learning. *Neural Comput.* **2001**, *13*. [CrossRef] [PubMed]

54. Van Den Oord, A.; Dieleman, S.; Zen, H.; Simonyan, K.; Vinyals, O.; Graves, A.; Kalchbrenner, N.; Senior, A.W.; Kavukcuoglu, K. WaveNet: A Generative Model for Raw Audio. *arXiv* **2016**, arXiv:1609.03499. Available online: https://arxiv.org/abs/1609.03499 (accessed on 7 September 2020).
55. Vincent, P.; Larochelle, H.; Bengio, Y.; Manzagol, P.A. Extracting and composing robust features with denoising autoencoders. In Proceedings of the 25th International Conference on Machine Learning, Helsinki, Finland, 5–9 July 2008; pp. 1096–1103.

 © 2020 by the authors. Licensee MDPI, Basel, Switzerland. This article is an open access article distributed under the terms and conditions of the Creative Commons Attribution (CC BY) license (http://creativecommons.org/licenses/by/4.0/).

Article

CEB Improves Model Robustness

Ian Fischer * and Alexander A. Alemi

Google Research, Mountain View, CA 94043, USA; alemi@google.com
* Correspondence: iansf@google.com

Received: 31 July 2020; Accepted: 21 September 2020; Published: 25 September 2020

Abstract: Intuitively, one way to make classifiers more robust to their input is to have them depend less sensitively on their input. The Information Bottleneck (IB) tries to learn compressed representations of input that are still predictive. Scaling up IB approaches to large scale image classification tasks has proved difficult. We demonstrate that the Conditional Entropy Bottleneck (CEB) can not only scale up to large scale image classification tasks, but can additionally improve model robustness. CEB is an easy strategy to implement and works in tandem with data augmentation procedures. We report results of a large scale adversarial robustness study on CIFAR-10, as well as the ImageNet-C Common Corruptions Benchmark, ImageNet-A, and PGD attacks.

Keywords: information theory; information bottleneck; machine learning

1. Introduction

We aim to learn models that make meaningful predictions beyond the data they were trained on. Generally we want our models to be robust. Broadly, robustness is the ability of a model to continue making valid predictions as the distribution the model is tested on moves away from the empirical training set distribution. The most commonly reported robustness metric is simply test set performance, where we verify that our model continues to make valid predictions on what we hope represents valid draws from the same data generating procedure as the training set.

Adversarial attacks test robustness in a worst case setting, where an attacker [1] makes limited targeted modifications to the input that are as fooling as possible. Many adversarial attacks have been proposed and studied (e.g., Szegedy et al. [1], Carlini and Wagner [2,3], Kurakin et al. [4], Madry et al. [5]). Most machine-learned systems appear to be vulnerable to adversarialexamples. Many defenses have been proposed, but few have demonstrated robustness against a powerful, general-purpose adversary [3,6]. Recent discussions have emphasized the need to consider forms of robustness besides adversarial [7]. The Common Corruptions Benchmark [8] measures image models' robustness to more mild real-world perturbations. Even these modest perturbations can fool traditional architectures.

One general-purpose strategy that has been shown to improve model robustness is data augmentation [9–11]. Intuitively, by performing modifications of the inputs at training time, the model is prevented from being too sensitive to particular features of the inputs that do not survive the augmentation procedure. We would like to identify complementary techniques for further improving robustness.

One approach is to try to make our models more robust by making them less sensitive to the inputs in the first place. The goal of this work is to experimentally investigate whether, by systematically limiting the complexity of the extracted representation using the Conditional Entropy Bottleneck (CEB), we can make our models more robust in all three of these senses: test set generalization (e.g., classification accuracy on "clean" test inputs), worst-case robustness, and typical-case robustness.

This paper is primarily empirical. We demonstrate:

- CEB models are easy to implement and train.

- CEB models show improved generalization performance over deterministic baselines on CIFAR10 and ImageNet.
- CEB models show improved robustness to untargeted Projected Gradient Descent (PGD) attacks on CIFAR10.
- CEB models trained on ImageNet show improved robustness on the ImageNet-C Common Corruptions Benchmark, the ImageNet-A Benchmark, and targeted PGD attacks.
- CEB models trained on ImageNet show improved calibration on the ImageNet validation set and on ImageNet-C.

We also show that adversarially-trained models *fail* to generalize to attacks they were not trained on, by comparing the results on L_2 PGD attacks from Madry et al. [5] to our results on the same baseline architecture. This result underscores the importance of finding ways to make models robust that do not rely on knowing the form of the attack ahead of time. Finally, for readers who are curious about theoretical and philosophical perspectives that may give insights into why CEB improves robustness, we recommend Fischer [12], which introduced CEB, as well as Achille and Soatto [13], Achille and Soatto [14], and Pensia et al. [15].

2. Materials and Methods

2.1. Information Bottlenecks

The Information Bottleneck (IB) objective [16] aims to learn a stochastic representation $Z \sim p(z|x)$ of some input X that retains as much information about a target variable Y while being as compressed as possible. The objective:

$$IB \equiv \max_{p(z|x)} I(Z;Y) - \sigma(-\rho) I(Z;X), \tag{1}$$

uses a Lagrange multiplier $\sigma(-\rho)$ to trade off between the relevant information ($I(Z;Y)$) and complexity of the representation ($I(Z;X)$). The IB objective is ordinarily written with a Lagrange multiplier $\beta \equiv \sigma(-\rho)$ with a natural range from 0 to 1. Here we use the sigmoid function: $\sigma(-\rho) \equiv \frac{1}{1+e^\rho}$ to reparameterize in terms of a control parameter ρ on the whole real line. As $\rho \to \infty$ the bottleneck turns off.

Because Z depends only on X ($Z \leftarrow X \leftrightarrow Y$), Z and Y are independent given X:

$$\begin{aligned} I(Z;X,Y) &= I(Z;X) + \cancel{I(Z;Y|X)} \\ &= I(Z;Y) + I(Z;X|Y). \end{aligned} \tag{2}$$

This allows us to write Equation (1) in an equivalent form:

$$\max_Z I(Z;Y) - e^{-\rho} I(Z;X|Y). \tag{3}$$

Just as the original IB objective (Equation (1)) admits a natural variational lower bound [17], so does this form. We can variationally lower bound the mutual information between our representation and the targets with a variational decoder $q(y|z)$:

$$\begin{aligned} I(Z;Y) &= \mathbb{E}_{p(x,y)p(z|x)} \left[\log \frac{p(y|z)}{p(y)} \right] \\ &\geq H(Y) + \mathbb{E}_{p(x,y)p(z|x)} \left[\log q(y|z) \right]. \end{aligned} \tag{4}$$

While we may not know $H(Y)$ exactly for real world datasets, in the IB formulation it is a constant outside of our control and so can be dropped in our objective. We can variationally upper bound our residual information:

$$I(Z;X|Y) = \mathbb{E}_{p(x,y)p(z|x)}\left[\log\frac{p(z|x,y)}{p(z|y)}\right]$$
$$\leq \mathbb{E}_{p(x,y)p(z|x)}\left[\log\frac{p(z|x)}{q(z|y)}\right], \qquad (5)$$

with a variational class conditional marginal $q(z|y)$ that approximates $\int dx\, p(z|x)p(x|y)$. Putting both bounds together gives us the Conditional Entropy Bottleneck objective [12]:

$$\min_{p(z|x)} \mathbb{E}_{p(x,y)p(z|x)}\left[-\log q(y|z) + e^{-\rho}\log\frac{p(z|x)}{q(z|y)}\right]. \qquad (6)$$

Compare this with the Variational Information Bottleneck (VIB) objective [17]:

$$\min_{p(z|x)} \mathbb{E}_{p(x,y)p(z|x)}\left[-\log q(y|z) - \sigma(-\rho)\log\frac{p(z|x)}{q(z)}\right]. \qquad (7)$$

The difference between CEB and VIB is the presence of a class conditional versus unconditional variational marginal. As can be seen in Equation (5), using an unconditional marginal provides a looser variational upper bound on $I(Z;X|Y)$. CEB (Equation (6)) can be thought of as a tighter variational approximation than VIB (Equation (7)) to Equation (3). Since Equation (3) is equivalent to the IB objective (Equation (1)), CEB can be thought of as a tighter variational approximation to the IB objective than VIB.

2.2. Implementing a CEB Model

In practice, turning an existing classifier architecture into a CEB model is very simple. For the stochastic representation $p(z|x)$ we simply use the original architecture, replacing the final softmax layer with a dense layer with d outputs. These outputs are then used to specify the means of a d-dimensional Gaussian distribution with unit diagonal covariance. That is, to form the stochastic representation, independent standard normal noise is simply added to the output of the network ($z = x + \epsilon$). For every input, this stochastic encoder will generate a random d-dimensional output vector. For the variational classifier $q(y|z)$ any classifier network can be used, including just a linear softmax classifier as done in these experiments. For the variational conditional marginal $q(z|y)$ it helps to use the same distribution as output by the classifier. For the simple unit variance Gaussian encoding we used in these experiments, this requires learning just d parameters per class. For ease of implementation, this can be represented as a single dense linear layer mapping from a one-hot representation of the labels to the d-dimensional output, interpreted as the mean of the corresponding class marginal.

In this setup the CEB loss takes a particularly simple form:

$$\mathbb{E}\left[w_y \cdot (f(x)+\epsilon) - \log\sum_{y'} e^{w_{y'}\cdot(f(x)+\epsilon)} - \frac{e^{-\rho}}{2}(f(x)-\mu_y)(f(x)-\mu_y+2\epsilon)\right]. \qquad (8)$$

The first two terms of Equation (8) are the usual softmax classifier loss, but acting on our stochastic representation $z = f(x) + \epsilon$, which is simply the output of our encoder network $f(x)$ with additive Gaussian noise. The w_y is the yth row of weights in the final linear layer outputting the logits. μ_y are the learned class conditional means for our marginal. ϵ are standard normal draws from an isotropic unit variance Gaussian with the same dimension as our encoding $f(x)$. The last term of Equation (8) is a stochastic sampling of the KL divergence between our encoder likelihood and the class conditional marginal likelihood. ρ controls the strength of the bottleneck and can vary on the whole real line. As $\rho \to \infty$ the bottleneck is turned off. In practice we find that ρ values near but above 0 tend to work best for modest size models, with the tendency for the best ρ to approach 0 as the model capacity increases. Notice that in expectation the second term in the loss is $(f(x)-\mu_y)^2$, which encourages the learned means μ_y to converge to the average of the representations of each element in the class. During testing we use the mean encodings and remove the stochasticity.

In its simplest form, training a CEB classifier amounts to injecting Gaussian random noise in the penultimate layer and learning estimates of the class-averaged output of that layer. In Appendix B we show simple modifications to the TPU-compatible ResNet implementation available on GitHub from the Google TensorFlow Team [18] that produce the same core ResNet50 models we use for our ImageNet experiments.

2.3. Consistent Classifier

An alternative classifier to the standard linear layer described in Section 2.2 performs the Bayesian inversion on the true class-conditional marginal:

$$p(y|z) = \frac{p(z|y)p(y)}{\sum_{y'} p(z|y')p(y')}. \tag{9}$$

Substituting $q(z|y)$ and using the empirical distribution over labels, we can define our variational classifier as:

$$q(y|z) \equiv \text{softmax}_y(q(z|y)p(y)) \tag{10}$$

In the case that the labels are uniformly distributed, that further simplifies to $q(y|z) \equiv \text{softmax}_y(q(z|y))$. We call this the *consistent classifier* because it is Bayes-consistent with the variational conditional marginal. This is in contrast to the standard feed-forward classifier, which may choose to classify a region of the latent space differently from the highest density class given by the conditional marginal.

2.4. Adversarial Attacks and Defenses

2.4.1. Attacks

The first adversarial attacks were proposed in Szegedy et al. [1], Goodfellow et al. [19]. Since those seminal works, an enormous variety of attacks has been proposed (Carlini and Wagner [2], Kurakin et al. [4], Madry et al. [5], Kurakin et al. [20], Moosavi-Dezfooli et al. [21], Eykholt et al. [22], Baluja and Fischer [23], etc.). In this work, we will primarily consider the Projected Gradient Descent (PGD) attack [5], which is a multi-step variant of the early Fast Gradient Method [19]. The attack can be viewed as having four parameters—p, the norm of the attack (typically 2 or ∞), ϵ, the radius the the p-norm ball within which the attack is permitted to make changes to an input, n, the number of gradient steps the adversary is permitted to take, and ϵ_i, the per-step limit to modifications of the current input. In this work, we consider L_2 and L_∞ attacks of varying ϵ and n, and with $\epsilon_i = \frac{4}{3}\frac{\epsilon}{n}$.

2.4.2. Defenses

A common defense for adversarial examples is adversarial training. Adversarial training was originally proposed in Szegedy et al. [1], but was not practical until the Fast Gradient Method was introduced. It has been studied in detail, with varied techniques [5,20,24,25]. Adversarial training can clearly be viewed as a form of data augmentation [26], where instead of using some fixed set of functions to modify the training examples, we use the model itself in combination with one or more adversarial attacks to modify the training examples. As the model changes, the distribution of modifications changes as well. However, unlike with non-adversarial data augmentation techniques, such as AutoAugment (AutoAug) [9], adversarial training techniques considered in the literature so far cause substantial reductions in accuracy on clean test sets. For example, the CIFAR10 model described in Madry et al. [5] gets 95.5% accuracy when trained normally, but only 87.3% when trained on L_∞ adversarial examples. More recently, Xie et al. [25] adversarially trains ImageNet models with impressive robustness to targeted PGD L_∞ attacks, but at only 62.32% accuracy on the non-adversarial test set, compared to 78.81% accuracy for the same model trained only on clean images.

2.5. Common Corruptions

The Common Corruptions Benchmark [8] offers a test of model robustness to common image processing pipeline corruptions. ImageNet-C modifies the ImageNet test set with the 15 corruptions applied at five different strengths. Within each corruption type we evaluate the average error at each of the five levels ($E_c = \frac{1}{5}\sum_{s=1}^{5} E_{cs}$). To summarize the performance across all corruptions, we report both the average corruption error (avg = $\frac{1}{15}\sum_c E_c$) and the *Mean Corruption Error* (mCE) [8]:

$$\text{mCE} = \frac{1}{15}\sum_c \frac{\sum_{s=1}^{5} E_{cs}}{\sum_{s=1}^{5} E_{cs}^{\text{AlexNet}}}. \quad (11)$$

The mCE weights the errors on each task against the performance of a baseline AlexNet model. Slightly different pipelines have been used for the ImageNet-C task [10]. In this work we used the AlexNet normalization numbers and data formulation from Yin et al. [11].

2.6. Natural Adversarial Examples

The ImageNet-A Benchmark [27] is a dataset of 7500 naturally-occurring "adversarial" examples across 200 ImageNet classes. The images exploit commonly-occurring weaknesses in ImageNet models, such as relying on textures often seen with certain class labels.

2.7. Calibration

One approach to estimating a model's robustness is to look at how well *calibrated* the model is. The *Expected Calibration Error* (ECE) [28] gives an intuitive metric of calibration:

$$ECE \equiv \sum_{s=1}^{S} \frac{|B_s|}{N} |\text{acc}(B_s) - \text{conf}(B_s)|, \quad (12)$$

where S is the number of confidence bins (30 in our experiments), N is the number of examples (50,000 for ImageNet and for each ImageNet-C corruption), $|B_s|$ is the number of examples in the sth bin, $\text{acc}(B_s)$ is the mean accuracy in the sth bin, and $\text{conf}(B_s)$ is the mean confidence of the model's predictions in the sth bin. The ECE ranges between 0 and 1. A perfectly calibrated model would have an ECE of 0. See Ovadia et al. [29] for further details.

3. Results

3.1. CIFAR10 Experiments

We trained a set of 25 28 × 10 Wide ResNet (WRN) CEB models on CIFAR10 at $\rho \in [-1, -0.75, \ldots, 5]$, as well as a deterministic baseline. They trained for 1500 epochs, lowering the learning rate by a factor of 0.3 after 500, 1000, and 1250 epochs. This long training regime was due to our use of the original AutoAug policies, which requires longer training. The only additional modification we made to the basic 28 × 10 WRN architecture was the removal of all Batch Normalization [30] layers. Every small CIFAR10 model we have trained with Batch Normalization enabled has had substantially worse robustness to L_∞ PGD adversaries, even though typically the accuracy is much higher. For example, 28 × 10 WRN CEB models rarely exceeded more than 10% adversarial accuracy. However, it was always still the case that lower values of ρ gave higher robustness. As a baseline comparison, a deterministic 28 × 10 WRN with BatchNorm, trained with AutoAug reaches 97.3% accuracy on clean images, but 0% accuracy on L_∞ PGD attacks at $\epsilon = 8$ and $n = 20$. Interestingly, that model was noticeably more robust to L_2 PGD attacks than the deterministic baseline without BatchNorm, getting 73% accuracy compared to 66%. However, it was still much weaker than the CEB models, which get over 80% accuracy on the same attack (Figure 1). Additional training details are in Appendix A.1.

Figure 1 demonstrates the adversarial robustness of CEB models to both targeted L_2 and L_∞ attacks. The CEB models show a marked improvement in robustness to L_2 attacks compared to an adversarially-trained baseline from Madry et al. [5] (denoted Madry). The attack parameters were selected to be about equally difficult for the adversarially-trained WRN 28 × 10 model from Madry et al. [5] (grey dashed and dotted lines in Figure 1). The deterministic baseline (Det.) only gets 8% accuracy on the L_∞ attacks, but gets 66% on the L_2 attack, substantially better than the 45.7% of the adversarially-trained model, which makes it clear that the adversarially-trained model failed to generalize in any reasonable way to the L_2 attack. The CEB models are always substantially more robust than Det., and many of them outperform Madry even on the L_∞ attack the Madry model was trained on, but for both attacks there is a clear general trend toward more robustness as ρ decreases. Finally, the CEB and Det. models all reach about the same accuracy, ranging from 93.9% to 95.1%, with Det. at 94.4%. In comparison, Madry only gets 87.3%.

Figure 2 shows the robustness of five of those models to PGD attacks as ϵ is varied. We selected the four CEB models to represent the most robust models across most of the range of ρ we trained. All values in the figure are collected at 20 steps of PGD. The Madry model [5] was trained with 7 steps of L_∞ PGD at $\varepsilon = 8$ (grey dashed line in the figure). All of the CEB models with $\rho \leq 4$ outperform Madry across most of the values of ϵ, even though they were not adversarially-trained. It is interesting to note that the Det. model eventually outperforms the CEB_5 model on L_2 attacks at relatively high accuracies. This result indicates that the CEB_5 model may be under-compressed.

Of the 25 CEB models we trained, only the models with $\rho \geq 1$ successfully trained. The remainder collapsed to chance performance. This is something we observe on all datasets when training models that are too low capacity. Only by increasing model capacity does it become possible to train at low ρ. Note that this result is predicted by the theory of the onset of learning in IB and its relationship to model capacity from Wu et al. [31].

We additionally tested two models ($\rho = 0$ and $\rho = 5$) on the CIFAR10 Common Corruptions test sets. At the time of training, we were unaware that AutoAug's default policies for CIFAR10 contain brightness and contrast augmentations that amount to training on those two corruptions from Common Corruptions (as mentioned in Yin et al. [11]), so our results are not appropriate for direct comparison with other results in the literature. However, they still allow us to compare the effect of bottlenecking the information between the two models. The $\rho = 5$ model reached an mCE of 61.2. The $\rho = 0$ model reached an mCE of 52.0, which is a dramatic relative improvement. Note that the mCE is computed relative to a baseline model. We use the baseline model from Yin et al. [11].

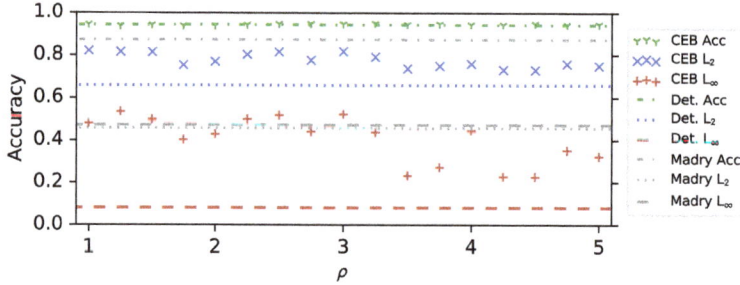

Figure 1. Conditional Entropy Bottleneck (CEB) ρ vs. test set accuracy, and L_2 and L_∞ Projected Gradient Descent (PGD) adversarial attacks on CIFAR10. *None of the CEB models is adversarially trained.*

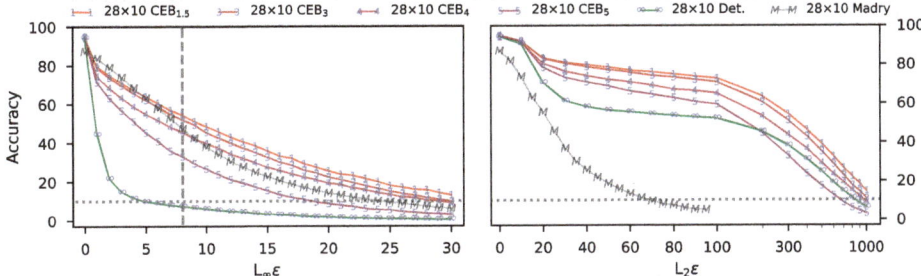

Figure 2. Untargeted adversarial attacks on CIFAR10 models showing both strong robustness to PGD L_2 and L_∞ attacks, as well as good test accuracy of up to 95.1%. (**Left**): Accuracy on untargeted L_∞ attacks at different values of ε for all 10,000 test set examples. (**Right**): Accuracy on untargeted L_2 attacks at different values of ε. Note the switch to log scale on the x axis at $L_2\varepsilon = 100$. 28×10 indicates the Wide ResNet size. CEB_x indicates a CEB model trained at $\rho = x$. Madry is the adversarially-trained model from Madry et al. [5] (values provided by Aleksander Madry). *None of the CEB models is adversarially-trained.*

3.2. ImageNet Experiments

To demonstrate CEB's ability to improve robustness, we trained four different ResNet architectures on ImageNet at 224×224 resolution, with and without AutoAug, using three different objective functions, and then tested them on ImageNet-C, ImageNet-A, and targeted PGD attacks.

As a simple baseline we trained ResNet50 with no data augmentation using the standard cross-entropy loss (XEnt). We then trained the same network with CEB at ten different values of $\rho = (1, 2, \ldots, 10)$. AutoAug [9] has previously been demonstrated to improve robustness markedly on ImageNet-C, so next we trained ResNet50 with AutoAug using XEnt. We similarly trained these AutoAug ResNet50 networks using CEB at the same ten values of ρ. ImageNet-C numbers are also sensitive to the model capacity. To assess whether CEB can benefit larger models, we repeated the experiments with a modified ResNet50 network where every layer was made twice as wide, training an XEnt model and ten CEB models, all with AutoAug. To see if there is any additional benefit or cost to using the consistent classifier (Section 2.3), we took the same wide architecture using AutoAug and trained ten consistent classifier CEB (cCEB) models. Finally, we repeated all of the previous experiments using ResNet152: XEnt and CEB models without AutoAug; with AutoAug; with AutoAug and twice as wide; and cCEB with AutoAug and twice as wide. All other hyperparameters (learning rate schedule, L_2 weight decay scale, etc.) remained the same across all models. All of those hyperparameters where taken from the ResNet hyperparameters given in the AutoAug paper. In total we trained 86 ImageNet models: 6 deterministic XEnt models varying augmentation, width, and depth; 60 CEB models additionally varying ρ; and 20 cCEB models also varying ρ. The results for the ResNet50 models are summarized in Figure 3. For ResNet152, see Figure 4. See Table 1 for detailed results across the matrix of experiments. Additional experimental details are given in Appendix A.2.

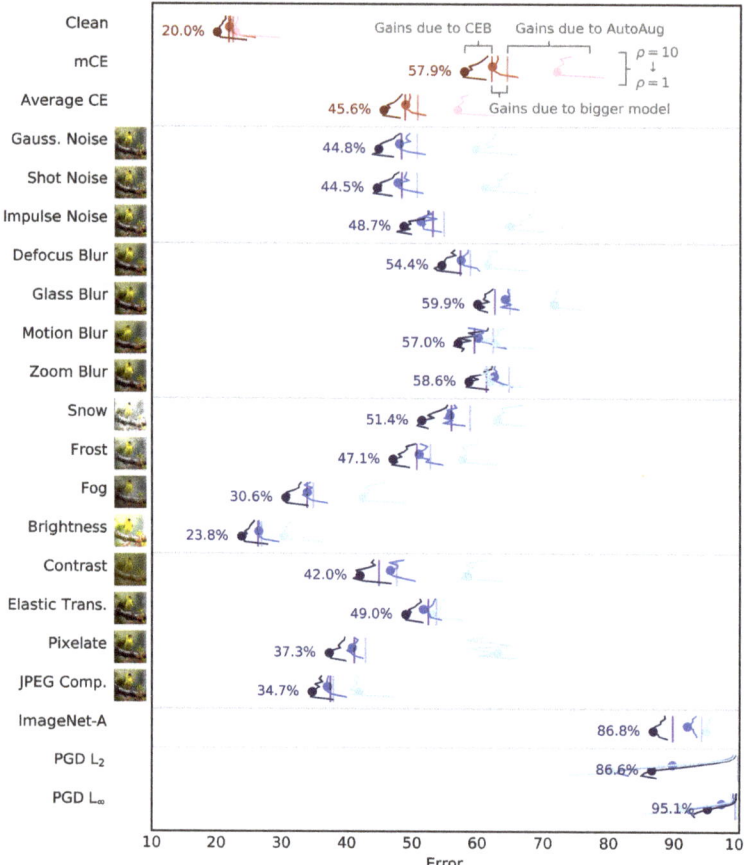

Figure 3. Summary of the ResNet50 ImageNet-C experiments. Lower is better in all cases. In the main part of the figure (in blue), the average errors across corruption magnitude are shown for 33 different networks for each of the labeled Common Corruptions, ImageNet-A, and targeted PGD attacks. The networks come in paired sets, with the vertical lines denoting the baseline XEnt network's performance, and then in the corresponding color the errors for each of 10 different CEB networks are shown with varying $\rho = [1, 2, \ldots, 10]$, arranged from 10 at the top to 1 at the bottom. The light blue lines indicate ResNet50 models trained without AutoAug. The blue lines show the same network trained with AutoAug. The dark blue lines show ResNet50 AutoAug networks that were made twice as wide. For these models, we display cCEB rather than CEB, which gave qualitatively similar but slightly weaker performance. The figure separately shows the effects of data augmentation, enlarging the model, and the additive effect of CEB on each model. At the top in red are shown the same data for three summary statistics. `clean` denotes the clean top-1 errors of each of the networks. `mCE` denotes the AlexNet regularized average corruption errors. `avg` shows an equally-weighted average error across all common corruptions. The dots denote the value for each CEB network and each corruption at ρ^*, the optimum ρ for the network as measured in terms of clean error. The values at these dots and the baseline values are given in detail in Table 1. Figure 4 show the same data for the ResNet152 models.

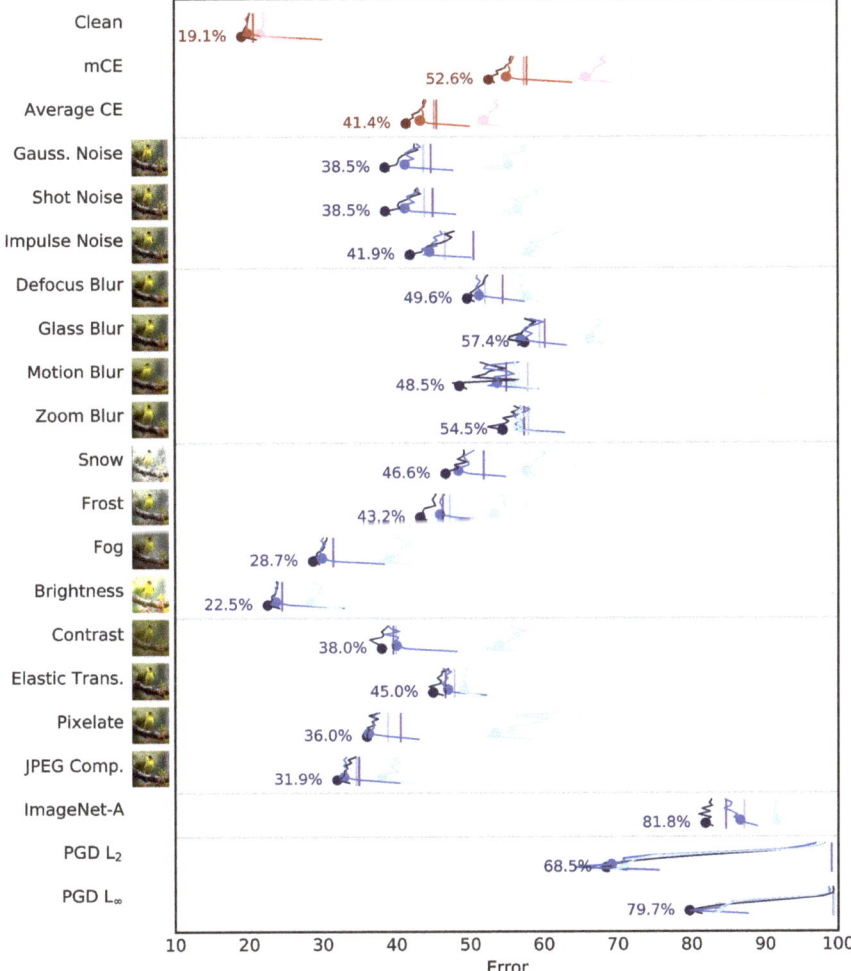

Figure 4. Replication of Figure 3 but for ResNet152. Lower is better in all cases. The light blue lines indicate ResNet152 models trained without AutoAug. The blue lines show the same network trained with AutoAug. The dark blue lines show ResNet152 AutoAug networks that were made twice as wide. As in Figure 3, we show the cCEB models for the largest network to reduce visual clutter. The deeper model shows marked improvement across the board compared to ResNet50, but the improvements due to CEB and cCEB are even more striking. Notice in particular the adversarial robustness to L_∞ and L_2 PGD attacks for the CEB models over the XEnt baselines. The L_∞ baselines all have error rates above 99%, so they are to the right edge of the figure. See Table 1 for details of the best-performing models, which correspond to the dots in this figure.

The CEB models highlighted in Figures 3 and 4 and Table 1 were selected by cross validation. These were values of ρ that gave the best *clean* test set accuracy. Despite being selected for classical generalization, these models also demonstrate a high degree of robustness on both average- and worst-case perturbations. In the case that more than one model gets the same test set accuracy, we choose the model with the lower ρ, since we know that lower ρ correlates with higher robustness.

The only model where we had to make this decision was for ResNet152 with AutoAug, where five models all were within 0.1% of each other, so we chose the $\rho = 3$ model, rather than $\rho \in \{5\ldots8\}$.

Table 1. Baseline and cross-validated CEB values for the ImageNet experiments. cCEB uses the consistent classifier. **XEnt** is the baseline cross entropy objective. "-aa" indicates AutoAug is not used during training. "x2" indicates the ResNet architecture is twice as wide. The CEB values reported here are denoted with the dots in Figures 3 and 4. Lower values are better in all cases, and the lowest value for each architecture is shown in bold. All values are percentages.

Architecture	ResNet152x2			ResNet152		ResNet152-aa		ResNet50x2			ResNet50		ResNet50-aa	
Objective	cCEB	CEB	XEnt	CEB	XEnt	CEB	XEnt	cCEB	CEB	XEnt	CEB	XEnt	CEB	XEnt
ρ	2	2	NA	3	NA	3	NA	4	3	NA	6	NA	4	NA
Clean	19.1	19.3	20.7	19.9	20.7	21.6	22.4	20.0	20.2	21.8	21.9	22.5	22.8	24.0
mCE	52.6	53.2	57.8	55.0	57.4	65.7	71.9	57.9	57.8	62.0	62.1	64.4	72.0	77.0
Average CE	41.4	41.8	45.5	43.3	45.2	51.9	56.8	45.6	45.5	48.9	49.0	50.8	56.9	60.9
Gauss. Noise	38.5	40.1	44.7	41.2	43.7	55.3	62.5	44.8	43.9	48.3	48.0	50.7	59.6	67.3
Shot Noise	38.5	40.3	45.0	41.2	43.8	56.5	63.7	44.5	43.9	48.4	47.8	50.7	61.2	68.8
Impulse Noise	41.9	43.6	50.5	44.5	46.6	57.9	66.8	48.7	48.1	53.1	51.3	54.8	64.8	72.7
Defocus Blur	49.6	48.8	54.5	51.3	52.1	57.7	58.3	54.4	54.2	57.3	57.4	58.8	61.5	62.7
Glass Blur	57.4	56.7	60.1	56.9	59.4	66.2	67.7	59.9	61.0	62.6	64.2	64.9	71.5	72.3
Motion Blur	48.5	51.4	55.0	53.7	57.8	55.6	59.7	57.0	56.6	59.5	60.0	62.3	62.7	68.1
Zoom Blur	54.5	54.7	57.3	56.8	57.9	56.6	59.8	58.6	58.0	61.3	62.5	64.8	61.8	63.7
Snow	46.6	46.6	51.9	48.4	51.8	57.6	64.2	51.4	50.9	55.9	55.7	58.8	63.1	68.7
Frost	43.2	43.9	46.3	45.9	47.2	53.2	57.6	47.1	47.1	50.7	51.0	52.7	57.6	61.7
Fog	28.7	28.7	31.4	29.9	31.5	39.1	43.3	30.6	30.2	33.9	33.9	34.8	42.3	47.0
Brightness	22.5	22.6	24.5	23.6	24.4	28.4	30.8	23.8	24.1	26.3	26.4	26.8	30.3	33.4
Contrast	38.0	37.7	39.5	40.0	39.9	54.0	58.7	42.0	42.4	44.9	46.7	47.6	58.5	62.8
Elastic Trans.	45.0	45.3	46.6	46.9	47.8	49.2	51.4	49.0	48.8	52.4	51.7	53.7	53.0	56.0
Pixelate	36.0	35.2	40.5	36.2	38.8	53.4	64.9	37.3	37.9	41.1	40.8	42.8	63.1	64.6
JPEG Comp.	31.9	31.8	34.9	32.9	34.5	38.0	43.0	34.7	35.1	37.4	37.0	37.9	41.8	43.5
ImageNet-A	81.8	82.0	84.6	86.5	87.1	91.5	93.4	86.8	88.1	89.8	92.0	94.2	94.9	96.8
PGD L_2	68.5	68.0	99.1	69.1	99.2	70.9	99.4	86.6	84.5	99.8	89.7	99.7	80.2	99.7
PGD L_∞	79.7	79.3	99.3	83.8	99.4	83.8	99.4	95.1	93.2	99.4	97.3	99.4	91.0	99.5

3.2.1. Accuracy, ImageNet-C, and ImageNet-A

Increasing model capacity and using AutoAug have positive effects on classification accuracy, as well as on robustness to ImageNet-C and ImageNet-A, but for all three classes of models CEB gives substantial additional improvements. cCEB gives a small but noticeable additional gain for all three cases (except indistinguishable performance compared to CEB on ImageNet-A with the wide ResNet152 architecture), indicating that enforcing variational consistency is a reasonable modification to the CEB objective. In Table 1 we can see that CEB's relative accuracy gains increase as the architecture gets larger, from gains of 1.2% for ResNet50 and ResNet152 without AutoAug, to 1.6% and 1.8% for the consistent wide models with AutoAug. This indicates that even larger relative gains may be possible when using CEB to train larger architectures than those considered here. We can also see that for the XEnt 152x2 and 152 models, the smaller model (152) actually has better mCE and equally good top-1 accuracy, indicating that the wider model may be overfitting, but the 152x2 CEB and cCEB models substantially outperform both of them across the board. cCEB gives a noticeable boost over CEB for clean accuracy and mCE in both wide architectures.

3.2.2. Targeted PGD Attacks

We tested on the random-target version of the PGD L_2 and L_∞ attacks [4]. The L_∞ attack used $\epsilon = 16$, $n = 10$, and $\epsilon_i = 2$, which is considered to be a strong attack still [25]. The L_2 attack used $\epsilon = 200$, $n = 10$, and $\epsilon_i = 220$. Those parameters were chosen by attempting to match the baseline XEnt ResNet50 without AutoAug model's performance on the L_∞ attack—the performance of the CEB models were not considered when selecting the L_2 attack strength. Interestingly, for the PGD attacks, AutoAug was detrimental—the ResNet50 models without AutoAug were substantially more robust than those with AutoAug, and the ResNet152 models without AutoAug were nearly as robust as the AutoAug and wide models, in spite of having much worse test set accuracy. The ResNet50

CEB models show a dramatic improvement over the XEnt model, with top-1 accuracy increasing from 0.3% to 19.8% between the XEnt baseline without AutoAug and the corresponding $\rho = 4$ CEB model, a relative increase of 66 times. Interestingly, the CEB ResNet50 models *without* AutoAug are much more robust to the adversarial attacks than the AutoAug and wide ResNet50 models. As with the accuracy results above, the robustness gains due to CEB increase as model capacity increases, indicating that further gains are possible.

3.2.3. Calibration and ImageNet-C

Following the experimental setup in Reference [29], in Figure 5 we compare accuracy and ECE on ResNet models for both the clean ImageNet test set and the collection of 15 ImageNet-C corruptions at each of the five different corruption intensities. It is easy to see in the figure that the CEB models always have superior mean accuracy and ECE for all six different sets of test sets.

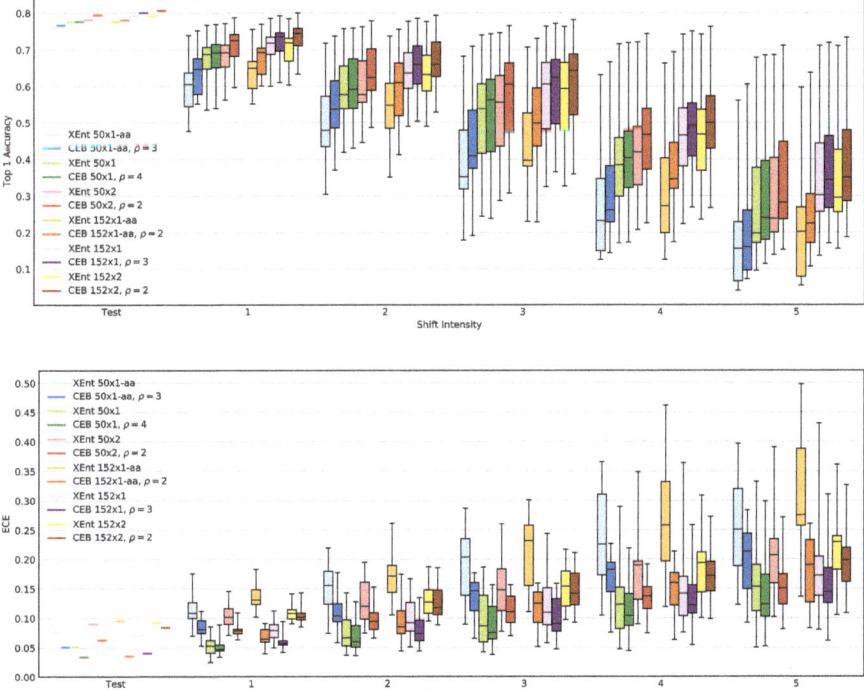

Figure 5. Comparison of accuracy and Expected Calibration Error (ECE) between Xent baseline models and corresponding CEB models at the value of ρ that gives the closest accuracy to the XEnt baseline. Higher is better for accuracy; lower is better for ECE. The box plots show the minimum, 25th percentile, mean, 75th percentile, and maximum values across the 15 different ImageNet-C corruptions for the given shift intensity. XEnt baseline models are always the lighter color, with the corresponding CEB model having the darker color.

Because accuracy can have a strong impact on ECE, we use a different model selection procedure than in the previous experiments. Rather than selecting the CEB model with the highest accuracy, we instead select the CEB model with with the *closest* accuracy to the corresponding XEnt model. This resulted in selecting models with lower ρ than in the previous experiments for four out of the six CEB model classes. We note that by selecting models with lower ρ (which are more compressed), we see more dramatic differences in ECE, but even if we select the CEB models with highest accuracy as in the previous experiments, all six CEB models outperform the corresponding XEnt baselines on all six different sets of test sets.

4. Conclusions

The Conditional Entropy Bottleneck (CEB) provides a simple mechanism to improve robustness of image classifiers. We have shown a strong trend toward increased robustness as ρ decreases in the standard 28×10 Wide ResNet model on CIFAR10, and that this increased robustness does not come at the expense of accuracy relative to the deterministic baseline. We have shown that CEB models at a range of ρ outperform an adversarially-trained baseline model, even on the attack the adversarial model was trained on, and have incidentally shown that the adversarially-trained model generalizes to at least one other attack *less well* than a deterministic baseline. Finally, we have shown that on ImageNet, CEB provides substantial gains over deterministic baselines in validation set accuracy, robustness to Common Corruptions, Natural Adversarial Examples, and targeted Projected Gradient Descent attacks, and gives large improvements to model calibration, all without any change to the inference architecture. We hope these empirical demonstrations inspire further theoretical and practical study of the use of bottlenecking techniques to encourage improvements to both classical generalization and robustness.

Author Contributions: Conceptualization, I.F.; methodology, I.F. and A.A.A.; software, I.F.; validation, I.F. and A.A.A.; formal analysis, A.A.A. and I.F.; investigation, I.F.; writing—original draft preparation, I.F. and A.A.A.; writing—review and editing, I.F. and A.A.A.; visualization, I.F. and A.A.A. All authors have read and agreed to the published version of the manuscript.

Funding: This research received no external funding.

Acknowledgments: We would like to thank Justin Gilmer for helpful conversations on the use of ImageNet-C.

Conflicts of Interest: The authors declare no conflict of interest.

Appendix A. Experiment Details

Here we give additional technical details for the CIFAR10 and ImageNet experiments.

Appendix A.1. CIFAR10 Experiment Details

We trained all of the models using Adam [32] at a base learning rate of 10^{-3}. We lowered the learning rate three times by a factor of 0.3 each time. The only additional trick to train the CIFAR10 models was to start with $\rho = 100$, anneal down to $\rho = 10$ over 2 epochs, and then anneal to the target ρ over one epoch once training exceeded a threshold of 20%. This jump-start method is inspired by experiments on VIB in Wu et al. [31]. It makes it much easier to train models at low ρ, and appears to not negatively impact final performance.

Appendix A.2. ImageNet Experiment Details

We follow the learning rate schedule for the ResNet 50 from Cubuk et al. [9], which has a top learning rate of 1.6, trains for 270 epochs, and drops the learning rate by a factor of 10 at 90, 180, and 240 epochs. The only difference for all of our models is that we train at a batch size of 8192 rather than 4096. Similar to the CIFAR10 models, in order to ensure that the ImageNet models train at low ρ, we employ a simple jump-start. We start at $\rho = 100$ and anneal down to the target ρ over 12,000 steps. The first learning rate drop occurs a bit after 14,000 steps. Also similar to the CIFAR10 28×10 WRN experiments, none of the models we trained at $\rho = 0$ succeeded, indicating that ResNet50 and wide ResNet50 both have insufficient capacity to fully learn ImageNet. We were able to train ResNet152 at $\rho = 0$, but only by disabling L_2 weight decay and using a slightly lower learning rate. Since that involved additional hyperparameter tuning, we do not report those results here, beyond noting that it is possible, and that those models reached top-1 accuracy around 72%.

Appendix B. CEB Example Code

In Listings 1 to 3 we give annotated code changes needed to make ResNet CEB models, based on the TPU-compatible ResNet implementation from the Google TensorFlow Team [18].

Listing 1. Modifications to the `model.py` file.

```python
# In model.py:
def resnet_v1_generator(block_fn, layers, num_classes, ...):
  def model(inputs, is_training):
    # Build the ResNet model as normal up to the following lines:
    inputs = tf.reshape(
        inputs, [-1, 2048 if block_fn is bottleneck_block else 512])
    # Now, instead of the final dense layer, just return inputs,
    # which for ResNet50 models is a [batch_size, 2048] tensor.
    return inputs
```

Listing 2. Modification to the head of `resnet_main.py`.

```python
# In resnet_main.py add the following imports and functions:
import tensorflow_probability as tfp
tfd = tfp.distributions

def ezx_dist(x):
  """Builds the encoder distribution, e(z|x)."""
  dist = tfd.MultivariateNormalDiag(loc=x)
  return dist

def bzy_dist(y, num_classes=1000, z_dims=2048):
  """Builds the backwards distribution, b(z|y)."""
  y_onehot = tf.one_hot(y, num_classes)
  mus = tf.layers.dense(y_onehot, z_dims, activation=None)
  dist = tfd.MultivariateNormalDiag(loc=mus)
  return dist

def cyz_dist(z, num_classes=1000):
  """Builds the classifier distribution, c(y|z)."""
  # For the classifier, we are using exactly the same dense layer
  # initialization as was used for the final layer that we removed
  # from model.py.
  logits = tf.layers.dense(
      z, num_classes, activation=None,
      kernel_initializer=tf.random_normal_initializer(stddev=.01))
  return tfd.Categorical(logits=logits)

def lerp(global_step, start_step, end_step, start_val, end_val):
  """Utility function to linearly interpolate two values."""
  interp = (tf.cast(global_step - start_step, tf.float32)
            / tf.cast(end_step - start_step, tf.float32))
  interp = tf.maximum(0.0, tf.minimum(1.0, interp))
  return start_val * (1.0 - interp) + end_val * interp
```

Listing 3. Modifications to `resnet_model_fn` in `resnet_main.py`.

```python
# Still in resnet_main.py, modify resnet_model_fn as follows:
def resnet_model_fn(features, labels, mode, params):
# Nothing changes until after the definition of build_network:
def build_network():
# Elided, unchanged implementation of build_network.

if params['precision'] == 'bfloat16':
# build_network now returns the pre-logits, so we'll change
# the variable name from logits to net.
with tf.contrib.tpu.bfloat16_scope():
net = build_network()
net = tf.cast(net, tf.float32)
elif params['precision'] == 'float32':
net = build_network()

# Get the encoder, e(z|x):
with tf.variable_scope('ezx', reuse=tf.AUTO_REUSE):
ezx = ezx_dist(net)
# Get the backwards encoder, b(z|y):
with tf.variable_scope('bzy', reuse=tf.AUTO_REUSE):
bzy = bzy_dist(labels)

# Only sample z during training. Otherwise, just pass through
# the mean value of the encoder.
if mode == tf.estimator.ModeKeys.TRAIN:
z = ezx.sample()
else:
z = ezx.mean()

# Get the classifier, c(y|z):
with tf.variable_scope('cyz', reuse=tf.AUTO_REUSE):
cyz = cyz_dist(z, params)

# cyz.logits is the same as what the unmodified ResNet model would return.
logits = cyz.logits

# Compute the individual conditional entropies:
hzx = -ezx.log_prob(z)       # H(Z|X)
hzy = -bzy.log_prob(z)       # H(Z|Y) (upper bound)
hyz = -cyz.log_prob(labels)  # H(Y|Z) (upper bound)

# I(X;Z|Y) = -H(Z|X) + H(Z|Y)
#          >= -hzx + hzy =: Rex, the residual information.
rex = -hzx + hzy

rho = 3.0  # You should make this a hyperparameter.
rho_to_gamma = lambda rho: 1.0 / np.exp(rho)
gamma = tf.cast(rho_to_gamma(rho), tf.float32)

# Get the global step now, so that we can adjust rho dynamically.
global_step = tf.train.get_global_step()

anneal_rho = 12000  # You should make this a hyperparameter.
if anneal_rho > 0:
# Anneal rho from 100 down to the target rho
# over the first anneal_rho steps.
gamma = lerp(global_step, 0, aneal_rho,
rho_to_gamma(100.0), gamma)

# Replace all the softmax cross-entropy loss computation with the following line:
loss = tf.reduce_mean(gamma * rex + hyz)
# The rest of resnet_model_fn can remain unchanged.
```

References

1. Szegedy, C.; Zaremba, W.; Sutskever, I.; Bruna, J.; Erhan, D.; Goodfellow, I.; Fergus, R. Intriguing properties of neural networks. *arXiv* **2013**, arXiv:1312.6199.
2. Carlini, N.; Wagner, D. Towards evaluating the robustness of neural networks. In Proceedings of the 2017 IEEE Symposium on Security and Privacy (SP), San Jose, CA, USA, 25 May 2017; pp. 39–57.

3. Carlini, N.; Wagner, D. Adversarial examples are not easily detected: Bypassing ten detection methods. In Proceedings of the 10th ACM Workshop on Artificial Intelligence and Security, Dallas, TX, USA, 3 November 2017; pp. 3–14.
4. Kurakin, A.; Goodfellow, I.; Bengio, S. Adversarial examples in the physical world. In Proceedings of the ICLR Workshop, International Conference on Learning Representations, Toulon, France, 24–26 April 2017.
5. Madry, A.; Makelov, A.; Schmidt, L.; Tsipras, D.; Vladu, A. Towards Deep Learning Models Resistant to Adversarial Attacks. In Proceedings of the International Conference on Learning Representations, Vancouver, BC, Canada, 30 April–3 May 2018.
6. Athalye, A.; Carlini, N.; Wagner, D. Obfuscated Gradients Give a False Sense of Security: Circumventing Defenses to Adversarial Examples. In Proceedings of the 35th International Conference on Machine Learning, ICML 2018, Stockholm Sweden, 10–15 July 2018.
7. Engstrom, L.; Gilmer, J.; Goh, G.; Hendrycks, D.; Ilyas, A.; Madry, A.; Nakano, R.; Nakkiran, P.; Santurkar, S.; Tran, B.; et al. A Discussion of 'Adversarial Examples Are Not Bugs, They Are Features'. *Distill* **2019**. Available online: https://distill.pub/2019/advex-bugs-discussion (accessed on 24 September 2020). [CrossRef]
8. Hendrycks, D.; Dietterich, T. Benchmarking Neural Network Robustness to Common Corruptions and Perturbations. In Proceedings of the International Conference on Learning Representations, New Orleans, LA, USA, 6–9 May 2019.
9. Cubuk, E.D.; Zoph, B.; Mane, D.; Vasudevan, V.; Le, Q.V. AutoAugment: Learning Augmentation Strategies From Data. In Proceedings of the IEEE/CVF Conference on Computer Vision and Pattern Recognition (CVPR), Long Beach, CA, USA, 15–20 June 2019.
10. Lopes, R.G.; Yin, D.; Poole, B.; Gilmer, J.; Cubuk, E.D. Improving Robustness Without Sacrificing Accuracy with Patch Gaussian Augmentation. *arXiv* **2019**, arXiv:1906.02611.
11. Yin, D.; Gontijo Lopes, R.; Shlens, J.; Cubuk, E.D.; Gilmer, J. A Fourier Perspective on Model Robustness in Computer Vision. In *Advances in Neural Information Processing Systems 32*; Wallach, H., Larochelle, H., Beygelzimer, A., d'Alché-Buc, F., Fox, E., Garnett, R., Eds.; Curran Associates, Inc.: Red Hook, NY, USA, 2019; pp. 13276–13286.
12. Fischer, I. The Conditional Entropy Bottleneck. *Entropy* **2020**, *22*, 999. [CrossRef]
13. Achille, A.; Soatto, S. Emergence of Invariance and Disentanglement in Deep Representations. *J. Mach. Learn. Res.* **2018**, *19*, 1–34.
14. Achille, A.; Soatto, S. Information dropout: Learning optimal representations through noisy computation. *IEEE Trans. Pattern Anal. Mach. Intell.* **2018**, *40*, 2897–2905. [CrossRef] [PubMed]
15. Pensia, A.; Jog, V.; Loh, P.L. Extracting robust and accurate features via a robust information bottleneck. *IEEE J. Select. Areas Inf. Theory* **2020**. [CrossRef]
16. Tishby, N.; Pereira, F.C.; Bialek, W. The information bottleneck method. In Proceedings of the 37th annual Allerton Conference on Communication, Control, and Computing, Monticello, IL, USA, 22–24 September 1999; pp. 368–377.
17. Alemi, A.A.; Fischer, I.; Dillon, J.V.; Murphy, K. Deep Variational Information Bottleneck. In Proceedings of the International Conference on Learning Representations, Toulon, France, 24–26 April 2017.
18. Google TensorFlow Team. Cloud TPU ResNet Implementation. 2019. Available online: https://github.com/tensorflow/tpu/tree/master/models/official/resnet (accessed on 30 September 2019).
19. Goodfellow, I.J.; Shlens, J.; Szegedy, C. *Explaining and Harnessing Adversarial Examples*; ICLR, 2015; Available online: http://arxiv.org/abs/1412.6572 (accessed 7 September 2020).
20. Kurakin, A.; Goodfellow, I.; Bengio, S. Adversarial machine learning at scale. In Proceedings of the International Conference on Learning Representations, Toulon, France, 24–26 April 2017.
21. Moosavi-Dezfooli, S.M.; Fawzi, A.; Frossard, P. Deepfool: A simple and accurate method to fool deep neural networks. In Proceedings of the IEEE Conference on Computer Vision and Pattern Recognition, Las Vegas, NV, USA, 27–30 June 2016, pp. 2574–2582.
22. Eykholt, K.; Evtimov, I.; Fernandes, E.; Li, B.; Rahmati, A.; Xiao, C.; Prakash, A.; Kohno, T.; Song, D. Robust physical-world attacks on deep learning models. *arXiv* **2017**, arXiv:1707.08945.
23. Baluja, S.; Fischer, I. Learning to Attack: Adversarial Transformation Networks. In Proceedings of the AAAI Conference on Artificial Intelligence. Association for the Advancement of Artificial Intelligence, New Orleans, LA, USA, 2–7 February 2018.

24. Ilyas, A.; Santurkar, S.; Tsipras, D.; Engstrom, L.; Tran, B.; Madry, A. Adversarial Examples Are Not Bugs, They Are Features. In *Advances in Neural Information Processing Systems 32*; Wallach, H., Larochelle, H., Beygelzimer, A., d'Alché-Buc, F., Fox, E., Garnett, R., Eds.; Curran Associates, Inc.: Red Hook, NY, USA, 2019; pp. 125–136.
25. Xie, C.; Wu, Y.; Maaten, L.V.D.; Yuille, A.L.; He, K. Feature denoising for improving adversarial robustness. In Proceedings of the IEEE Conference on Computer Vision and Pattern Recognition, Long Beach, CA, USA, 16–20 June 2019; pp. 501–509.
26. Tsipras, D.; Santurkar, S.; Engstrom, L.; Turner, A.; Madry, A. Robustness May Be at Odds with Accuracy. In Proceedings of the International Conference on Learning Representations, New Orleans, LA, USA, 6–9 May 2019.
27. Hendrycks, D.; Zhao, K.; Basart, S.; Steinhardt, J.; Song, D. Natural adversarial examples. *arXiv* **2019**, arXiv:1907.07174.
28. Naeini, M.P.; Cooper, G.; Hauskrecht, M. Obtaining Well Calibrated Probabilities Using Bayesian Binning. In *AAAI Conference on Artificial Intelligence*; Association for the Advancement of Artificial Intelligence: Menlo Park, CA, USA, 2015.
29. Ovadia, Y.; Fertig, E.; Ren, J.; Nado, Z.; Sculley, D.; Nowozin, S.; Dillon, J.; Lakshminarayanan, B.; Snoek, J. Can you trust your model's uncertainty? Evaluating predictive uncertainty under dataset shift. In *Advances in Neural Information Processing Systems 32*; Wallach, H., Larochelle, H., Beygelzimer, A., d'Alché-Buc, F., Fox, E., Garnett, R., Eds.; Curran Associates, Inc.: Red Hook, NY, USA, 2019; pp. 13991–14002.
30. Ioffe, S.; Szegedy, C. Batch Normalization: Accelerating Deep Network Training by Reducing Internal Covariate Shift. In *Proceedings of Machine Learning Research*; Bach, F., Blei, D., Eds.; PMLR: Lille, France, 2015; Volume 37, pp. 448–456.
31. Wu, T.; Fischer, I.; Chuang, I.L.; Tegmark, M. Learnability for the Information Bottleneck. *Entropy* **2019**, *21*, 924. [CrossRef]
32. Kingma, D.; Ba, J. Adam: A method for stochastic optimization. In Proceedings of the International Conference on Learning Representations, San Diego, CA, USA, 7–9 May 2015.

© 2020 by the authors. Licensee MDPI, Basel, Switzerland. This article is an open access article distributed under the terms and conditions of the Creative Commons Attribution (CC BY) license (http://creativecommons.org/licenses/by/4.0/).

Article
A Comparison of Variational Bounds for the Information Bottleneck Functional

Bernhard C. Geiger [1,*] **and Ian S. Fischer** [2]

[1] Know-Center GmbH, Inffeldgasse 13/6, 8010 Graz, Austria
[2] Google Research, Mountain View, CA 94043, USA; iansf@google.com
* Correspondence: geiger@ieee.org

Received: 24 September 2020; Accepted: 20 October 2020; Published: 29 October 2020

Abstract: In this short note, we relate the variational bounds proposed in Alemi et al. (2017) and Fischer (2020) for the information bottleneck (IB) and the conditional entropy bottleneck (CEB) functional, respectively. Although the two functionals were shown to be equivalent, it was empirically observed that optimizing bounds on the CEB functional achieves better generalization performance and adversarial robustness than optimizing those on the IB functional. This work tries to shed light on this issue by showing that, in the most general setting, no ordering can be established between these variational bounds, while such an ordering can be enforced by restricting the feasible sets over which the optimizations take place. The absence of such an ordering in the general setup suggests that the variational bound on the CEB functional is either more amenable to optimization or a relevant cost function for optimization in its own regard, i.e., without justification from the IB or CEB functionals.

Keywords: information bottleneck; deep learning; neural networks

1. Introduction

The celebrated information bottleneck (IB) functional [1] is a cost function for supervised lossy compression. More specifically, if X is an observation and Y a stochastically related random variable (RV) that we associate with relevance, then the IB problem aims to find an encoder $e_{Z|X}$, i.e., a conditional distribution of Z given X, that minimizes

$$\mathcal{L}_{\text{IB}} := I(X;Z) - \beta I(Y;Z). \tag{1}$$

In (1), $I(X;Z)$ and $I(Y;Z)$ denote the mutual information between observation X and representation Z and between relevant variable Y and representation Z, respectively, and β is a Lagrangian parameter. The aim is to obtain a representation Z that is simultaneously compressed (small $I(X;Z)$) and informative about the relevant variable Y (large $I(Y;Z)$), and the parameter β trades between these two goals.

Recently, Fischer proposed an equivalent formulation, termed the conditional entropy bottleneck (CEB) [2]. While the IB functional inherently assumes the Markov condition $Y - X - Z$, the CEB is motivated from the principle of Minimum Necessary Information, which lacks this Markov condition and which aims to find a representation Z that compresses a bi-variate dataset $(X;Y)$ while still being useful for a given task. Instantiating the principle of Minimum Necessary Information induces then a Markov condition. For example, the task of finding a representation Z that makes X and Y conditionally independent induces the Markov condition $X - Z - Y$, and the representation optimal w.r.t. the principle of Minimum Necessary Information turns out to be $\arg\inf_{X-Z-Y} I(X,Y;Z)$, i.e., it is related to Wyner's common information [3]. The task relevant in this work—estimating Y from a representation Z that is obtained exclusively from X—induces the Markov condition $Y - X - Z$ and the constraint $I(Y;Z) \geq I(X;Y)$. A Lagrangian formulation of the constrained optimization problem $\inf_{I(Y;Z) \geq I(X;Y)} I(X;Z)$, where the infimum is taken over all encoders $e_{Z|X}$ that take only X as input, yields the CEB functional (see Section 2.3 of [2])

$$\mathcal{L}_{\text{CEB}} := I(X;Z|Y) - \gamma I(Y;Z). \tag{2}$$

Due to the chain rule of mutual information [4] (Theorem 2.5.2), (2) is equivalent to (1) for $\gamma = \beta - 1$. Nevertheless, (2) has additional appeals. To this end, note that $I(X;Z|Y)$ captures the information about X contained in the representation Z that is redundant for the task of predicting the class variable Y. In the language of [5], which essentially also proposed (2), $I(X;Z|Y)$ thus quantifies class-conditional compression. Minimizing this class-conditional compression term $I(X;Z|Y)$ is not in conflict with maximizing $I(Y;Z)$, whereas minimizing $I(X;Z)$ is (see Figure 2 in [2] and Section 2 in [5]). At the same time, as stated in [2] (p. 6), $I(X;Z|Y)$ allows to "measure in absolute terms how much more we could compress our representation at the same predictive performance", i.e., by how much $I(X;Z|Y)$ could potentially be further reduced without simultaneously reducing $I(Y;Z)$.

Aside from these theoretical considerations that make the CEB functional preferable over the equivalent IB functional, it has been shown that minimizing variational bounds on the former achieve better performance than minimizing variational bounds on the latter [2,6]. More specifically, it was shown that variational CEB (VCEB) achieves higher classification accuracy and better robustness against adversarial attacks than variational IB (VIB) proposed in [7].

The exact underlying reason why VCEB outperforms VIB is currently still being investigated. Comparing these two bounds at $\beta - 1 = \gamma = 1$, Fischer suggests that "we may expect VIB to converge to a looser approximation of $I(X;Z) = I(Y;Z) = I(X;Y)$", where the later equation corresponds to the Minimum Necessary Information point (see Section 2.5.1 of [2]). Furthermore, Fischer and Alemi claim that VCEB "can be thought of as a tighter variational approximation to the IB objective than VIB" (see Section 2.1 of [6]). Nevertheless, the following question remains: Does VCEB outperform VIB because the variational bound of VCEB is tighter, or because VCEB is more amenable to optimization than VIB?

To partly answer this question, we compare the optimization problems corresponding to VCEB and VIB. Rather than focusing on actual (commonly neural network-based) implementations of these problems, we keep an entirely mathematical perspective and discuss the problem of finding minimizers within well-defined feasible sets (see Section 3). Our main result in Section 4 shows that the optimization problems corresponding to VCEB and VIB are indeed ordered if additional constraints are added: If VCEB is constrained to use a consistent classifier-backward encoder pair (see Definition 1 below), then (unconstrained) VIB yields a tighter approximation of the IB functional. In contrast, if VIB is constrained to use a consistent classifier-marginal pair, then (constrained and unconstrained) VCEB yields a tighter approximation. If neither VCEB nor VIB are constrained, then no ordering can be shown between the resulting optimal variational bounds. Taken together, these results indicate that the superiority of VCEB over VIB observed in [2,6] cannot be due to VCEB better approximating the IB functional. Rather, we conclude in Section 5 that the variational bound provided in [2] is either more amenable to optimization, at least when the variational terms in VCEB and VIB are implemented using neural networks (NNs), or a successful cost function for optimization in its own regard, i.e., without justification from the IB or Minimum Necessary Information principles.

Related Work and Scope. Many variational bounds for mutual information have been proposed [8], and many of these bounds can be applied to the IB functional. Both the VIB and VCEB variational bounds belong to the class of Barber & Agakov bounds, cf. Section 2.1 of [8]. As an alternative example, the authors of [9] bounded the IB functional using the Donsker–Varadhan representation of mutual information. Aside from that, the IB functional has been used for NN training also without resorting to purely variational approaches. For example, the authors of [10] applied the Barber & Agakov bound to replace $I(Y;Z)$ by the standard cross-entropy loss of a trained classifier, but used a non-parametric estimator for $I(X;Z)$. Rather than comparing multiple variational bounds with each other, in this work we focus exclusively on the VIB [7] and VCEB [2] bounds. The structural similarity of these bounds allows a direct comparison and still yields interesting insights that can potentially carry over to other variational approaches.

We finally want to mention two works that draw conclusions similar to ours. First, Achille and Soatto [11] pointed to the fact that their choice of injecting multiplicative noise to neuron activations is not only a restriction of the feasible set over which the optimization is performed, but it can also be interpreted as a means of regularization or as an approach to perform optimization. Thus, the authors claim, there is an intricate connection between regularization (i.e., the cost function), the feasible set, and the method of optimization (see Section 9 of [11]); this claim resonates with our Section 5. Second, Wieczorek and Roth [12] investigate the difference between IB and VIB: While IB implicitly assumes the Markov condition $Y - X - Z$, the variational approach taken in VIB assumes that an estimate of Y is obtained from the representation Z, i.e., $X - Z - Y$. Dropping the former assumption allows to express the difference between the VIB bound and the IB functional via mutual and lautum information, which, taken together, measure the violation of the condition $Y - X - Z$. The authors thus argue that dropping this condition enables VIB and similar variants to optimize over larger sets of joint distributions of X, Y, and Z. In this work, we take a slightly different approach and argue that the posterior distribution of Y given Z is approximated by a classifier with input Z that responds with a class estimate $\hat{Y}$. Thus, we stick to the Markov condition inherent to IB and extend it by an additional variable, resulting in $Y - X - Z - \hat{Y}$. As a consequence, our variational approach does not assume that $X - Z - Y$ holds, which also leads to a larger set of joint distributions of X, Y, and Z. Finally, while [12] compares the IB functional with the VIB bound, in our work we compare two variational bounds on the IB functional with each other.

Notation. We consider a classification task with a feature RV X on $\mathbb{R}^m$ and a class RV Y on the finite set $\mathcal{Y}$ of classes. We assume that the joint distribution of X and Y is denoted by p_{XY}. In this work we are interested in representations Z of the feature RV X. This (typically real-valued) representation Z is obtained by feeding X to a stochastic encoder $e_{Z|X}$, and the representation Z can be used to infer the class label by feeding it to a classifier $c_{\hat{Y}|Z}$. Note that this classifier yields a class *estimate* $\hat{Y}$ that need not coincide with the class RV Y. Thus, the setup of encoder, representation, and classifier yields the following Markov condition: $Y - X - Z - \hat{Y}$. We abuse notation and abbreviate the conditional probability (density) $p_{W|V=v}(\cdot)$ of a RV W given that another RV V assumes a certain value v as $p_{W|V}(\cdot|v)$. For example, the probability density of the representation Z for an input $X = x$ is induced by the encoder $e_{Z|X}$ and is given as $e_{Z|X}(\cdot|x)$.

We obtain encoder, classifier, and eventual variational distributions via solving a constrained optimization problem. For example, $\min_{e_{Z|X} \in \mathcal{E}} \mathcal{J}$ minimizes the objective $\mathcal{J}$ over all encoders $e_{Z|X}$ from a given family $\mathcal{E}$. In practice, encoder, classifier, and variational distributions are parameterized by (stochastic) feed-forward NNs. The chosen architecture has a certain influence on the feasible set; e.g., $\mathcal{E}$ may denote the set of encoders that can be parameterized by a NN of a given architecture.

We assume that the reader is familiar with information-theoretic quantities. More specifically, we let $I(\cdot;\cdot)$ and $D(\cdot\|\cdot)$ denote mutual information and Kullback–Leibler divergence, respectively. The expectation w.r.t. to a RV W drawn from a distribution p_W is denoted as $E_{W \sim p_W}[\cdot]$.

2. Variational Bounds on the Information Bottleneck Functional

We consider the IB principle for NN training. Specifically, we are interested in a (real-valued) representation Z, obtained directly from X, that minimizes the following functional:

$$\mathcal{L}_{\text{IB}}(\beta) := I(X;Z) - \beta I(Y;Z) = I(X;Z|Y) - (\beta - 1)I(Y;Z) =: \mathcal{L}_{\text{CEB}}(\beta - 1) \qquad (3)$$

Rather than optimizing (3) directly (which was shown to be ill-advised at least for deterministic NNs in [13]), we rely on minimizing variational upper bounds. More specifically, the authors of [7] introduced the following variational bound on $\mathcal{L}_{\text{IB}}$:

$$\mathcal{L}_{\text{VIB}}(\beta) := E_{X\sim p_X}\left[D\left(e_{Z|X}(\cdot|X)\|q_Z\right)\right] - \beta H(Y) - \beta E_{XYZ\sim p_{XY}e_{Z|X}}\left[\log c_{\hat{Y}|Z}(Y|Z)\right] \quad (4)$$

where $e_{Z|X}$, $c_{\hat{Y}|Z}$, and q_Z are called the encoder, classifier, and marginal. The classifier is used as a variational approximation to the distribution $p_{Y|Z}$. The marginal q_Z is a learned distribution that aims to marginalize out the encoder $e_{Z|X}$. As such, this distribution is conceptually different from a fixed (unlearned) prior distribution in a Bayesian framework as in, e.g., the variational auto-encoder [14].

As an alternative and motivated by the principle of Minimum Necessary Information, the author of [2] proposed the variational bound on the CEB functional:

$$\mathcal{L}_{\text{VCEB}}(\beta) := E_{XY\sim p_{XY}}\left[D\left(e_{Z|X}(\cdot|X)\|b_{Z|Y}(\cdot|Y)\right)\right] - \beta H(Y) - \beta E_{XYZ\sim p_{XY}e_{Z|X}}\left[\log c_{\hat{Y}|Z}(Y|Z)\right] \quad (5)$$

where $b_{Z|Y}$ is called the backward encoder, which is a variational approximation to the distribution $p_{Z|Y}$.

3. Variational IB and Variational CEB as Optimization Problems

While it is known that $\mathcal{L}_{\text{IB}}(\beta) \le \mathcal{L}_{\text{VIB}}(\beta)$ and $\mathcal{L}_{\text{IB}}(\beta) \le \mathcal{L}_{\text{VCEB}}(\beta - 1)$ for all possible p_{XY} and all choices of $e_{Z|X}$, $b_{Z|Y}$, $c_{\hat{Y}|Z}$, and q_Z, it is not obvious how $\mathcal{L}_{\text{IB}}(\beta)$ and $\mathcal{L}_{\text{VCEB}}(\beta - 1)$ compare during optimization. In other words, we are interested in determining whether there is an ordering between

$$\min_{e_{Z|X}, c_{\hat{Y}|Z}, q_Z} \mathcal{L}_{\text{VIB}}(\beta) \quad (6a)$$

and

$$\min_{e_{Z|X}, c_{\hat{Y}|Z}, b_{Z|Y}} \mathcal{L}_{\text{VCEB}}(\beta - 1). \quad (6b)$$

Since we will always compare variational bounds for equivalent parameterization, i.e., compare $\mathcal{L}_{\text{VIB}}(\beta)$ with $\mathcal{L}_{\text{VCEB}}(\beta - 1)$, we will drop the arguments β and $\beta - 1$ for the sake of readability.

For a fair comparison, we need to ensure that both cost functions are optimized over comparable feasible sets $\mathcal{E}$, $\mathcal{C}$, $\mathcal{B}$, and $\mathcal{Q}$ for the encoder, classifier, the backward encoder, and the marginal. We make this explicit in the following assumption.

Assumption 1. *The optimizations of VCEB and VIB are performed over equivalent feasible sets. Specifically, the families $\mathcal{E}$ and $\mathcal{C}$ from which VCEB and VIB can choose encoder $e_{Z|X}$ and classifier $c_{\hat{Y}|Z}$ shall be the same. Depending on the scenario, we may require that the optimization over the marginal q_Z is able to choose from the same mixture models as are induced by VCEB. I.e., if $b_{Z|Y}(\cdot|y)$ is a feasible solution of $\mathcal{L}_{\text{VCEB}}$, then $q_Z(\cdot) = \sum_y b_{Z|Y}(\cdot|y)p_Y(y)$ shall also be a feasible solution for $\mathcal{L}_{\text{VIB}}$; we thus require that $\mathcal{Q} \supseteq \{q_Z : q_Z(z) = \sum_y b_{Z|Y}(z|y)p_Y(y), b_{Z|Y} \in \mathcal{B}\}$. Depending on the scenario, we may require that every feasible solution for the marginal q_Z shall be achievable by selecting feasible backward encoders; we thus require that $\mathcal{B} \supseteq \{b_{Z|Y} : q_Z(z) = \sum_y b_{Z|Y}(z|y)p_Y(y), q_Z \in \mathcal{Q}\}$. If both conditions are fulfilled, then we write that $\mathcal{B} \leftrightarrow \mathcal{Q}$.*

We furthermore need the following definition:

Definition 1. *In the optimization of $\mathcal{L}_{\text{VCEB}}$, we say that backward encoder $b_{Z|Y}$ and classifier $c_{\hat{Y}|Z}$ are a consistent pair if*

$$c_{\hat{Y}|Z}(y|z) = \frac{p_Y(y)b_{Z|Y}(z|y)}{\sum_{y'} p_Y(y')b_{Z|Y}(z|y')} = \frac{p_Y(y)b_{Z|Y}(z|y)}{q'_Z(z)} \quad (7)$$

holds. In the optimization of $\mathcal{L}_{\text{VIB}}$, we say that marginal q_Z and classifier $c_{\hat{Y}|Z}$ are a consistent pair if

$$p_Y(y) = \sum_z c_{\hat{Y}|Z}(y|z) q_Z(z) \qquad (8)$$

holds.

The restriction to consistent pairs restricts the feasible sets. For example, for VCEB, if $\mathcal{C}$ is large enough to contain all classifiers consistent with backward encoders in $\mathcal{B}$, i.e., if $\mathcal{C} \supseteq \{c_{\hat{Y}|Z} : c_{\hat{Y}|Z}(y|z) \propto p_Y(y) b_{Z|Y}(z|y), b_{Z|Y}(\cdot|y) \in \mathcal{B}\}$, then the triple minimization

$$\min_{\substack{e_{Z|X} \in \mathcal{E}, c_{\hat{Y}|Z} \in \mathcal{C}, b_{Z|Y} \in \mathcal{B} \\ (c_{\hat{Y}|Z}, b_{Z|Y}) \text{ consistent}}} \mathcal{L}_{\text{VCEB}} \qquad (9)$$

is reduced to the double minimization

$$\min_{e_{Z|X} \in \mathcal{E}, b_{Z|Y} \in \mathcal{B}} \mathcal{L}_{\text{VCEB}}. \qquad (10)$$

Equivalently, one can write the joint triple minimization as a consecutive double minimization and a single minimization, where the inner minimization runs over all backwards encoders consistent with the classifier chosen in the outer minimization (where the minimization over an empty set returns infinity):

$$\min_{e_{Z|X} \in \mathcal{E}, c_{\hat{Y}|Z} \in \mathcal{C}} \left[\min_{b_{Z|Y} \in \mathcal{B} \cap \left\{ b'_{Z|Y} : \frac{p_Y(y) b'_{Z|Y}(z|y)}{\sum_{y'} p_Y(y') b'_{Z|Y}(z|y')} = c_{\hat{Y}|Z}(y|z) \right\}} \mathcal{L}_{\text{VCEB}} \right]. \qquad (11)$$

Similar considerations hold for VIB.

4. Main Results

Our first main result is negative in the sense that it shows $\mathcal{L}_{\text{VIB}}$ and $\mathcal{L}_{\text{VCEB}}$ cannot be ordered in general. To this end, consider the following two examples.

Example 1 (VIB < VCEB). *In this example, let $\mathcal{B} \leftrightarrow \mathcal{Q}$, where $\mathcal{B}$ and $\mathcal{Q}$ are constrained, and let $\mathcal{C}$ be unconstrained, thus $\min_{c_{\hat{Y}|Z} \in \mathcal{C}} -E_{XYZ \sim p_{XY} e_{Z|X}} \left[\log c_{\hat{Y}|Z}(Y|Z) \right] = H(Y|Z)$. Suppose further that we have selected a fixed encoder $e_{Z|X}$ that induces the marginal and conditional distributions p_Z and $p_{Z|Y}$, respectively. With this, we can write*

$$E_{XY \sim p_{XY}} \left[D \left(e_{Z|X}(\cdot|X) \| b_{Z|Y}(\cdot|Y) \right) \right] = I(X;Z|Y) + E_{Y \sim p_Y} \left[D \left(p_{Z|Y}(\cdot|Y) \| b_{Z|Y}(\cdot|Y) \right) \right] \qquad (12a)$$

and

$$E_{X \sim p_X} \left[D \left(e_{Z|X}(\cdot|X) \| q_Z \right) \right] = I(X;Z) + D \left(p_Z \| q_Z \right). \qquad (12b)$$

Suppose that $b_{Z|Y}^{\text{VCEB}}$ is a minimizer of (12a) over $\mathcal{B}$ and that $q_Z^{\text{VCEB}}(z) = \sum_y p_Y(y) b_{Z|Y}^{\text{VCEB}}(z|y)$. By the chain rule of of Kullback–Leibler divergence [4] (Th. 2.5.3) and with $b_{Y|Z}^{\text{VCEB}}(y|z) = p_Y(y) b_{Z|Y}^{\text{VCEB}}(z|y) / q_Z^{\text{VCEB}}(z)$, we can expand

$$D\left(p_{ZY}\|b_{Z|Y}^{\text{VCEB}}p_Y\right) = D\left(p_Z\|q_Z^{\text{VCEB}}\right) + \underbrace{E_{Z\sim p_Z}\left[D\left(p_{Y|Z}(\cdot|Z)\|b_{Y|Z}^{\text{VCEB}}(\cdot|Z)\right)\right]}_{\geq 0}$$

$$= \underbrace{D\left(p_Y\|p_Y\right)}_{=0} + E_{Y\sim p_Y}\left[D\left(p_{Z|Y}(\cdot|Y)\|b_{Z|Y}^{\text{VCEB}}(\cdot|Y)\right)\right]$$

thus

$$E_{Y\sim p_Y}\left[D\left(p_{Z|Y}(\cdot|Y)\|b_{Z|Y}^{\text{VCEB}}(\cdot|Y)\right)\right] \geq D\left(p_Z\|q_Z^{\text{VCEB}}\right).$$

Suppose that $e_{Z|X}$ is such that the inequality above is strict. Then,

$$\min_{e_{Z|X}\in\mathcal{E}, c_{\hat{Y}|Z}\in\mathcal{C}, b_{Z|Y}\in\mathcal{B}} \mathcal{L}_{\text{VCEB}}(\beta-1)$$

$$= I(X;Z|Y) + E_{Y\sim p_Y}\left[D\left(p_{Z|Y}(\cdot|Y)\|b_{Z|Y}^{\text{VCEB}}(\cdot|Y)\right)\right] - (\beta-1)I(Y;Z)$$

$$> I(X;Z|Y) + D\left(p_Z\|q_Z^{\text{VCEB}}\right) - (\beta-1)I(Y;Z)$$

$$= I(X;Z) + D\left(p_Z\|q_Z^{\text{VCEB}}\right) - \beta I(Y;Z)$$

$$\geq \min_{e_{Z|X}\in\mathcal{E}, c_{\hat{Y}|Z}\in\mathcal{C}, q_Z\in\mathcal{Q}} \mathcal{L}_{\text{VIB}}(\beta)$$

where the last inequality follows because q_Z^{VCEB} may not be optimal for the VIB cost function.

Example 2 (VIB > VCEB). *Let $\mathcal{B} \leftrightarrow \mathcal{Q}$, where $\mathcal{Q}$ and $\mathcal{B}$ are unconstrained, thus with (12) we have*

$$\min_{b_{Z|Y}\in\mathcal{B}} E_{XY\sim p_{XY}}\left[D\left(e_{Z|X}(\cdot|X)\|b_{Z|Y}(\cdot|Y)\right)\right] = I(X;Z|Y)$$

and

$$\min_{q_Z\in\mathcal{Q}} E_{X\sim p_X}\left[D\left(e_{Z|X}(\cdot|X)\|q_Z\right)\right] = I(X;Z).$$

Suppose further that $\mathcal{C}$ is such that $\min_{c_{\hat{Y}|Z}\in\mathcal{C}} -E_{XYZ\sim p_{XY}e_{Z|X}}\left[\log c_{\hat{Y}|Z}(Y|Z)\right] = H(Y|Z) + \varepsilon$, where $\varepsilon > 0$. It then follows that

$$\min_{e_{Z|X}\in\mathcal{E}, c_{\hat{Y}|Z}\in\mathcal{C}, q_Z\in\mathcal{Q}} \mathcal{L}_{\text{VIB}}(\beta) = I(X;Z) - \beta I(Y;Z) + \beta\varepsilon$$

$$> I(X;Z|Y) - (\beta-1)I(Y;Z) + (\beta-1)\varepsilon = \min_{e_{Z|X}\in\mathcal{E}, c_{\hat{Y}|Z}\in\mathcal{C}, b_{Z|Y}\in\mathcal{B}} \mathcal{L}_{\text{VCEB}}(\beta-1).$$

In both of these examples we have ensured that the comparison is fair in the sense of Assumption 1. Aside from showing that VIB and VCEB in general allow no ordering, additional interesting insights can be gleaned from Examples 1 and 2. First, whether VIB or VCEB yield tighter approximations of the IB and CEB functionals for a fixed encoder depends largely on the feasible sets $\mathcal{C}$ and $\mathcal{B}$: Constraints on $\mathcal{C}$ cause disadvantages for VIB, while constraints on $\mathcal{B}$ lead to the VCEB bound becoming looser. Second, for fixed encoders, the tightness of the respective bounds and the question which of the bounds is tighter do not depend on how well the IB and CEB objectives are met: These objectives are functions only of the encoder $e_{Z|X}$, whereas the tightness of the variational bounds depends on $\mathcal{C}$, $\mathcal{B}$, and $\mathcal{Q}$. (Of course, the tightness of the respective bounds after the triple optimization in (6) depends also on $\mathcal{E}$, as the optimization over $\mathcal{B}$ and $\mathcal{Q}$ in Example 1 and over $\mathcal{C}$ in Example 2 interacts with the optimization over $\mathcal{E}$ in a non-trivial manner.)

Our second main result, in contrast, shows that the variational bounds can indeed be ordered if additional constraints are introduced. More specifically, if the variational bounds are restricted to consistent pairs as in Definition 1, then the following ordering can be shown. The proof of Theorem 1 is deferred to Section 6.

Theorem 1. *If VCEB is constrained to a consistent classifier-backward encoder pair, and if $\mathcal{Q} \supseteq \{q_Z : q_Z(z) = \sum_y b_{Z|Y}(z|y) p_Y(y), b_{Z|Y} \in \mathcal{B}\}$, then*

$$\min_{\substack{e_{Z|X} \in \mathcal{E}, c_{\hat{Y}|Z} \in \mathcal{C}, q_Z \in \mathcal{Q}}} \mathcal{L}_{\text{VIB}} \leq \min_{\substack{e_{Z|X} \in \mathcal{E}, c_{\hat{Y}|Z} \in \mathcal{C}, b_{Z|Y} \in \mathcal{B} \\ (c_{\hat{Y}|Z}, b_{Z|Y}) \text{ consistent}}} \mathcal{L}_{\text{VCEB}}. \tag{13a}$$

If VIB and VCEB are constrained to a consistent classifier–marginal and classifier-backward encoder pair, respectively, and if $\mathcal{B} \supseteq \{b_{Z|Y} : b_{Z|Y}(z|y) = c_{\hat{Y}|Z}(y|z) q_Z(z)/p_Y(y), q_Z \in \mathcal{Q}, c_{\hat{Y}|Z} \in \mathcal{C}\}$, then

$$\min_{\substack{e_{Z|X} \in \mathcal{E}, c_{\hat{Y}|Z} \in \mathcal{C}, q_Z \in \mathcal{Q} \\ (c_{\hat{Y}|Z}, q_Z) \text{ consistent}}} \mathcal{L}_{\text{VIB}} \geq \min_{\substack{e_{Z|X} \in \mathcal{E}, c_{\hat{Y}|Z} \in \mathcal{C}, b_{Z|Y} \in \mathcal{B} \\ (c_{\hat{Y}|Z}, b_{Z|Y}) \text{ consistent}}} \mathcal{L}_{\text{VCEB}}. \tag{13b}$$

A fortiori, (13b) *continues to hold if VCEB is not constrained to a consistent classifier-backward encoder pair.*

Theorem 1 thus relates the cost functions of VIB and VCEB in certain well-defined scenarios, contingent on the size of the feasible sets $\mathcal{B}$ and $\mathcal{Q}$. If the variational approximations are implemented using NNs, then these bounds are thus contingent on the capacity of the NNs trained to represent the backward encoder in case of VCEB and the marginal in the case of VIB. A few clarifying statements are now in order.

First, it is easy to imagine scenarios in which the inequalities are strict. Trivially, this is the case for (13a) if $\mathcal{C}$ and $\mathcal{B}$, and for (13b) if $\mathcal{C}$ and $\mathcal{Q}$ do not contain a consistent pair. Furthermore, if the set relations in the respective conditions do not hold with equality, the optimization over the strictly larger set of, e.g., marginals in (13a), may yield strictly smaller values for the cost function $\mathcal{L}_{\text{VIB}}$.

Second, the condition that $\mathcal{B} \supseteq \{b_{Z|Y} : b_{Z|Y}(z|y) = c_{\hat{Y}|Z}(y|z) q_Z(z)/p_Y(y), q_Z \in \mathcal{Q}, c_{\hat{Y}|Z} \in \mathcal{C}\}$ is less restrictive than the condition stated in Assumption 1. This is because every backward encoder that is written as $b_{Z|Y}(z|y) = c'_{\hat{Y}|Z}(y|z) q'_Z(z)/p_Y(y)$ for $q'_Z \in \mathcal{Q}$ and $c'_{\hat{Y}|Z} \in \mathcal{C}$ satisfies trivially that $\sum_y b'_{Z|Y}(z|y) p_Y(y) = q'_Z(z)$. Thus, if one accepts Assumption 1 as reasonable for a fair comparison between VCEB and VIB, then one must also accept that the ordering provided in the theorem is mainly a consequence of the restriction to consistent pairs, and not to one of the optimization problems having access to a significantly larger feasible set.

Finally, if $\mathcal{C}$, $\mathcal{B}$, and $\mathcal{Q}$ are sufficiently large, i.e., if the NNs implementing the classifier, backward encoder, and marginal are sufficiently powerful, then both VCEB and VIB can be assumed to yield equally good approximations of the IB functional. To see this, let p_Z, $p_{Z|Y}$, and $p_{Y|Z}$ denote the marginal and conditional distributions induced by $e_{Z|X}$ and note that with (12) we get

$$\mathcal{L}_{\text{VIB}}(\beta) = \mathcal{L}_{\text{IB}}(\beta) + D(p_Z \| q_Z) + \beta E_{Z \sim p_Z} \left[D\left(p_{Y|Z}(\cdot|Z) \| c_{\hat{Y}|Z}(\cdot|Z)\right) \right] \tag{14a}$$

and

$$\mathcal{L}_{\text{VCEB}}(\beta - 1)$$
$$= \mathcal{L}_{\text{IB}}(\beta) + E_{Y \sim p_Y} \left[D\left(p_{Z|Y}(\cdot|Y) \| b_{Z|Y}(\cdot|Y)\right) \right] + (\beta - 1) E_{Z \sim p_Z} \left[D\left(p_{Y|Z}(\cdot|Z) \| c_{\hat{Y}|Z}(\cdot|Z)\right) \right]. \tag{14b}$$

Large $\mathcal{B}$ and $\mathcal{Q}$ render the second terms in both equations close to zero for all choices of $e_{Z|X}$ (see Example 2), while large $\mathcal{C}$ renders the last terms close to zero (see Example 1). Thus, in this case not only do we have $\mathcal{L}_{\text{VIB}}(\beta) \approx \mathcal{L}_{\text{VCEB}}(\beta - 1) \approx \mathcal{L}_{\text{IB}}(\beta)$, but we also have that VCEB employs a consistent classifier-backward encoder pair by the fact that $b_{Z|Y} \approx p_{Z|Y}$ and $c_{\hat{Y}|Z} \approx p_{Y|Z}$. Thus, one may argue that if the feasible sets are sufficiently large, the restriction to consistent pairs may not lead to significantly looser bounds.

5. Discussion

In this note we have compared the IB and CEB functionals and their respective variational approximations. While IB and CEB are shown to be equivalent, the variational approximations VIB and VCEB yield different results after optimization. Specifically, it was observed that using VCEB as a training objective for stochastic NNs outperforms VIB in terms of accuracy, adversarial robustness, and out-of-distribution detection (see Section 3.1 of [2]). In our analysis we have observed that, although in general there is no ordering between VIB and VCEB (Examples 1 and 2), the optimal values of the cost functions can be ordered if additional restrictions are imposed (Theorem 1). Specifically, if VCEB is constrained to a consistent classifier-backward encoder pair, then its optimal value cannot fall below the optimal value of VIB. If, in contrast, VIB is constrained, then the optimal value of VIB cannot fall below the optimal value of VCEB (constrained or unconstrained). Thus, as expected, adding restrictions weakens the optimization problem w.r.t. the unconstrained counterpart.

These results imply that the superiority of VCEB is not caused by enabling a tighter bound on the IB functional than VIB does. Furthermore, it was shown in Table 1 of [6] that VCEB, constrained to a consistent classifier-backward encoder pair, yields better classification accuracy and robustness against corruptions than the unconstrained VCEB objective. Since obviously

$$\min_{\substack{e_{Z|X}\in\mathcal{E}, c_{\hat{Y}|Z}\in\mathcal{C}, b_{Z|Y}\in\mathcal{B} \\ (c_{\hat{Y}|Z}, b_{Z|Y}) \text{ consistent}}} \mathcal{L}_{\text{VCEB}} \geq \min_{e_{Z|X}\in\mathcal{E}, c_{\hat{Y}|Z}\in\mathcal{C}, b_{Z|Y}\in\mathcal{B}} \mathcal{L}_{\text{VCEB}} \qquad (15)$$

the achievable tightness of a variational bound on the IB functional appears to be even negatively correlated with generalization performance in this set of experiments. (We note that [6] only reports constrained VCEB results for the largest NN models, and the constrained models perform slightly worse on robustness to adversarial examples than the unconstrained VCEB models of the same size.)

One may hypothesize, though, that VCEB is more amenable to optimization, in the sense that it achieves a tighter bound on the IB functional when encoder, classifier, and variational distributions are implemented and optimized using NNs. However, optimizing VCEB and VIB was shown to yield very similar results in terms of a lower bound on $I(X;Z)$ for several values of β, cf. Figure 4 of [2], which seems not to support above hypothesis.

We therefore conclude that the superiority of (constrained) VCEB is not due to it better approximating the IB functional. While the hypothesis that the optimized VCEB functional approximates the optimized IB functional better cannot be ruled out, we will now formulate an alternative hypothesis. Namely, that the VCEB cost function itself instills desirable properties in the encoder that would otherwise not be instilled when relying exclusively on the IB functional, cf. Section 5.4 of [13]. For example, neither IB nor the Minimum Necessary Information principle include a classifier $c_{\hat{Y}|Z}$ in their formulations. Thus, by the invariance of mutual information under bijections, there may be many encoders $e_{Z|X}$ in the feasible set $\mathcal{E}$ that lead to representations Z equivalent in terms of (1) and (2). Only few of these representations are useful in the sense that the information about the class Y can be extracted "easily". The variational approach of using a classifier to approximate $I(Y;Z)$, however, ensures that, among all encoders $e_{Z|X}$ that are equivalent under the IB principle, one is chosen such that there exists a classifier $c_{\hat{Y}|Z}$ in $\mathcal{C}$ that allows inferring the class variable Y from Z with low entropy: While the IB and Minimum Necessary Information principles ensure that Z is informative about Y, the variational approaches of VIB and VCEB ensure that this information can be accessed in practice. Regarding the observed superiority of VCEB over VIB, one may argue that a variational bound relying on a backward encoder instills properties in the latent representation Z that are preferable over those that are achieved by optimizing a variational bound relying on a marginal only.

In other words, VCEB and VIB are justified as cost functions for NN training even without recourse to the IB and Minimum Necessary Information principles. This does not say that the concept of compression, inherent in both of these principles, is not a useful guidance—whether compression and generalization are causally related is the topic of an ongoing debate to which we do not want to contribute in this work. Rather, we claim that variational approaches may yield desirable properties that go beyond compression and that may be overlooked when too much focus is put on the functionals that are approximated with these variational bounds.

In combination with the variational approach, the selection of feasible sets can also have profound impact on the properties of the representation Z. A representation Z is called disentangled if its distribution p_Z factorizes. Disentanglement can thus be measured by total correlation, i.e., the Kullback–Leibler divergence between p_Z and the product of its marginals Section 5 of the [11]. Achille and Soatto have shown that selecting $\mathcal{Q}$ in the optimization of VIB as a family of factorized marginals is equivalent to adding a total correlation term to the IB functional, effectively encouraging disentanglement, cf. Proposition 1 in [11]. Similarly, Amjad and Geiger note that selecting $\mathcal{B}$ in the optimization of VCEB as a family of factorized backward encoders encourages class-conditional disentanglement; i.e., it enforces a Naive Bayes structure on the representation Z, cf. Corollary 1 & Section 3.1 of [5]. To understand the implications of these observations, it is important to note that neither disentanglement nor class-conditional disentanglement are encouraged by the IB or CEB functionals. However, by appropriately selecting the feasible sets of VIB or VCEB, disentanglement and class-conditional disentanglement can be achieved. While we leave it to the discretion of the reader to decide whether disentanglement is desirable or not, we believe that it is vital to understand that disentanglement is an achievement of optimizing a variational bound over an appropriately selected feasible set, and not one of the principles based on which these variational approaches are motivated.

6. Proof of Theorem 1

We start with the first assertion. Assume that $e_{Z|X}^{\text{VCEB}}$, $b_{Z|Y}^{\text{VCEB}}$, and $c_{\hat{Y}|Z}^{\text{VCEB}}$ are the optimal encoder, backward encoder, and classifier in terms of the VCEB cost function under the assumption of consistency, i.e.,

$$\min_{\substack{e_{Z|X}\in\mathcal{E}, c_{\hat{Y}|Z}\in\mathcal{C}, b_{Z|Y}\in\mathcal{B} \\ (c_{\hat{Y}|Z}, b_{Z|Y}) \text{ consistent}}} \mathcal{L}_{\text{VCEB}} = E_{XY\sim p_{XY}}\left[D\left(e_{Z|X}^{\text{VCEB}}(\cdot|X)\|b_{Z|Y}^{\text{VCEB}}(\cdot|Y)\right)\right]$$

$$- (\beta-1)H(Y) - (\beta-1)E_{XYZ\sim p_{XY}e_{Z|X}^{\text{VCEB}}}\left[\log c_{\hat{Y}|Z}^{\text{VCEB}}(Y|Z)\right] \quad (16)$$

where

$$c_{\hat{Y}|Z}^{\text{VCEB}}(y|z) = \frac{p_Y(y)b_{Z|Y}^{\text{VCEB}}(z|y)}{\sum_{y'} p_Y(y')b_{Z|Y}^{\text{VCEB}}(z|y')} = \frac{p_Y(y)b_{Z|Y}^{\text{VCEB}}(z|y)}{q'_Z(z)}. \quad (17)$$

Certainly, if $\mathcal{C}$ and $\mathcal{B}$ are such that they do not admit a consistent pair, then this minimum is infinity and the inequality holds trivially.

For the VIB optimization problem, we obtain

$$\min_{e_{Z|X}\in\mathcal{E}, c_{\hat{Y}|Z}\in\mathcal{C}, q_Z\in\mathcal{Q}} \mathcal{L}_{\text{VIB}}$$

$$= \min_{e_{Z|X}\in\mathcal{E}, c_{\hat{Y}|Z}\in\mathcal{C}, q_Z\in\mathcal{Q}} E_{X\sim p_X}\left[D\left(e_{Z|X}(\cdot|X)\|q_Z\right)\right] - \beta H(Y) - \beta E_{XYZ\sim p_{XY}e_{Z|X}}\left[\log c_{\hat{Y}|Z}(Y|Z)\right]$$

$$\stackrel{(a)}{=} \min_{e_{Z|X}\in\mathcal{E}, c_{\hat{Y}|Z}\in\mathcal{C}, q_Z\in\mathcal{Q}} E_{XZ\sim p_X e_{Z|X}}\left[\log \frac{e_{Z|X}(Z|X)}{q_Z(Z)}\right] - \beta H(Y) - \beta E_{XYZ\sim p_{XY}e_{Z|X}}\left[\log c_{\hat{Y}|Z}(Y|Z)\right]$$

$$\stackrel{(b)}{=} \min_{e_{Z|X}\in\mathcal{E}, c_{\hat{Y}|Z}\in\mathcal{C}, q_Z\in\mathcal{Q}} E_{XYZ\sim p_{XY}e_{Z|X}}\left[\log \frac{e_{Z|X}(\cdot|X)c_{\hat{Y}|Z}^{\text{VCEB}}(Y|Z)}{q_Z(Z)c_{\hat{Y}|Z}^{\text{VCEB}}(Y|Z)}\right] - \beta H(Y)$$
$$\quad - \beta E_{XYZ\sim p_{XY}e_{Z|X}}\left[\log c_{\hat{Y}|Z}(Y|Z)\right]$$

$$\stackrel{(c)}{\leq} \min_{e_{Z|X}\in\mathcal{E}, q_Z\in\mathcal{Q}} E_{XYZ\sim p_{XY}e_{Z|X}}\left[\log \frac{e_{Z|X}(\cdot|X)c_{\hat{Y}|Z}^{\text{VCEB}}(Y|Z)}{q_Z(Z)c_{\hat{Y}|Z}^{\text{VCEB}}(Y|Z)}\right] - \beta H(Y) - \beta E_{XYZ\sim p_{XY}e_{Z|X}}\left[\log c_{\hat{Y}|Z}^{\text{VCEB}}(Y|Z)\right]$$

$$\stackrel{(d)}{\leq} \min_{e_{Z|X}\in\mathcal{E}} E_{XYZ\sim p_{XY}e_{Z|X}}\left[\log \frac{e_{Z|X}(Z|X)c_{\hat{Y}|Z}^{\text{VCEB}}(Y|Z)}{p_Y(Y)b_{Z|Y}^{\text{VCEB}}(Z|Y)}\right] - \beta H(Y) - \beta E_{XYZ\sim p_{XY}e_{Z|X}}\left[\log c_{\hat{Y}|Z}^{\text{VCEB}}(Y|Z)\right]$$

$$\stackrel{(e)}{=} \min_{e_{Z|X}\in\mathcal{E}} E_{XY\sim p_{XY}}\left[D\left(e_{Z|X}(\cdot|X)\|b_{Z|Y}^{\text{VCEB}}(\cdot|Y)\right)\right] + E_{XYZ\sim p_{XY}e_{Z|X}}\left[\log \frac{c_{\hat{Y}|Z}^{\text{VCEB}}(Y|Z)}{p_Y(Y)}\right]$$
$$\quad - \beta H(Y) - \beta E_{XYZ\sim p_{XY}e_{Z|X}}\left[\log c_{\hat{Y}|Z}^{\text{VCEB}}(Y|Z)\right]$$

$$\stackrel{(f)}{=} \min_{e_{Z|X}\in\mathcal{E}} E_{XY\sim p_{XY}}\left[D\left(e_{Z|X}(\cdot|X)\|b_{Z|Y}^{\text{VCEB}}(\cdot|Y)\right)\right] - (\beta-1)H(Y)$$
$$\quad - (\beta-1)E_{XYZ\sim p_{XY}e_{Z|X}}\left[\log c_{\hat{Y}|Z}^{\text{VCEB}}(Y|Z)\right]$$

$$\stackrel{(g)}{\leq} E_{XY\sim p_{XY}}\left[D\left(e_{Z|X}^{\text{VCEB}}(\cdot|X)\|b_{Z|Y}^{\text{VCEB}}(\cdot|Y)\right)\right] - (\beta-1)H(Y) - (\beta-1)E_{XYZ\sim p_{XY}e_{Z|X}^{\text{VCEB}}}\left[\log c_{\hat{Y}|Z}^{\text{VCEB}}(Y|Z)\right]$$

$$= \min_{e_{Z|X}\in\mathcal{E}, c_{\hat{Y}|Z}\in\mathcal{C}, b_{Z|Y}\in\mathcal{B}} \mathcal{L}_{\text{CEB}}$$

where

- (a) follows by writing the KL divergence as an expectation of the logarithm of a ratio;
- (b) follows by multiplying both numerator and denominator in the first term with $c_{\hat{Y}|Z}^{\text{VCEB}}$;
- (c) is because of the (potential) suboptimality of $c_{\hat{Y}|Z}^{\text{VCEB}}$ for the VIB cost function;
- (d) is because $\mathcal{Q} \supseteq \{q_Z: q_Z(z) = \sum_y b_{Z|Y}(z|y)p_Y(y), b_{Z|Y} \in \mathcal{B}\}$, thus we may choose $q_Z = q'_Z$ where q'_Z is defined in (17); and because this particular choice may be suboptimal for the VIB cost function;
- (e) follows by splitting the logarithm
- (f) follows by noticing that $E_{XYZ\sim p_{XY}e_{Z|X}}[\log p_Y(Y)] = -H(Y)$
- (g) follows because $e_{Z|X}^{\text{VCEB}}$ may be suboptimal for the VIB cost function.

Comparing the last line with (16) completes the proof of the first assertion.

We next turn to the second assertion. Assume that $e_{Z|X}^{\text{VIB}}$, $c_{\hat{Y}|Z}^{\text{VIB}}$, and q_Z^{VIB} are the optimal encoder, classifier, and marginal in terms of the VIB cost function under the assumption of consistency, i.e.,

$$\min_{\substack{e_{Z|X}\in\mathcal{E}, c_{\hat{Y}|Z}\in\mathcal{C}, q_Z\in\mathcal{Q} \\ (c_{\hat{Y}|Z}, q_Z) \text{ consistent}}} \mathcal{L}_{\text{VIB}} := E_{X\sim p_X}\left[D\left(e_{Z|X}^{\text{VIB}}(\cdot|X)\|q_Z^{\text{VIB}}\right)\right] - \beta H(Y) - \beta E_{XYZ\sim p_{XY}e_{Z|X}^{\text{VIB}}}\left[\log c_{\hat{Y}|Z}^{\text{VIB}}(Y|Z)\right] \quad (18)$$

where
$$p_Y(y) = \sum_z c_{\hat{Y}|Z}^{\text{VIB}}(y|z) q_Z^{\text{VIB}}(z). \tag{19}$$

Again, if $\mathcal{C}$ and $\mathcal{Q}$ are such that they do not admit a consistent pair, then this minimum is infinity and the inequality holds trivially.

For the VCEB optimization problem, we obtain

$$\min_{\substack{e_{Z|X} \in \mathcal{E}, c_{\hat{Y}|Z} \in \mathcal{C}, b_{Z|Y} \in \mathcal{B} \\ (c_{\hat{Y}|Z}, b_{Z|Y}) \text{ consistent}}} \mathcal{L}_{\text{VCEB}} + (\beta - 1) H(Y)$$

$$= \min_{\substack{e_{Z|X} \in \mathcal{E}, c_{\hat{Y}|Z} \in \mathcal{C}, b_{Z|Y} \in \mathcal{B} \\ (c_{\hat{Y}|Z}, b_{Z|Y}) \text{ consistent}}} E_{XY \sim p_{XY}} \left[D\left(e_{Z|X}(\cdot|X) \| b_{Z|Y}(\cdot|Y) \right) \right] - (\beta - 1) E_{XYZ \sim p_{XY} e_{Z|X}} \left[\log c_{\hat{Y}|Z}(Y|Z) \right]$$

$$\stackrel{(a)}{=} \min_{\substack{e_{Z|X} \in \mathcal{E}, c_{\hat{Y}|Z} \in \mathcal{C}, b_{Z|Y} \in \mathcal{B} \\ (c_{\hat{Y}|Z}, b_{Z|Y}) \text{ consistent}}} E_{XYZ \sim p_{XY} e_{Z|X}} \left[\log \frac{e_{Z|X}(Z|X)}{b_{Z|Y}(Y|Z)} \right] - (\beta - 1) E_{XYZ \sim p_{XY} e_{Z|X}} \left[\log c_{\hat{Y}|Z}(Y|Z) \right]$$

$$\stackrel{(b)}{=} \min_{e_{Z|X} \in \mathcal{E}, c_{\hat{Y}|Z} \in \mathcal{C}} \min_{b_{Z|Y} \in \mathcal{B} \cap \left\{ b'_{Z|Y} : \frac{p_Y(y) b'_{Z|Y}(z|y)}{\sum_{y'} p_Y(y') b'_{Z|Y}(z|y')} = c_{\hat{Y}|Z}(y|z) \right\}} E_{XYZ \sim p_{XY} e_{Z|X}} \left[\log \frac{e_{Z|X}(Z|X)}{b_{Z|Y}(Y|Z)} \right]$$

$$- (\beta - 1) E_{XYZ \sim p_{XY} e_{Z|X}} \left[\log c_{\hat{Y}|Z}(Y|Z) \right]$$

$$\stackrel{(c)}{\leq} \min_{e_{Z|X} \in \mathcal{E}} \min_{b_{Z|Y} \in \mathcal{B} \cap \left\{ b'_{Z|Y} : \frac{p_Y(y) b'_{Z|Y}(z|y)}{\sum_{y'} p_Y(y') b'_{Z|Y}(z|y')} = c_{\hat{Y}|Z}^{\text{VIB}}(y|z) \right\}} E_{XYZ \sim p_{XY} e_{Z|X}} \left[\log \frac{e_{Z|X}(Z|X)}{b_{Z|Y}(Z|Y)} \right]$$

$$- (\beta - 1) E_{XYZ \sim p_{XY} e_{Z|X}} \left[\log c_{\hat{Y}|Z}^{\text{VIB}}(Y|Z) \right]$$

$$\stackrel{(d)}{\leq} \min_{e_{Z|X} \in \mathcal{E}} E_{XYZ \sim p_{XY} e_{Z|X}} \left[\log \frac{e_{Z|X}(Z|X)}{c_{\hat{Y}|Z}^{\text{VIB}}(Y|Z) q_Z^{\text{VIB}}(Z)} \right] - H(Y) - (\beta - 1) E_{XYZ \sim p_{XY} e_{Z|X}} \left[\log c_{\hat{Y}|Z}^{\text{VIB}}(Y|Z) \right]$$

$$= \min_{e_{Z|X} \in \mathcal{E}} E_{XZ \sim p_X e_{Z|X}} \left[\log \frac{e_{Z|X}(Z|X)}{q_Z^{\text{VIB}}(Z)} \right] - H(Y) - \beta E_{XYZ \sim p_{XY} e_{Z|X}} \left[\log c_{\hat{Y}|Z}^{\text{VIB}}(Y|Z) \right]$$

$$\stackrel{(e)}{\leq} E_{XZ \sim p_X e_{Z|X}^{\text{VIB}}} \left[\log \frac{e_{Z|X}^{\text{VIB}}(Z|X)}{q_Z^{\text{VIB}}(Z)} \right] - H(Y) - \beta E_{XYZ \sim p_{XY} e_{Z|X}^{\text{VIB}}} \left[\log c_{\hat{Y}|Z}^{\text{VIB}}(Y|Z) \right]$$

$$= \min_{\substack{e_{Z|X} \in \mathcal{E}, c_{\hat{Y}|Z} \in \mathcal{C}, q_Z \in \mathcal{Q} \\ (c_{\hat{Y}|Z}, q_Z) \text{ consistent}}} \mathcal{L}_{\text{VIB}} + (\beta - 1) H(Y)$$

where

- (a) follows by writing the KL divergence as an expectation of the logarithm of a ratio;
- (b) follows by the assumption that the VCEB problem is constrained to a consistent classifier-backward encoder pair, and from (11);
- (c) is because of the (potential) suboptimality of $c_{\hat{Y}|Z}^{\text{VIB}}$ for the VCEB cost function;
- (d) follows by adding and subtracting $H(Y)$; by choosing $b_{Z|Y}^{\text{VIB}} = c_{\hat{Y}|Z}^{\text{VIB}} q_Z^{\text{VIB}} / p_Y$, which is possible because $\mathcal{B} \supseteq \{ b_{Z|Y} : b_{Z|Y}(z|y) = c_{\hat{Y}|Z}(y|z) q_Z(z) / p_Y(y), \ q_Z \in \mathcal{Q}, c_{\hat{Y}|Z} \in \mathcal{C} \}$; and by the fact that this particular choice may be suboptimal for the VCEB cost function;
- (e) follows because $e_{Z|X}^{\text{VIB}}$ may be suboptimal for the VCEB cost function.

This completes the proof. □

Author Contributions: Conceptualization, formal analysis, validation, writing: B.C.G. and I.S.F.; Proof of Theorem 1: B.C.G. Both authors have read and agreed to the published version of the manuscript. All authors have read and agreed to the published version of the manuscript.

Funding: The work of Bernhard C. Geiger was supported by the iDev40 project. The iDev40 project has received funding from the ECSEL Joint Undertaking (JU) under grant agreement No 783163. The JU receives support from the European Union's Horizon 2020 research and innovation programme. It is co-funded by the consortium members, grants from Austria, Germany, Belgium, Italy, Spain and Romania. The information and results set out in this publication are those of the authors and do not necessarily reflect the opinion of the ECSEL Joint Undertaking. The Know-Center is funded within the Austrian COMET Program-Competence Centers for Excellent Technologies - under the auspices of the Austrian Federal Ministry for Climate Action, Environment, Energy, Mobility, Innovation and Technology, the Austrian Federal Ministry of Digital and Economic Affairs, and by the State of Styria. COMET is managed by the Austrian Research Promotion Agency FFG.

Conflicts of Interest: The authors declare no conflict of interest.

References

1. Tishby, N.; Pereira, F.C.; Bialek, W. The Information Bottleneck Method. In Proceedings of the Allerton Conference on Communication, Control, and Computing, Monticello, IL, USA, 22–24 September 1999; pp. 368–377.
2. Fischer, I. The Conditional Entropy Bottleneck. *Entropy* **2020**, *22*, 999. [CrossRef]
3. Wyner, A. The common information of two dependent random variables. *IEEE Trans. Inf. Theory* **1975**, *21*, 163–179. [CrossRef]
4. Cover, T.M.; Thomas, J.A. *Elements of Information Theory*, 1st ed.; John Wiley & Sons, Inc.: New York, NY, USA, 1991.
5. Amjad, R.A.; Geiger, B.C. Class-Conditional Compression and Disentanglement: Bridging the Gap between Neural Networks and Naive Bayes Classifiers. *arXiv* **2019**, arXiv:1906.02576.
6. Fischer, I.; Alemi, A.A. CEB Improves Model Robustness. *Entropy* **2020**, *22*, 1081. [CrossRef]
7. Alemi, A.A.; Fischer, I.; Dillon, J.V.; Murphy, K. Deep Variational Information Bottleneck. In Proceedings of the International Conference on Learning Representations (ICLR), Toulon, France, 24–26 April 2017.
8. Poole, B.; Ozari, S.; van den Oord, A.; Alemi, A.A.; Tucker, G. On Variational Bounds of Mutual Information. In Proceedings of the International Conference on Machine Learning (ICML), Long Beach, CA, USA, 10–15 June 2019; pp. 5171–5180.
9. Belghazi, M.I.; Baratin, A.; Rajeshwar, S.; Ozair, S.; Bengio, Y.; Courville, A.; Hjelm, D. Mutual Information Neural Estimation. In Proceedings of the International Conference on Machine Learning (ICML), Stockholm, Sweden, 10–15 July 2018; pp. 531–540.
10. Kolchinsky, A.; Tracey, B.D.; Wolpert, D.H. Nonlinear Information Bottleneck. *Entropy* **2019**, *21*, 1181. [CrossRef]
11. Achille, A.; Soatto, S. Information Dropout: Learning Optimal Representations Through Noisy Computation. *IEEE Trans. Pattern Anal. Mach. Intell.* **2018**, *40*, 2897–2905. [CrossRef] [PubMed]
12. Wieczorek, A.; Roth, V. On the difference between the Information Bottleneck and the Deep Information Bottleneck. *Entropy* **2020**, *22*, 131. [CrossRef]
13. Amjad, R.A.; Geiger, B.C. Learning Representations for Neural Network-Based Classification Using the Information Bottleneck Principle. *IEEE Trans. Pattern Anal. Mach. Intell.* **2020**, *42*, 2225–2239. [CrossRef] [PubMed]
14. Kingma, D.P.; Welling, M. Auto-encoding variational Bayes. In Proceedings of the International Conference on Learning Representations (ICLR), Banff, AB, Canada, 14–16 April 2014.

Publisher's Note: MDPI stays neutral with regard to jurisdictional claims in published maps and institutional affiliations.

© 2020 by the authors. Licensee MDPI, Basel, Switzerland. This article is an open access article distributed under the terms and conditions of the Creative Commons Attribution (CC BY) license (http://creativecommons.org/licenses/by/4.0/).

MDPI
St. Alban-Anlage 66
4052 Basel
Switzerland
Tel. +41 61 683 77 34
Fax +41 61 302 89 18
www.mdpi.com

Entropy Editorial Office
E-mail: entropy@mdpi.com
www.mdpi.com/journal/entropy

www.ingramcontent.com/pod-product-compliance
Lightning Source LLC
LaVergne TN
LVHW070507100526
838202LV00014B/1805